The Decarbonization Delusion

The Decarbonization Delusion

What 3.5 Billion Years of Biological Sustainability Can Teach Us

ANDREW MOORE

OXFORD
UNIVERSITY PRESS

Oxford University Press is a department of the University of Oxford. It furthers the University's objective of excellence in research, scholarship, and education by publishing worldwide. Oxford is a registered trade mark of Oxford University Press in the UK and certain other countries.

Published in the United States of America by Oxford University Press
198 Madison Avenue, New York, NY 10016, United States of America.

© Oxford University Press 2023

All rights reserved. No part of this publication may be reproduced, stored in a retrieval system, or transmitted, in any form or by any means, without the prior permission in writing of Oxford University Press, or as expressly permitted by law, by license, or under terms agreed with the appropriate reproduction rights organization. Inquiries concerning reproduction outside the scope of the above should be sent to the Rights Department, Oxford University Press, at the address above.

You must not circulate this work in any other form
and you must impose this same condition on any acquirer.

Library of Congress Cataloging-in-Publication Data
Names: Moore, Andrew (Science writer), author.
Title: The decarbonization delusion : what 3.5 billion years of biological sustainability can teach us / Andrew Moore.
Description: New York, NY : Oxford University Press, 2023. | Includes bibliographical references and index.
Identifiers: LCCN 2023006074 (print) | LCCN 2023006075 (ebook) | ISBN 9780197664834 (hardback) | ISBN 9780197664858 (epub) | ISBN 9780197664841 | ISBN 9780197664865
Subjects: LCSH: Carbon dioxide mitigation. | Atmospheric carbon dioxide. | Renewable energy sources. | Sustainability.
Classification: LCC TD885.5.C3 M66 2023 (print) | LCC TD885.5.C3 (ebook) | DDC 363.738/746—dc23/eng/20230412
LC record available at https://lccn.loc.gov/2023006074
LC ebook record available at https://lccn.loc.gov/2023006075

DOI: 10.1093/oso/9780197664834.001.0001

Printed by Sheridan Books, Inc., United States of America

I dedicate this book to my parents, who made me curious about the world and supported me so much during my scientific education and training, and life in general.

Contents

Foreword	xix
Acknowledgments	xxi

1. What carbon "does" in the universe: From the first stars to life on Earth ... 1

   Where did carbon come from? ... 1

      The foundries of creation: How the elements arose ... 1

         Understanding elements by protons, neutrons, and electrons ... 1
         Back to the beginning: The birth of matter, energy, and the universe ... 3
         The first atoms appear out of the plasma soup, and the universe becomes structured ... 4
         The first stars, re-emergence of nuclear fusion, and creation of heavier elements ... 5
         Elements move and concentrate in stellar/planetary systems ... 6

   What does carbon do in the universe? ... 8

      AGBs, the first hydrocarbons, molecular matryoshkas, and more ... 8
      Incomparable chemistry: How can carbon make so many compounds? ... 11

         Why electrons don't really "orbit" the nucleus: The weird world of quantum mechanics ... 13
         Carbon's "unsatisfying" electronic configuration is its greatest virtue ... 15
         Carbon's chemical "cousins," and why silicon could never be the element of life ... 18

      A range of strengths and stabilities: Carbon's bonds make it dynamically flexible ... 18

         Diamonds are forever, but only above 100,000 atmospheres pressure ... 19
         Hopping over activation energy barriers with help and creatively channeling energy fluxes ... 21

viii CONTENTS

The element that life "chose" — 22

  Primordial chemical soup on Earth: A nice idea at the time — 23
  A better idea, based on different chemistry and energy flux as the driver — 25

    Some things Miller and Urey could not have known, but others they could have — 25
    Origin in oceanic abysses: A new theory based on alkaline water and electrochemistry — 26
    Chemistry on ancient membranes: H^+ gradients may have worked with "green rust" as the crucial catalyst — 27
    How did life graduate from rusty membranes to the first membrane-bound cells? — 29

  As life evolved, it became more and more carbon-dependent — 30

Photosynthesis: The beginning of the shift away from mineral energy — 32

  How photosynthesizing plants convert the Sun's energy into chemically reduced compounds — 32
  Photosynthesis is intimately integrated with material and energy metabolism — 34
  Photosynthesis indirectly caused life on Earth to shift gear and explode in complexity — 37

Conclusion: The creation of carbon and its special chemistry—
A toolbox of interconvertible compounds waiting for life to use — 41

2. The carbon economy of nutrition and food production: Getting out of control in most respects — 44

Human nutrition: What we consume and how our bodies use it — 44

  What are we made of, and what do we need? — 45
  Carbohydrates: Most efficient use of land, easy to store, mobilize, and interconvert — 47
  A bi-fuel metabolic machine: What happens when carbohydrates run low — 49

Energetic specifics of carbohydrates and fats: It's all in the chemical bonds — 51

  What are biological fats and oils in terms of chemistry? — 51
  Why fats and oils are denser in energy than sugars — 52
  Metabolic economies and health: Why biological hydrocarbons have the edge over carbohydrates — 57

    What is "normal" human fuel intake? — 57

Metabolism is highly sensitive to the "handedness" of molecules	58
Signs that ketogenesis really can extend life span as well	59
So why don't we do ketosis all the time?	59

The place of protein in the larger picture of metabolism and environmental impact — 61

    Linkages between energy metabolism and protein metabolism — 61

Some important health-related qualifications	62
We evolved biologically and culturally as mere opportunistic meat-eaters	62
But we are continuing to eat more and more meat per capita	63

    Energy inefficiency and GHG footprints of animal products: The cases of milk and meat — 65

A sobering analysis of milk consumption in fossil-fuel equivalence	65
Livestock competes with transportation in terms of GHG emissions	67
Methane and nitrous oxide are the most worrying components	69
Concerned scientists emphasize the urgent need for more vegetarian diets	70
How wasting food exacerbates an already terrifying trend	71

Food in flames: How humans' long relationship with fire formed tolerance to many carbon compounds — 71

    Cooking directly with fire put many "unusual" carbon compounds into our diet (and into our lungs) — 73

Some are mutagenic, but the link with human cancer is not simple	74
What happens in our liver makes a big difference	74
Why modeling humans with rats is not ideal in cancer research	75
The smoking that came before smoking: Fumes from the ancient barbecue	77
Probably produced selection pressure for "genetic tolerance"	77
It seems that humans have higher tolerance than other animals to organic combustion products	78

Parallels between human technological evolution and biological evolution: Perfection is relative — 79

Conclusion: Countering further increases in agriculture's enormous GHG impact, getting human nutrition back on track, and learning from biology — 81

3. Sources and sinks: Where carbon compounds accumulate on Earth, and what they do there   83

In the atmosphere: Anthropogenic CO_2 inputs to the carbon cycle continue to rise at around 2.5 ppm per year   83

    Why the truth about CO_2-driven global warming is complicated   84
    CO_2 absorbs infrared radiation in particular wavelength bands   86
    Atmospheric H_2O dominates IR absorption, but CO_2 cycles much more slowly and absorbs strongly at two crucial wavelengths   90
    CO_2's main absorption bands "close windows" in the H_2O spectrum   93

The climate-changing effects of CO_2 started to be researched around 150 years ago   96

    CO_2 exchanges energy with its environment in ways that produce a surprising effect   99

        Heat capacity of gas is only a small part of the picture   99
        Why the saturation argument against CO_2-driven global warming doesn't work   100
        And air currents don't disrupt layers on fast enough timescales   101
        Ultimately, thermal energy, not just radiation, is transferred, but how?   102
        CO_2 as a thermal catalyst   105

    Why methane is a more potent GHG than CO_2 in the shorter term   106

CO_2 cycles and sinks: How the inorganic and organic Earth work together   108

    Big blue: The largely mysterious workings of Earth's oceans   109
    Where does carbon go in geological cycles, and for how long?   113
    Creating coal: Of ancient plants and parallels to cosmic carbon   114

        Brown coal, or "lignite": We can still see the wood   116
        Sub-bituminous and bituminous: On the way to high-grade coal   117
        Anthracite: The hardest, densest, and highest-quality coal   117
        Still some mysteries surrounding the chemistry of coal formation   118

    The origins of oil: Of cell membranes and the largest wax deposits ever identified   120

        Theory 1: The biogenic origin of oil from ancient cell membranes   120
        Theory 2: The abiogenic origin of oil from geochemical processes and meteorites   123
        Is peak oil now really in sight?   125

The genesis of graphite: Carbonaceous sediments that got hotter and denser	125
Conclusion: Thermally catalytic CO_2 is overwhelming Earth's recycling and buffering capacity, but much more than climate is at stake	126
4. Fuels, efficiency, and emissions: Understanding carbon-based energy carriers in the larger picture of sustainability	128
We must fully understand energy density and its consequences	128
Portable fuels: Applications depend on energy density, energy balance, and politics	129
Methanol, the first alcohol, and a promising auto fuel	129
A mere peculiarity of motor sport?	130
The Open Fuel Standard Act and why it failed to stimulate technology diversity	131
China's hesitancy to switch from diesel to methanol trucks: A matter of cost?	132
Global demand for methanol as an energy carrier is growing fastest, but renewable methanol is still scant	133
More alcohols to consider: A large assortment, with large potential pitfalls	134
Replacing gasoline with bioethanol: How much could we achieve?	134
And what does the standard (crop material fermentation) energy balance for bioethanol look like?	135
An alcoholic muddle: If you you're not confused, you're underinformed!	136
A struggle to compare: Policymakers can't judge different technologies properly, and industry can take advantage	138
Dramatically accelerating (photo)synthesis of carbon-based energy carriers: E-fuels	139
E-methanol from photovoltaic energy, H_2 generation and CO_2 capture	139
Power-to-X fuels in general and a much cleaner diesel	141
Portable energy versus stationary energy: No universal solution	144
Weight, transient power, and heat dissipation: Why hydrocarbons are the aviation fuel of choice	144
Why methanol or ethanol could also not replace aviation fuel	146
What are the high-energy-density options for aviation fuel?	146
Many alternatives, but which are environmentally sustainable?	146
And which are, in addition, economically sustainable?	148

Energy carriers for stationary applications: Somewhat
different considerations, but no less important — 149

   *Industrial and domestic requirements: Big challenges
but immediate opportunities* — 150
   *Why burning wood is not an ecologically sound large-
scale heating solution* — 151
   *If most households burned wood for heating, our forests
would be gone in a few years* — 152

Efficiency must be viewed from several perspectives, from
fuel production to use — 153

   Why *expending* energy to *store* energy is necessary: A parallel
with biology — 153

      *A reversible power station and chemical plant in one: Central
metabolism unites material with energetic fluxes* — 155
      How thermally efficient are living organisms? — 155
      *Thought experiments reveal quite modest energy efficiency
in humans* — 156

   Minimizing avoidable losses but making sensible compromises:
How much heat can we reasonably recover? — 158

      Direct harnessing of thermal energy into mechanical energy — 158
      Thermoelectric mechanisms of energy recovery — 159

   Combustion engines: Not a technology at the end of the road — 160

      Much achieved, but still great scope for improving the ICE — 160
      Jet engines: Heading for 20% reduction in consumption by 2040 — 162

From emissions to healthcare: Mismatches between public perception
and the realities of challenges and progress — 163

   The transport sector shows greater success than most others — 163

      *The iconic catalytic converter: Now achieving 95%-plus reduction
of by-product gases* — 163
      Particulate filters: Removing solids formed during combustion — 164
      Non–ICE-related particulates are now taking center stage — 166
      *Political developments versus the truth of the present
road-transport situation* — 167
      Emissions from jet engines: A very hard nut to crack — 168

   Healthcare-related aspects of carbon emissions: We might be
surprised by the realities — 169

      Stationary sources: Your kitchen might be quite a significant one — 169
      PM2.5s: The smallest routinely measured particulates — 171

| | CONTENTS | xiii |

Domestic heating: One of the largest current challenges	172
There's more to emissions than meets the microscope:	
Complex particulates	172

Reaching a realistic perception of risks and potential for improvement   174

Conclusion: Increasingly feasible and acceptable replacements for
fossil fuels, more efficiently produced and burned   175

5. The call to "decarbonize": Public perception, hard-to-abate carbon
positives, and hard-to-achieve carbon negatives   178

Decarbonizing public perception: Of idealistic politics, bizarre
consumer behavior, and harsh reality   178

Killer tomatoes on the rampage: A public relations experience to
learn from   179
Psychological games with carbon: How we are persuaded
and persuade ourselves   181
Waste incineration: A sizeable percentage of anthropogenic
CO_2 emissions   183

The CO_2 footprint of plastics: An ever-increasing challenge   183
Domestic waste incineration produces at least 2% of
  CO_2 *emissions*   184
Are there environmental arguments for waste incineration?   184
Tying us into a new, unsustainable energy system:
  Waste supports both legal and illegal economies   185

How risk-blindness and complacency develop from ubiquity   187

No precautionary principle here!   187
The ubiquity of synthetic carbon-based products in our lives   188
How can we reduce and replace the fossil origin of raw
  ingredients?   189

Hard-to-abate areas: Sectors where fossil fuels are very
difficult to replace   192

Steel manufacture: The quest for an affordable energy source
and chemical reductant in one   192
Cement manufacture: Transitioning toward waste as fuel?   194
Transport sectors: The larger the vehicle, the harder it is to abate   196

Jetting around the world: A disproportionate environmental
  impact   196
Giants of the sea: The most efficient means of transport
  per unit of energy consumed   199
Giants of the land: An awkward combination of
  weight, distance, and cost   201

Are we helping hard-to-abate sectors by reducing consumption?	202
Tiny steps toward reducing per capita consumption-related CO_2 emissions	203
If one can't reduce, one can "compensate," but how impactful can that be?	205

Carbon-negative: Projects for taking CO_2 out of the atmosphere (semi-permanently) ... 208

Deep sequestration of CO_2: The further down, the better, but the greater the energy needed ... 209

Into rock, cavities, and pseudo-cavities ... 209
Into water, from saline aquifers to oceans ... 212
Pilot projects are promising, but progress is far too slow ... 213

Conclusion: Recognizing the absurdity of "decarbonizing" whole economies and the scale of necessary reductions in CO_2 emissions ... 214

6. Decarbonizing the car: Trading off CO_2 against larger environmental problems? ... 217

Sustainable personal transport: How can we reduce the current impact? ... 218

"Inconvenient" results lead curious minds to investigate: EVs are not nearly as sustainable as we are led to believe ... 218
Implications for my personal car: Consider driving much less and sticking with an economical ICEV ... 221
Approaching 10-year ownership with a BEV is much more worrisome (personally and environmentally) than with an ICEV ... 225
Doing the "right" thing when buying a new car: Modest ICEVs used little are probably the best option at present ... 226

Painting a fuller picture of the impact of global car manufacture and driving ... 228

Energy impact: It's much more for EVs than usually reported ... 228

Full production chain energies for battery manufacture appear to be greatly underestimated ... 228
An average BEV without battery and large parts of drive-train and electronics is at least as energetically costly to produce as an ICEV ready to drive ... 229
A model for comparing the incomparable: Electrons in a BEV's battery with e-fuel in an ICEV's tank ... 231
The BEV and ICEV economies break even on energy consumption within the range of 6,700 to 9,800 km per year ... 232

Environmental impact: The BEV economy is several times as bad as the ICEV economy at average mileages ... 234

Human health impact: The BEV economy is 10–20 times as bad
  as the ICEV economy at average mileages ... 236
Matching the ADEME and Fraunhofer conclusions: ICEVs
  running on e-fuel have lower full-cycle impacts than BEVs ... 239
Car life spans and mileages: Consumer habits and culture do not
  bode well for the sustainability of EVs ... 241
Manufacture and recycling: ICEVs have many more "moving"
  parts than BEVs, but that argument is weak and deceptive ... 243

European-style zero emissions in exchange for environmental
  degradation elsewhere ... 245

High-level politics is supporting an unsustainable consumption
  culture, conscious of the false underlying logic ... 245
And not heeding studies that show how important it is to
  recognize environmental impacts beyond CO_2 emissions ... 246

Car makers under pressure, the failure of the world's most
  economical car, and the road to madness ... 248

The heat is on for automobile manufacturers because the
  temperature is rising ... 248
Cars are becoming bigger and heavier despite the need to
  reduce environmental burden ... 249
And range increases are largely made simply by increasing
  battery mass ... 250

The electric mobility revolution has been financially supported
  enough: Now it's the turn of e-fuels ... 251

How e-fuel technology could help us reach net-zero CO_2 as fast
  as possible ... 252

*Driving industry harder with newer technologies: A good way
  to reach net zero CO_2?* ... 252
*Could greatly reducing the use of what we already have be
  a better immediate solution?* ... 253
*Choosing areas in which to move quickly with least
  environmental impact: Driving less and adding
  e-fuels to the mix* ... 254
Cost of e-fuel to the customer is now "right" ... 255
*And we desperately need to develop more geopolitical
  independence in energy* ... 255

Conclusion: Understanding full environmental impact,
  true recycling, and the need for openness to technology diversity ... 257

7. A carbonaceous, biology-inspired recipe for sensible and environmentally conscious energy economies — 261

Reduction in consumption of almost everything is mandatory — 262
We must strive for true recyclability — 264
Integration and economy of metabolism are crucial for making the most of natural energy and energy carriers — 264

- The basic concept — 265
- We must distinguish between fuels for different uses — 265
- Natural energy must be stored sensibly when abundant — 266
  - *Should we mainly use batteries for renewable electricity storage?* — 266
  - *A variety of options for energy storage: From physical to chemical, from stationary to mobile* — 269
  - *Methane and methanol can form technically feasible cyclical systems of synthesis and use* — 273
  - *We must critically compare storage media, their advantages, disadvantages, and acceptable tradeoffs* — 275
  - *Interconnecting energy and material transformations is essential for minimizing waste and facilitating flexibility: How does biology do this?* — 277
- Toward an industrial metabolism with high similarity to biology: From theory to practice — 282
  - *Ideas for practical facilities are emerging* — 282
  - *There are striking parallels with biology* — 282
  - *More explicitly, with the most prominent biochemical pathway on Earth* — 284
  - *The facility mimics the biological interface between energy and material economies* — 285
  - *But what are the hard figures surrounding efficiency?* — 286
  - *We must not shy away from complexity: It is probably the key to sustainability* — 287
  - *The exact "contraption" is less important than the principle of integration* — 287

We must apply opportunism to prevent useful materials and energy becoming environmental problems — 288
Economy of scale and decentralized production are needed in appropriate measures — 290
Diversity of technology must be supported politically — 291
Experimentation, close monitoring, and adaptation are prerequisites for success — 292
We must be prepared to pay more for almost everything — 293

It's time for a new kind of materialism: Valuing and caring more for
    the things we have   295
    We need to derive more satisfaction from preserving what we have   295
    We must develop sustainable personal behavior   296
    Does this mean a slow-down for life and industry?   297
    On the one hand yes, on the other no!   298
        *Cradle-to-cradle, in contrast to cradle-to-grave, must become
            the norm*   298
        *Modular design, construction, and "component-exchangeability"
            must replace monoblock culture*   299
        *Negative feedback must be incorporated into economic systems to
            prevent damaging trends becoming self-reinforcing*   300
Conclusion: Thinking more biologically, developing closed-loop
    economies, and getting back into Earth's "buffering" zone   301

List of figures, tables and information boxes   305
References   309
Index of topics   341

Foreword

Carbon has gotten a very bad name, not because it's bad, but because we have taken it from sources and used it in ways that are clearly unsustainable. However, life on Earth is a massive, sustainable carbon-based mechanism of energy and material transitions: recycling at its best. My motivation for writing this book was a growing concern that human technology development and consumption habits are taking us further and further away from the principles that have enabled life on Earth to exist sustainably for around 3.5 billion years. Instead of seeking mineral-based energy solutions—the human equivalents of which create massive environmental degradation and long-term contamination of land and water—life exists in equilibrium with the inorganic Earth in a relatively thin layer at the interface between the atmosphere and the crust: the biosphere. Here it cycles carbon in integrated networks of energy and material transformations powered ultimately by the Sun.

We now have a very small window of opportunity to learn from biology in this respect and devise carbon-based economies on similar principles. Interestingly, this is the exact opposite of the way in which many people interpret the concept of decarbonization. We arguably have more knowledge about carbon compounds and the monitoring of carbon-based gases in our atmosphere than any other realm of chemistry. We can use this to embrace rather than reject carbon, but it—or any other "solution" for that matter—will only work if we greatly reduce consumption and become sensitive to feedback from our biotic and abiotic environment: just as life does. In my analysis, if we continue to distance ourselves technologically yet further from the natural world, we will reduce our chances of sustainable existence, not increase them. The largest mismatches are, indeed, in our economies of energy and material. The consequences of our fossil fuel–dependent culture have just started to manifest themselves: there's much worse to come. However, if we try to solve this problem by creating new technologies that take us yet further away from natural biological economies of energy and material, we will almost certainly fail.

It took me three years of intensive literature research and thinking to write this book, and I would like to help you to take in its most important insights quickly. Some parts are easy to read, others are inevitably considerably harder because I must make certain points with calculations, numbers, and specialist science. In the interest of economy of resources in the print edition of this book, the calculations and details of the modeling on which I base certain lines of argumentation, conclusions, and insights are available online at https://doi.org/10.1093/oso/9780197664834.001.0001, grouped according to chapter.

I have given the chapters, their sections, and subsections very explicit and meaningful titles in order to help readers identify and grasp high-level concepts as quickly as possible. Chapters 1 and 7 mirror each other in important ways and are the easiest to read: they essentially describe how carbon became such an important element in the universe and in life, and how human civilization can harness it sustainably. Beyond that, feel free to scan the table of contents, identify a topic that particularly interests you, and simply start there. I hope that you enjoy your journey through some of the history, science, and present-day relevance of a truly fascinating element.

—Andrew Moore, PhD

Acknowledgments

This book would not have been possible without the unfailing support of my partner, Gerlind Wallon, who encouraged me; kept me on track; read, understood, and corrected my writing; and gave me invaluable guidance and constructive criticism. I am indebted to Guiyan Zang (Argonne National Laboratory) for explaining the particulars of a model synthetic methanol plant and for checking my calculations related to energy/exergy balances for synthetic methanol production from carbon dioxide and hydrogen; Sebastian Verhelst (University of Gent) for an extremely useful email exchange and conversation on the mechanical and thermodynamic considerations of using methanol as a fuel for internal combustion engines; Jannik Burre (Technical University of Aachen) for explaining energetic aspects of the synthetic manufacture of dimethylether; Richard Starling for reading and commenting on Chapter 6 from the perspective of an economist and financial analyst; Robin Upton for keeping me up to date with the latest reporting on developments in the electrical vehicle socioeconomic debate; and Martin Koš, also for reading and commenting on Chapter 6. I also thank Jeremy Lewis, Senior Editor at Oxford University Press, for realizing the potential of my book proposal, very efficiently organizing a peer review of my summary chapter, and discussing with me the final shape of the book; Michelle Kelley, Project Editor at Oxford University Press, for her support, valuable guidance, and capacity for solution-finding; and Bala Subramanian, Project Manager at Newgen, for his diligence and professionalism in supervising the production of this book.

1
What carbon "does" in the universe
From the first stars to life on Earth

Carbon was a relative latecomer in the formation of the universe as we know it (i.e., the stars and galaxies that we can see with the naked eye). Carbon arose long after the most abundant elements, but to understand its place in the order of things, I go back to the Big Bang. As we will see, carbon's birth started to change the universe in ways that no other element could. Though carbon is, indeed, the element of life, its compounds are not the "essence" of life, as many people believe. However, without carbon, or an element that can achieve a similar, astounding versatility of chemistry, life would not have got much further than simple chemical concentration gradients. It is with an exploration of carbon's incorporation into life that we end this chapter.

Where did carbon come from?

Carbon (C), nitrogen (N), and oxygen (O) are among the most abundant elements in the cosmos, but they are peanuts compared with hydrogen (H), which makes up 99% of the mass of the universe. Next in line is helium (He), at 1%. So where are C, N and O, the elements making up 86.5% of our bodies? Well, they're in the mathematical rounding error between hydrogen and helium! Most of the "detectable" matter in the universe consists of stars, intergalactic and galactic gas, and black holes, but most of it is not visible to the naked eye: neutron stars, for example, were detected via their radio wavelength pulses. Most stars that we can "see" consist largely of hydrogen, and the ones that are still emitting in the visible spectrum are burning hydrogen to helium via nuclear fusion.

The foundries of creation: How the elements arose

Understanding elements by protons, neutrons, and electrons
The "weight" of an element (correctly speaking, its mass) is usually given in terms of atomic mass, which is overwhelmingly constituted by the nucleus of positively charged protons and zero-charge neutrons. Electrons have a tiny mass: around

1,800 times less than that of a proton or neutron. Weighing in as the ultimate featherweight of elements is an atom of "normal" hydrogen (H) with an atomic mass of almost exactly 1 (one proton and one negligibly weighty electron). I wrote "normal" because hydrogen exists in two other variants or *isotopes*—this word deriving from Ancient Greek "same" (*iso*) "place" (*tópos*) (see Box 1.1).

Now back to our periodic system of elements: why, if they have nuclei of very different masses, do hydrogen, deuterium, and tritium (or, for that matter, any isotopes) occupy the same place in Mendeleev's systematic categorization of the elements? Because Mendeleev was a chemist, and so he studied how elements behave when they encounter atoms of their own type or atoms of different types,

---

Box 1.1 More on isotopes: from hydrogen to heavy elements

The three isotopes of hydrogen occupy the same "place" in the so-called periodic system of elements devised by Russian chemist Dimitry Ivanovic Mendeleev (1834–1907). They differ, however, in atomic mass: hydrogen is 1; deuterium is 2 (one proton and one neutron, one negligible electron); tritium is 3 (one proton, two neutrons, one negligible electron). Briefly, to complete the picture, most elements have naturally occurring isotopes, and some of them are unstable: the ratio of protons to neutrons in the atomic nucleus creates a bit of atomic agitation, and the nucleus prefers to settle down to a more stable configuration. This can happen in various ways, via particle/energy emissions and nuclear changes. Emissions comprise the types of radioactive radiation that we know: alpha (which is the ejection of a cluster of two protons and two neutrons from the decaying nucleus), beta (which is the emission of a high-energy electron), and gamma (which is the emission of a high-energy photon—like light, but of an energy that our eyes can't see). The instability of certain isotopes is part of a more general phenomenon related to the nuclear mass and the forces holding the nucleus together (essentially, nuclear glue): atoms with 84 protons or more in their nucleus are all unstable, and they decay to atoms (elements) of lower atomic mass. Two atomic nuclei with *fewer* than 84 protons are also unstable—but rather as a result of the *ratio of neutrons to protons*, which for their particular atomic numbers simply can't be balanced by adding or subtracting a neutron: technetium (atomic number 43) and promethium (61). They also undergo radioactive decay. Not surprisingly, technetium is so unstable that it doesn't occur naturally and can only be made by nuclear physicists via nuclear collisions. Promethium is somewhat less unstable and can be found in tiny amounts in uranium ore, as a by-product of uranium decay.

to put it crudely. That behavior is determined overwhelmingly by how many *electrons* they have, regardless of what is in their nucleus.

And so to electrons, the negatively charged subatomic particles that complement the positively charged protons and which have quite a complicated existence. Electrons don't just swarm around a nucleus of protons and neutrons. Yes, they are a bit like "hangers-on" because they are attracted to the nucleus by the force of electrostatic attraction between their negative charge and the positive charge of the nucleus. However, unlike the hangers-on at public appearances of celebrities, they are at once very ordered, but also very enigmatic. Their orderliness is their confinement to very well-defined spaces around the nucleus, having, for example, spherical, donut-, or hourglass-shaped volumes.

The more energetic electrons occupy spaces further from the nucleus, and, by absorbing a photon of light of high enough energy, any electron can jump into a higher-energy space. If no atomic collision occurs in the meantime, the electron will fall back to its original space, giving off another photon of light with the same energy as the absorbed one. Similar electronic transitions happen in molecules, including carbon dioxide (CO_2), as we will see in Chapter 3. But no nuclear security guard would ever be able keep track of the electrons because they just can't be pinned down with certainty inside their confinement spaces. The German physicist Werner Heisenberg (1901–1976) investigated the electron's strange behavior and discovered that if one tried to measure its energy accurately, one couldn't nail its exact location; conversely, if one discovered exactly where it was, one could only get a rough idea of its energy. This is known as the *Heisenberg Uncertainty Principle*, and it forms part of the theoretical framework for studying and understanding subatomic particles known as *quantum mechanics* and applies to all subatomic particles.

Back to the beginning: The birth of matter, energy, and the universe

Subatomic particles came into being before the universe "knew" what an atom was, around 1 second after the Big Bang. In the moment just before the Big Bang, the entire mass of the universe was contained in a point roughly the size of a subatomic particle. After 10 seconds, and for the following 3 minutes or so, the nuclei of "light" elements were produced, mostly hydrogen, a tiny amount of helium, and vanishingly small amounts of lithium and beryllium, the first and second metals in the Periodic System (known as *alkali earth metals* because they are found in the earth and react—vigorously—with water to produce an alkaline solution). Essentially, the temperatures and density at this time were so high that nuclear fusion happened spontaneously: protons and neutrons collided and "stuck" together to form nuclei of deuterium (the first isotope of hydrogen); these then collided and stuck together to form nuclei of helium, and other collisions (and splittings) resulted in lithium and beryllium, as well as tritium, the second isotope of hydrogen.

In the first 3 minutes of the universe, the temperatures were unimaginable: starting at around 10^{32} Kelvin (1 with 32 zeros behind it; basically the same figure in °C) and falling to around 10^8 Kelvin (K). The nuclei of the light elements started to form at around 10^9 K (a fraction under 1,000,000,000 °C). In comparison, the Joint European Torus, designed to test the feasibility of generating energy via nuclear fusion, operates at a meager 200,000,000 K (roughly five times cooler than the start of the first fusion epoch of our universe). At such temperatures (as during the first 3 minutes of the universe), atoms don't exist: rather nuclei float around in a "sea" of electrons, a state known as a *plasma*. This is the form of matter of which all stars consist. Atoms have to be made "bottom-up," as it were, from their nuclei, and so the first 3 minutes (minus the first 10 seconds) of the Big Bang are known as the *Big Bang nucleosynthesis*.

The first atoms appear out of the plasma soup, and the universe becomes structured

At around 380,000 years after the Big Bang, in the so-called *recombination phase*, the energy (basically temperature) of matter had fallen to a level where electrons associate with nuclei—in equal numbers to the protons in those nuclei, thus forming neutral atoms. Now we're at around 4,000 K, or roughly 2.5 times hotter than molten iron. But still no carbon! At this point, the energy and density of the universe, on average, was too low to allow further fusion reactions: there simply weren't enough nuclei bumping into each other frequently enough and with enough energy. So how did larger nuclei arise? On average the universe had cooled and "thinned out" too much for nuclear fusion to happen ever again. But something fascinating had happened in the first billionths of a second of the Big Bang that would literally have cosmic effects, ultimately producing the order of the universe that we see today and allowing life itself to evolve: the mini-universe became inhomogeneous—very slightly uneven in distribution of matter/energy.

Despite arising from a so-called *singularity* (an almost dimensionless point of unimaginable density and energy), there were tiny (probably immeasurably small) fluctuations in density and energy in the proto-universe, and, as the universe expanded after the Big Bang, these fluctuations became ever more evident. The mere fact that we humans are here to observe them means that matter and energy weren't distributed completely evenly; if they had been, the universe would have continued expanding as an equilibrated, uniform "soup" of a kind that we can't really imagine. Instead it is scattered with blobs of concentrated matter and energy here and there, from small particles through to stars and planets and up to galaxies: an order that arose from a kind of "magnification" of the infinitesimally small irregularities in the proto-universe. These irregularities are manifest in irregularities in the background radiation that is still coursing through space from the Big Bang (the "temperature of space," as it were). This radiation is in the microwave range of wavelengths, having been greatly "stretched" out by the expanding universe. Physicists had predicted that we would detect the irregularities if only we had sensitive enough instruments: that is, if we pointed a "detector" to different places in space and looked for the "right" wavelengths, we would see minute fluctuations in their strength. This technological feat was achieved in the early 1990s. In 2006, it resulted in the Nobel Prize in physics for Cosmic Background Explorer (COBE) scientists John Mather, at the NASA Goddard Space Flight Center, and George Smoot, at the University of California, Berkeley.

Basically, the irregularities in matter/energy density caused tiny regions of the universe that were particularly "dense" in hydrogen to start contracting under gravity—and against the larger-scale expansion of the universe. At around 150 million years after the Big Bang, the first stars were starting to form from these condensing clouds of hydrogen. So-called *stellar nurseries* became the first galaxies at around 300 to 400 million years post Big Bang.

The first stars, re-emergence of nuclear fusion, and creation of heavier elements

These first stars were the beginning of the second phase of nuclear fusion in the universe (i.e., *stellar nucleosynthesis*.[1]) This is the process whereby nuclear fusion reactions of increasing energy (temperature) and pressure create successively heavier elements. In doing so, they largely destroy lithium, beryllium, and boron (elements 3, 4, and 5 in the Periodic Table),[2,3] hence making these elements very rare in the present universe. But, at the same time, the larger stars in the developing universe (typically ones with greater than 10 times the mass of our Sun) were generating heavier elements that would allow the formation of planetary systems somewhat later.

Massive stars are not like nuclear *fission* reactors that humans build: even before exhausting their primary energy source, hydrogen (which fuses to form helium), the temperature and pressure within kick off nuclear fusion reactions of increasingly higher energy. In its old age, such a star—known as a *red giant*—has fusion reactions producing successively heavier elements as one descends toward its core. At its center, at around 4 billion K, silicon is being fused into nickel and iron. At around halfway between the core and the ever-thinning outer layer of hydrogen (which still makes up the majority of the star), carbon is being formed. This happens at a mere 1 billion K, essentially via the fusion of helium: three helium nuclei of atomic mass 4 fuse to produce one carbon nucleus of atomic mass 12. A tiny amount of carbon 13 (^{13}C)—containing an extra neutron—is produced by a side-reaction, and a minuscule quantity of carbon 14 (^{14}C)—containing two extra neutrons. These isotopes respectively constitute 1.1 and 0.0000000001% of carbon in the solar system, are very useful in a variety of research (Box 1.2), and, whereas ^{13}C is stable, ^{14}C is radioactive because its ratio of 8 neutrons to 6 neutrons is too much for the "nuclear glue" to keep together for long. In fact, it is the rapidly decreasing concentration of ^{14}C in atmospheric CO_2 that proves that the CO_2 increases are coming from burning fossil fuel—which has been underground for so long that all of its initial ^{14}C has decayed to ^{13}C.

At some point, the balance between the mounting pressure of the fusion reactions in the star and the gravity holding everything together tips, and the star explodes spectacularly as a *super nova*, spewing new elements out into the near vacuum of space at around 22 million kilometers per hour. The lighter elements that were formed in the Big Bang have almost all been fused into heavier elements by this point of the star's life. However, certain additional elements have also been formed because the fusion reactions that produce iron and nickel have minor side reactions that produce larger nuclei. Some of these are present in small quantities, such as gold; others in minuscule quantities (astatine is the rarest naturally occurring element on Earth, constituting a total mass of 25 grams). Some of the metals are known to us now as the "rare earth metals" and include elements such as promethium, gadolinium, and neodymium, which are, for instance, used in electronics and magnets (e.g., for motors).

Two or more super novae can occur in relative proximity, and hence such stellar explosions can generate swathes of matter known as *molecular clouds*. The speed of the initial explosion slows, and gravity ultimately starts to concentrate matter again into clumps. This time, of course, the matter is a mixture of elements, some of which have already reacted with each other to form compounds such as silicon dioxide (SiO_2), CO_2, water, and methane (CH_4). Because of the motion present in the original material, a slowly spinning disc of matter forms—an *accretion disc* containing dust and gas. At its center is a concentration of hydrogen, which, by gravity alone, will form a new, smaller star. The rest of the matter in the accretion disc will further condense into planets. This is, to a first approximation, how our solar system was formed around 4.6 billion years ago.

Elements move and concentrate in stellar/planetary systems

The unequal distribution of what astrophysicists generally refer to as metals and non-metals in the solar system has given Earth more than its fair share of lithium, beryllium, and boron—the small remnants that escaped the nuclear furnaces of the red giant. It is believed that during the formation of the solar system, the nascent Sun "blew" the lighter elements further away from itself, thus giving rise to the gas giants (Jupiter and Saturn)—largely consisting of hydrogen and helium—and leaving the heavier elements to form the rocky planets

Box 1.2 Carbon isotopes: one stable, one unstable, both useful

All isotopes of carbon have the same chemical properties because they all have the same number of electrons. However, their nuclei differ enough to give them very distinct *physical* properties. ^{13}C is particularly useful in Earth sciences because organisms that take up CO_2—mainly plants, algae, and photosynthetic bacteria—preferentially absorb ^{12}C, in the form of $^{12}CO_2$, and that leads to a higher ratio of ^{12}C to ^{13}C in these organisms compared with their environment. Animals that consume plants also have a skewed ^{12}C to ^{13}C ratio compared with non-living material. Hence soil and rock material that is of organic origin has a higher ratio of ^{12}C to ^{13}C than matter that is purely mineral in origin. But that's not all: ^{12}C and ^{13}C are differently absorbed by different types of plant. Two types of carbon fixation metabolism exist in different plant groups: C3 and C4 (named for the number of carbons in the first sugar derivative formed). Most C4 plants have long, thin leaves with parallel veins: maize, sorghum, sugarcane, millet, and switchgrass are examples. C4 metabolism is the more productive for crops, enabling them to produce more starch. However, many staple crop plants have C3 metabolism, examples being rice, wheat, cassava, and soybean. C3 plants include most of the plants with branched-veined leaves, such as classical shrubs and trees. Interestingly, C3 plants have a higher preference for ^{12}C than do C4 plants; hence, by measuring ratios of ^{12}C to ^{13}C in archaeological animal and human finds, we can make deductions about the kinds of plants that they ate. ^{14}C is the unstable (i.e., radioactive) isotope of carbon, having six protons and eight neutrons in its nucleus. Its occurrence in matter (though in infinitesimally small quantities) permits carbon dating: as with all radioactive isotopes, its bulk decay is very regular, corresponding to a half-life of 5,730 years, hence making it very useful for archaeological dating (up to around 60,000 years into the past). Most ^{14}C on Earth is produced via the action of thermal neutrons from cosmic rays on atmospheric nitrogen. A certain quantity is incorporated into all living organisms, and, after death, the clock starts ticking because no further incorporation can occur. By measuring the radioactivity of the tissue or organic material (e.g., papyrus bearing ancient scripts), the time since the ^{14}C was incorporated into it can be interpolated from the decay curve (representing the half-life). ^{14}C decays back to nitrogen by emitting a beta particle (high-energy electron) in the process of converting one of its nuclear neutrons into a proton (nitrogen is one element higher in the Periodic Table, having seven protons in its nucleus).

Mercury, Venus, Earth, and Mars, nearer to the Sun. That said, Earth has no more than 20 milligrams of lithium per kilogram of crust rock, placing lithium at position 33 in terms of elemental abundance. The two most common elements in Earth's crust are oxygen and silicon: rocks are largely silicates (derivatives of SiO_2). Carbon mainly reached us as carbon monoxide (CO), CO_2, and a variety of hydrocarbons and crystalline elemental carbon (diamond, graphite, and fullerites). Carbon is at position 17 in abundance,[4] comprising between 300 and 1,800 milligrams per kilogram of matter on Earth, depending on data source.

In addition to being spewed out from super novae, carbon is continuously leaching into the interstellar medium (IM) from so-called *asymptotic giant branch* (AGB) stars, which have a mass of between 0.6 and 10 solar masses. These stars have this strange name because, on the diagram of brightness against temperature (Hertzsprung-Russell diagram), they approach asymptotically (i.e., never quite reaching) the position of red giants. In their dotage, they start burning (fusing) helium to carbon and oxygen in their core. However, convection currents raise some of this mixture to the surface, where it is given off as a variety of forms and compounds of carbon. These drift off into space to form part of the molecular clouds from which stars and planetary systems are born.

Carbon is not the most abundant element in the universe or on Earth, but it is omnipresent in its myriad forms, and, even before it became part of living organisms, it was doing some things that no other element can do so well. The extraordinary chemistry of carbon, from inorganic beginnings through to its crucial role in life as we know it, is the topic of the next section.

What does carbon do in the universe?

Carbon is a major component of the *interstellar medium* (ISM) and fundamentally contributed to the evolution of the ISM and its properties. We now enter the realms of soot and dust as we contemplate the remarkable AGB stars mentioned earlier, which literally pump out "soot" into the ISM—soot without which life on Earth would not exist. This is basically like the incomplete combustion that you get in your coal or wood stove.

AGBs, the first hydrocarbons, molecular matryoshkas, and more

Carbon and oxygen rise up from the core of AGB stars, experience decreased pressure, and are surrounded by hydrogen in the thin remaining hydrogen shell of the star. Here some very interesting things start happening. Physicists

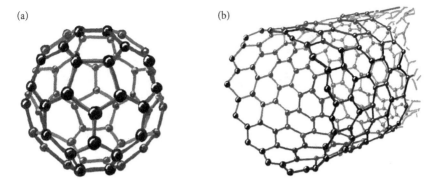

Figure 1.1 Carbon bucky ball (**a**) and bucky tube (**b**). The size (diameter) can vary greatly, but it is always dependent on a whole number of geodesic units (hexagons and pentagons). These structures are relatively stable because of the delocalized electrons in the C=C pi bonds.

can mimic in the laboratory the conditions of temperature, pressure, and elemental composition under which these processes occur and are hence quite confident that they also happen in aged medium-sized stars. As in a dirtily burning wood or coal stove (i.e., one starved of oxygen), carbon starts to combust with the available oxygen as it leaves the star and its temperature falls, thus mainly forming CO. This and the excess carbon (i.e., uncombusted) is free to react with whatever it will, and, in the outer layers of the star, there is plenty of hydrogen left: simple C-H compounds are formed—the first hydrocarbons in the universe—beginning with CH_4.

Carbon also reacts with itself: diamond and graphite are the best known examples—two elemental forms that could not be more different in properties (diamond being the hardest naturally occurring substance; graphite being soft and slippery). These properties are a result of carbon's unique electronic structure. But there are several other forms of carbon. You may, for example, have heard of things called "bucky balls," three-dimensional geodesic structures (made up of regular polygons) consisting only of carbon atoms. They are named after US architect, designer, and visionary Richard Buckminster Fuller (1895–1983), who built geodesic domes, realizing that these structures were very light and yet comparatively strong. The physical stability of Buckminster Fuller's beautiful buildings is mirrored in the chemical stability of the carbon form that is named after him.

AGBs put out some really bizarre bucky balls and bucky tubes (Figure 1.1), ranging from simple one-layer arrangements (rather like a traditional hexagon/pentagon football) to onion-like structures that are a bit like massive bucky balls within bucky balls (Figure 1.2)—perhaps "molecular matryoshka" would be a

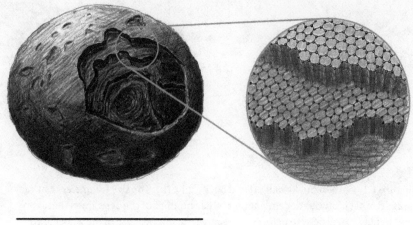

20–50 μm

Figure 1.2 Impression of a soot particle from an asymptotic giant branch (AGB) star. The particle formed via accretion of sheet-like carbon compounds (possibly containing small amounts of hydrogen) that also form bucky balls and bucky tubes (Figure 1.1). Many such particles cluster together to form agglomerations, forming a major component of interstellar dust that can be observed by astronomers because of its light-absorbing and -scattering properties. Scale bar indicates size range of approximately 20–50 micrometers (one micrometer is one thousandth of a mm).

fitting term. Bucky tubes are the essential constituent of synthetic carbon fiber, and it is the triangular bond configuration and "spread" double bonds that give carbon fiber its enormous strength but amazing lightness.

The variety of conditions through which the carbon passes on its way out of the star (essentially temperature and pressure gradients) is ideal for making relatively complex structures that are generically known to astrophysicists as *carbon grains*. These have a tendency to stick together to form larger particles; spectrographic measurements suggest that they are, indeed, present in the interstellar "dust." So, too, are numerous chain-like molecules consisting of C-C bonds—from a few, up to 18 or so. Researchers have also detected cyclic molecules, mainly five- and six-membered rings, sometimes in combination with each other. These carbon molecules also contain C-H bonds, given that H is the most abundant element with which carbon can react in the universe.

And the chemical party doesn't stop at the outskirts of the AGB; rather, it continues in the molecular cloud in places shielded from ultraviolet (UV) rays by interstellar dust (UV is good at splitting many carbon compounds). Some very large and complicated compounds can be formed (Figure 1.3).

WHAT CARBON "DOES" IN THE UNIVERSE   11

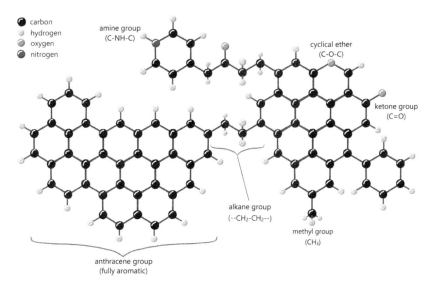

Figure 1.3 A large anthracene-based molecule with a variety of chemical groups. The cyclic (hexagonal) groups consisting only of carbons and hydrogens are "aromatic"; that is, each of the three pairs of carbon atoms is bound by a double bond (C=C), which is, for simplicity, not shown here.

The variety of carbon forms and compounds drifting around in space is worthy of a whole university course in itself. Ultimately, we can also assume that these compounds and grains of carbon were also present during the formation of the Earth and the other planets and their moons. If you've ever wondered where the seas of liquid CH_4 on Saturn's largest moon (Titan) come from, now you know. In fact, because small amounts of nitrogen (N_2) are also present in the molecular cloud, carbon had already reacted to form certain amino acids and cyclic compounds containing nitrogen that we find in protein and DNA in life today—billions of years before any life in the universe could be possible (Figure 1.4). In 2019, ribose (contained in DNA) and other sugars were detected in primitive meteorites.[5] The researchers conjecture that these organics might have helped life on Earth get going, though, as we will see, there are abiological (not linked with life) sources of sugars and many other biologically relevant molecules much closer to home.

Incomparable chemistry: How can carbon make so many compounds?

How can the element that also makes the core of a pencil do all of this varied chemistry so well? Though we are talking about the chemistry of carbon here,

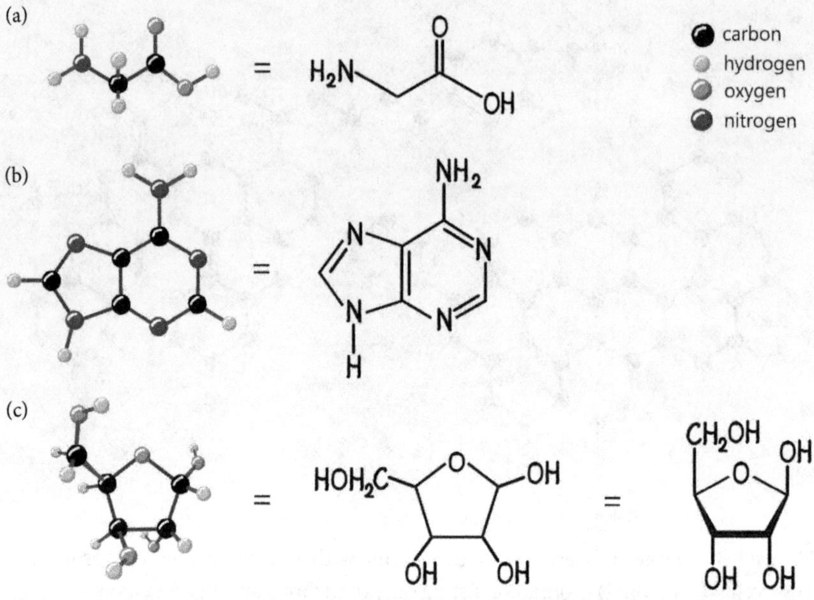

Figure 1.4. Examples of smaller organic compounds formed in the universe before the origin of life and also found in certain meteorites. (a) The simplest amino acid, glycine; (b) a nitrogenous base, adenine, that is found in DNA and RNA; (c) the sugar, ribose, which is found in RNA, and, missing the lower-right oxygen, in DNA as deoxyribose. At left are classical ball-and-stick representations (neglecting double-bonds), followed by Fischer projections in which only "functional" groups are given in letters and non-functional hydrogen atoms are omitted. For ribose, an additional 3D projection is shown (at right) to emphasize that ribose has a flat (planar) ring.

we should really term it the "electronics" of carbon. As mentioned earlier, the chemistry of an element is determined by the way in which the electrons buzzing around its nucleus behave. In this context, "behave" really means how much energy they have, where around the nucleus they "are" (remember, we can only approximate their positions if we want to know their energy, and vice versa), and how stable a particular configuration of electrons is compared with another. If that sounds a bit vague, it isn't: the amazing thing about electrons is that they "know" how to behave when they're whizzing around a given nucleus with a given energy. "Whizzing" is permissible, because they have speeds that are a significant fraction of the speed of light. The person who studied the electrons in atoms arguably more than anyone else was the quantum physicist Heisenberg (whom we met in "Understanding elements by protons, neutrons, and electrons"). The concept of "quanta" (permissible energy states, with no intermediate energy states) applies to all subatomic particles, but it is marvelously illustrated visually by the electron.

Why electrons don't really "orbit" the nucleus: The weird world of quantum mechanics

Electrons, being negatively charged, are attracted to the positively charged nucleus. But if they are above a certain energy (typically upward of 6,000 K, roughly the same in °C) they are moving so fast that that no nucleus can hold on to them for long, and we have a *plasma* (atomic nuclei in a sea of electrons). If we slow electrons down to a reasonable speed (below typical plasma temperatures), they associate spontaneously with nuclei: any nucleus will do—electrons (as all subatomic particles) are universally equal. Let's imagine an atom with many protons in its nucleus, say 30 (that would be zinc). The highest-energy electrons around the nucleus are, as one can intuit, the ones that are farthest away, to put it crudely. As we move closer to the nucleus, there is a step-reduction in the energy of the electrons, and we find the next group. Then, a little closer to the nucleus, and a further step in energy lower, we find the next group of electrons.

To understand the chemistry of any element, however, we need to refine this crude model. That was also part of Heisenberg's work. Using a branch of mathematics that was developed together with quantum mechanics, he determined the "shapes" of the volumes that the electrons inhabit. Why shapes—plural—you might ask? Aren't we talking about small objects orbiting the nucleus, rather like the planets orbit the sun? Surely it's enough to imagine a circular or ellipsoidal orbit and leave it at that. Well, that would be OK for an object that we can see with the naked eye—the behavior of which we can accurately predict and explain with Newtonian (classical) mechanics. But subatomic particles behave in a very special way because they obey different physical laws: the *laws of quantum mechanics*. They don't "orbit" the atomic nucleus in the classical sense at all, but rather exist in a certain space around the nucleus with a certain probability. These volumes are described as 95% probability volumes because an electron has a 95% chance of being found within a given volume at a given time (note, this is also a consequence of the Heisenberg Uncertainty Principle).

The lowest energy volume—the one closest to the nucleus—was calculated to be spherical and was named $1s$ (first spherical orbital—and, yes, you'd be right to criticize the use of "orbital" here, but that's what the physicists decided!). Incidentally, the letters denoting the orbitals were chosen according to the absorption lines in the alkali metal spectra, which had long since been dubbed "*s*harp," "*p*rincipal," "*d*iffuse," and "*f*undamental": each absorption line is caused by an electron jumping from its "home" orbital to a higher one. The $1s$ orbital can hold up to two electrons. One step higher in energy (i.e., one quantum more energetic) is a 2s orbital—also spherical—similarly containing a maximum of two electrons. But now for something quite bizarre: the calculations showed that this second energy level (95% probability volume) also contained three orbitals shaped like squashed hourglasses, or two satsumas almost touching each other, each pair oriented at 90 degrees to the other two. These were dubbed "*p* orbitals" (electrons jumping into a higher energy orbital from these orbitals produced the "*principal*" absorption lines in the alkali metal absorption spectra). The *p* orbitals lie along imaginary axes x, y, and z, and are thus termed $2p_{x,y,z}$ (Figure 1.5). Each pair of *p* orbitals can hold up to two electrons. Why the second-level electrons inhabit different shapes of orbital (one spherical and three hourglass-shaped) can only be explained by complicated mathematics that are outside the scope of this book: suffice it to say that, though they all have the same energy, two of the second-level electrons live in a spherical orbital and the remaining six (three *p* orbitals × two electrons) live in hourglass-shaped ones. *d* and *f* orbitals have even more surprising and complicated shapes, but we need not go into them here (carbon doesn't have them anyway).

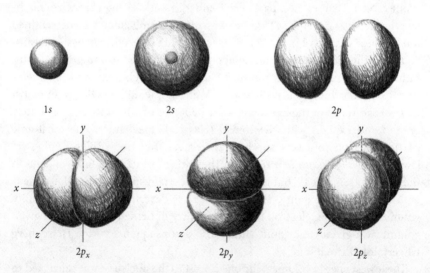

Figure 1.5 *s* and *p* orbitals represented as 95% probability volumes in which the respective electrons are to be found. The 2s orbital is calculated (using equations from the quantum mechanics framework) to contain a discontinuity of density near its core, hence appearing as a sphere within a sphere. Each orbital can hold up to two electrons. An atom that fills 1s, 2s, and all three 2p orbitals has 10 electrons (the element neon). This "electronic completeness" makes neon very stable and unreactive.

Carbon's "unsatisfying" electronic configuration is its greatest virtue

You might already have calculated that because carbon is element number 6 in the Periodic Table (i.e., has six protons and six electrons) it doesn't fill all of its available 2p orbitals. In fact, it only fills one of them, and that is written as $1s(2)$, $2s(2)$, $2p(2)$ (i.e., two electrons in the first [spherical] orbital; two electrons in the second [spherical] orbital; and two electrons in the second [hourglass-shaped] orbital). The element above carbon (nitrogen) has the electronic configuration $1s(2)$, $2s(2)$, $2p_x(2)_y(1)$. By deduction, therefore, element number 10 (neon) has $1s(2)$, $2s(2)$, $2p_x(2)_y(2)_z(2)$. In other words, neon has all of its available second-tier orbitals filled. Not surprisingly this is a very "satisfying" configuration, and its stability makes neon an extremely unreactive element (it's the first of the noble gases). Hence we can deduce that carbon, with only partly filled second-tier orbitals, might not be so "satisfied" with its electronic configuration and would therefore be more reactive than neon. We'd be right, of course, but there is something more interesting about the arrangement of electrons in carbon's second tier.

Having the same energy level, the electrons in the 2s and 2p orbitals can "mix" and form *sp* hybrid orbitals. This is what happens when carbon reacts with hydrogen to form methane (CH_4). The four C-H bonds are equally spaced and form a perfectly symmetrical tetrahedral arrangement (Figure 1.6). This electronic configuration is also "satisfying" because each bond (which is made with one of the hybrid *sp* orbitals) contains two electrons (one from the carbon and one from the hydrogen atom): a standard chemical bond. That is, each hybrid orbital is "full."

In a sense, carbon makes available this hybrid orbital configuration whenever it has the chance to react with another atom, including itself. In this way, it can (under the right conditions of temperature and pressure) react both with itself and hydrogen to form chains in which each carbon in the chain is bonded to two hydrogens, and the two end carbons are each bonded to three hydrogens. Such "straight chain" molecules of hydrogen and carbon are known as *alkanes*, and they are the simplest hydrocarbons. In fact, the chains are (despite the chemical nomenclature) not straight, but kinked, because of the tetrahedral arrangement of the bonds (Figure 1.7.), which is similar to that in methane. Butane (the gas often used in camping stoves) has a four-carbon chain. Chains can join ends to form rings, a common example being cyclohexane (the cyclized version of the six-carbon hexane). But where carbon becomes really special, indeed unique, is in the way it manages its *p* orbital when it reacts with itself or another atom to form a *double*-bond.

A single-bond consists of two electrons in a shared orbital between two atoms, and in carbon it is known as a *sigma bond* (Greek letter σ); a double bond consists of four electrons shared between two atoms in two shared orbitals. When carbon does this, it forms one bond with its 2s orbital and another with its 2p orbital.

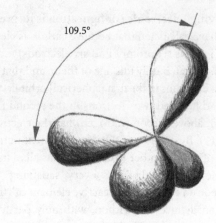

Figure 1.6 The *sp* hybrid orbitals in methane (CH_4). Each lobe is a kind of merger between the 1*s* orbital of hydrogen and the 2*p* orbitals of carbon. These orbitals are the bonds between hydrogen and carbon, and they are exactly equally spaced at 109.5 degrees to each other. Carbon achieves electronic stability by "inventing" more orbitals to its existing $2p_x$ orbital, hence filling all its level 2 orbitals (i.e., with eight electrons) albeit with four of those electrons coming from hydrogen. The electronic stability of an element can be increased by sharing of electrons with another element in the process of reacting with it to form bonds. However, that stability is never as great as in an element that naturally has its own electrons to fill all of its orbitals (e.g., neon).

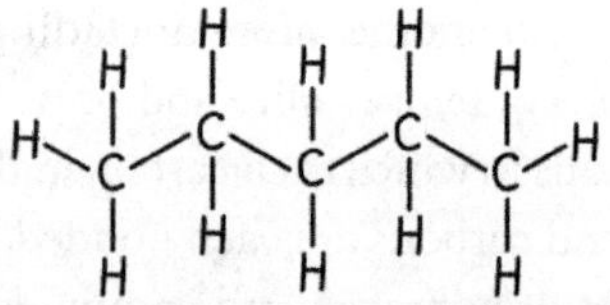

Figure 1.7 2D projection of a typical alkane (pentane), which is described as "straight chain" because it has an uninterrupted and non-branching chain of C-C bonds. The 3D geometry of the molecule is, however, kinked because of the formation of *sp* hybrid orbitals in the C-C and C-H bonds, as in the case of methane (Figure 1.6).

This second bond has a characteristic shape, known as *pi* because it resembles the Greek letter pi (Figure 1.8).

The pi bond has remarkable properties: if two of them are separated by a C-C single bond (such as in the cyclic six-carbon compound benzene), they

(a) (b)

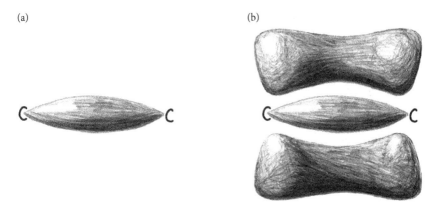

Figure 1.8 Sigma (a) and pi (b) bonds between carbon atoms. When carbon forms a single bond with itself (e.g., in a saturated alkane such as in Figure 1.7), that bond is made with *sp* hybrid orbitals, is cigar-shaped, and is referred to as a sigma bond (Greek letter σ). The second bond that carbon makes with itself (to form a double bond) is formed with the $2p_x$ orbital and can be thought of as a kind of merger of the two originally dumbbell-shaped orbitals. This is known as a pi bond (Greek letter π) because of its shape. For simplicity, only the orbitals involved in bonding between the two carbon atoms are shown.

"communicate" with each other such that one can't tell where the electrons really are. This is known as a *delocalized state*, and it leads to some effects that, though we take them for granted, are very special in the chemical world. One of them is the ability to "soak up" a highly reactive electron from a so-called radical. Examples are oxygen radicals (e.g., $O_2^{\bullet -}$ "superoxide") that are formed in our bodies when we burn food. Neutralizing radicals is how so-called antioxidants work. Oxygen radicals essentially react much more vigorously with biological material than simple oxygen does, and hence they are much more powerful oxidizing agents. Our body tries its best to keep us in a largely non-oxidized state that is compatible with life.

Another notable property of carbon compounds with pi bonds is their ability to conduct electrons from one place to another: photosynthesis and respiration in our cells' mitochondria (power plants) are examples. Lastly, many pi orbital carbon compounds that have double bonds in a ring structure fluoresce (i.e., emit light when "excited" by another light source [often UV]). The electronic flexibility of carbon, if I may quaintly summarize it thus, is so special that carbon in interstellar molecular clouds is the major electron donor and hence facilitator of reactions: even in the absence of any living organism, it produces a diversity of chemistry and compounds unmatched in the Periodic Table of the elements.

Carbon's chemical "cousins," and why silicon could never be the element of life

What about carbon's cousins? I turn next to a chemical conundrum that entered the popular realm via a science fiction series that started in the late 1960s. A few other elements can also form *sp* hybrids. Its position directly under carbon (i.e., also in Group 14 of the Periodic Table) suggests that silicon might also be quite electronically creative. So why don't we see such an abundance and complexity of silicon compounds? In a series of episodes of *Star Trek* in 1976, Captain Kirk's brave team of intergalactic explorers and peacemakers stumbled across a planet inhabited by a silicon-based life form: "It's life, Jim, but not as we know it," remarked Dr. McCoy. Could it be that carbon-based life forms on Earth outcompeted silicon-based ones, or that silicon life exists somewhere in the universe? The truth is more prosaic. Charming as it might be to imagine a silvery silicate worm slithering across your lawn, the types of molecule found in life on Earth, and indeed absolutely necessary for that life, cannot be made with silicon. The further down Group 14 of the Periodic Table we go, the less "life-facilitating" the elements: not even Captain Kirk would dream of coming across life based on germanium, tin, or lead, which, of course, are metals (the further down the Periodic Table one goes, the more metallic the elements).

Silicon does react with hydrogen to make silane (SiH_4), the equivalent of methane, and the *sp* hybrid orbitals are identical in shape and regular tetrahedral orientation to those in methane. However, SiH_4 doesn't form nearly as easily as CH_4, and longer-chain silanes (the equivalent of carbon's alkanes) are very limited in extent. Stable cyclic Si compounds consisting only of Si and H (e.g., an analog of carbon's cyclohexane) are unknown: it seems that silicon can only form closed rings in combination with oxygen as a bridging atom (cyclic silicones). Why such limited chemistry? The further away from the nucleus that the electrons available for making bonds with other atoms are (the "valence" electrons) the weaker the bonds that they make (simplistically put); these electrons in silicon are that much farther away from the nucleus than in carbon. Carbon's place in the order of the elements is "just right" in terms of its valence electrons and the types of orbitals that they can form—in particular, the hybrid *sp* and delocalized *pi* orbitals. This allows carbon to form relatively strong bonds but still have orbitals available to make other bonds with other carbon atoms.

A range of strengths and stabilities: Carbon's bonds make it dynamically flexible

The relatively strong and prolific bonding behavior of carbon makes carbon compounds relatively stable and long-lived, but not too stable and long-lived. This compromise is crucial for life and the "recycling" of carbon-based components of life because life is always in flux: the making, breaking, and recycling of organic molecules and macromolecules is an essential principle of life. But what, exactly, is "not too long-lived"? The stability of C-C and C-H bonds in most organic molecules, in our normal everyday ambient conditions, is relatively high. However, we must go into a little more detail to understand this fully. First, the thermodynamics: this tells us basically in which form a substance will ultimately come to rest in given conditions of temperature and pressure.

Diamonds are forever, but only above 100,000 atmospheres pressure

An intriguing carbon-based example of thermodynamic stability is diamond: under conditions of high pressure (e.g., in the molten core of the Earth), diamond is the stable form of carbon. At normal atmospheric pressure, it's graphite—considerably less valuable. This is rather disappointing for those who thought that "diamonds are forever" for diamond will, eventually, change into graphite under ambient conditions. The reason why the diamond industry is still around is that the process is so incredibly slow that, as calculated in one thought experiment online,[6] it would take more than a billion years for 1 cubic centimeter of diamond to "degrade" to graphite on your finger. The tetrahedral C-C bonds in diamond (Figure 1.9) are so strong (they are formed from sp hybrid orbitals) that an enormous amount of energy is needed to break them. That is, indeed, why diamond is the hardest naturally occurring substance. It is exceedingly rare, at ambient conditions, for a molecular collision of sufficient energy to exceed this so-called *activation energy barrier*. But if it does, a little bit of the diamond is converted to graphite and stays that way (its stable state) until it meets conditions of high temperature and pressure that favor diamond instead.

Both processes are painfully slow, but humans can speed them up greatly and make diamonds artificially: small ones for industrial use (and a little for jewelry) are made in enormous quantities, by far the largest producer being China.[7] The original production method was to mimic the conditions of natural diamond formation: high temperature and pressure (typically upward of 2,000 K and 10 Gigapascals [that's 100,000 times normal atmospheric pressure]). Basically a "diamond reactor" consists of a tiny chamber containing pure graphite, a tiny seed diamond, and steel (the steel becomes molten under electric heating). These are compressed by two types of "anvils," made respectively of hard steel and tungsten carbide.

The first artificial diamond made in this way came out of the press in 1953, the result of a collaboration between the United States, Sweden, and the Soviet Union. Since then, other methods have been developed. It seems that the most promising is a type of plasma deposition that also starts with a tiny seed diamond (this is a prerequisite for all types of diamond manufacture). Intriguingly, the

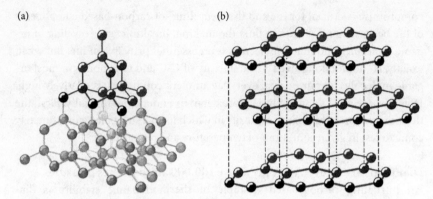

Figure 1.9 Diamond and graphite. **(a)** Diamond forms a crystal lattice with tetrahedral hybrid orbital bonds as in methane. This 3D triangular arrangement gives diamond its enormous hardness (10 on the Mohs scale); **(b)** graphite forms hexagonal sheets, the most common form of which is A-B-A, where the A layers are in register with each other, and the B layer is displaced by half a hexagon (as shown with the dotted lines). The weak associations *between* sheets (allowing sliding), but strong bonds *within* sheets, makes graphite a soft substance that is often used as a dry lubricant. A single layer is known as "graphene," and this material has remarkable properties that are extremely useful in a large range of technologies.

highly pure graphite in the reaction chamber is subjected to very *low* pressure, methane and hydrogen are added, and a plasma is created. The methane and hydrogen react on the surface of the seed crystal, forming diamond. So, if one uses chemical, as opposed to purely physical methods (the pressure reactors), one can deposit diamond using less energy. In 2017, this method was used to produce a truly enormous stone: the largest sheet of diamond ever made, at over 9 centimeters in diameter, a millimeter or so thick, and weighing in at 155 carats (an average engagement ring diamond is only 1 carat).[8]

The plasma deposition technique can also be used to introduce other elements into the growing diamond. So-called *doped diamond* is a semiconductor. Undoped diamond is a very good electric insulator but a very good thermal conductor: it doesn't have free electrons for conducting electricity, and its crystal structure transmits thermal vibrations very efficiently because it's so "stiff." Both doped and undoped diamonds are very promising for electronics and computing applications, but for different reasons. For completeness, the final method for artificial diamond formation is *explosion technology*, where graphite is mixed with an explosive charge in a very robust container and literally blown up. The reaction is very quick (and entirely physical), but it produces only nanometer-scale diamonds, for industrial abrasive dusts.

Possibly even more intriguing for industrial applications is *graphene*—a single layer of graphite. Graphene's properties include high flexibility but great mechanical strength, high surface area (hence well-suited for reaction chemistry and as a "carrier"), high electron mobility (relatively good conductor), high optical transparency, and high thermal conductivity. Arguably the most fascinating areas of graphene application are in biomedicine, for example as a carrier substrate for antibacterial agents or biosensors, in bioimaging, 3D bioprinting, tissue engineering, gene therapy (as a DNA carrier), drug delivery (as a drug carrier), photothermal therapy in cancer treatment (ablating a tumor by doping it with graphene and then exposing it to near-infrared light), and as interfaces between electronic devices and living tissue (e.g., in neurobiology and regenerative medicine).

Hopping over activation energy barriers with help and creatively channeling energy fluxes

Having discussed the *thermodynamic* stability of molecules, we now move on to the second determinant of reactivity: *kinetics*. This is the study of factors depending on the energy—that is, speed of movement—of particles: kinetic stability tells us how fast a reaction will occur. Diamond under ambient conditions is kinetically very stable. If we take oxidation of carbon compounds with ordinary oxygen in ambient air, organic substances—particularly those in living organisms—mostly react very slowly (at least on the life span scale of an organism), but much faster than diamond; the faster non-catalyzed reactions of carbon compounds, and those that are more detrimental to organisms, occur with radicals (molecules containing a free, unpaired electron), particularly oxygen radicals. Fast reactions can also happen as a result of catalysis, and this is one crucial distinguishing facet of the chemistry of life: reactions that would take years to reach completion via standard temperature/standard pressure chemistry can be accelerated massively by biological catalysts: *enzymes*. At the extreme end of the scale, we have enzymes that turn over at 200,000 reactions per second, somewhere in the region of 200 million times faster than the uncatalyzed reaction. Essentially, they are able to lower the activation barrier to a reaction so that molecules of moderate energy level (rather than very hot and under pressure, as in a chemical plant) can hop over it and participate. The aggregate energy change (i.e., the average including the molecules and their surroundings) is always in the inexorable direction of hot + cold → warm (the Second Law of Thermodynamics), but numerous chemical reactions can create relatively high-energy products at the expense of their surroundings becoming lower in energy (stealing from Peter to pay Paul).

It would be fair to say that the kinetics (momentary energy in relationship to activation energy barrier) and thermodynamics (final energy level) of carbon chemistry make it possible for life to create extraordinary ordering of matter. That is, life creates complicated, self-renewing systems that are very distinct from their inanimate surroundings. This suggests to us the defining characteristic of life: to channel energy fluxes to make complex ordered systems and structures by "subtly steering" the inexorable course of the Second Law of Thermodynamics in the surrounding universe. These processes require a flux of energy from high energy to low energy, or "hot" to "cold," as it were (Second Law). On Earth, we have four main energy sources: the Sun, the molten magma under the Earth's crust, heat from radioactive decay in rock, and chemical energy from reduced compounds in the Earth's geosphere. Life is not about reaching equilibrium— that is, the mere flux of energy from a hot object to a cold object, resulting in two lukewarm objects where nothing further happens. Rather, life is about using this energy flux to create new disequilibria (unbalanced thermodynamic situations) and maintain those disequilibria. Moreover, life then propagates these disequilibria to new offspring through an extraordinary process that makes a pristine new organism from an aging one (reproduction).

The order found in DNA and proteins (both long-chain polymers) is a good example. But both protein and DNA can be broken down by enzymes (other biological entities that essentially manipulate energy levels and probability of reaction). Most chain-like molecules in biology require the input of energy to make them (i.e., they take energy in doing so), and that is described as an *endothermic* process. Very good examples, in preparation for what follows later in this book, are carbohydrates, fats, and oils. We will look into their biosynthesis, biochemistry, natural turnover, and latest insights into the connections between their metabolism and aging. We will also examine industry's "artificial metabolism" of carbon compounds.

The element that life "chose"

Earlier in this chapter I briefly covered the formation of organic molecules "in space," but we should probably be more scientific in that respect and say "under abiotic conditions" (involving no living organism). As we now know, organic molecules form in the complete absence of life in all manner of situations in the universe: from interstellar dust clouds, through meteorites and comets, to planets and planetoids—our own Earth being a conspicuous example: yes, there is plenty of organic chemistry going on on Earth, completely oblivious to all the life forms that are doing the same thing! And it started long before primitive life. This is, as it transpires, very important for theories of the origin of life on Earth and one in particular: origin at *submarine alkaline vents* (SAV theory), a relatively recent postulation. Chemical theories of the origin of life on Earth and experiments to verify them started in the first half of the 20th century, and we will first explore the early hypotheses.

Primordial chemical soup on Earth: A nice idea at the time

The first informed thinking on the chemical origin of life led to key experiments for which we must go back in time to the early 1950s. Two American chemists, Stanley Miller and Harold Urey, his research supervisor, started building a kind of time machine. The experiment that it would perform would mimic the chemical and physical conditions believed at the time to have prevailed on early Earth (Box 1.3). The researchers' aim was to test a conjecture made by Alexander Oparin and J. B. S. Haldane in the early 1920s. Oparin and Haldane reckoned that life had begun in a watery broth that produced organic molecules typical of living organisms in general (in particular, amino acids and components of DNA and RNA). Indeed, this conjecture soon became fondly known as the "primordial soup hypothesis" (as well as the Oparin-Haldane theory). Miller and Urey's apparatus consisted of a reaction chamber containing water vapor, methane (CH_4), ammonia (NH_3), and hydrogen (H_2). CH_4 and NH_3 could be let into the system to maintain a relatively constant concentration of these gases in the reaction chamber, and the mixture was driven through a cycle via a receptacle (containing just water initially) heated by a laboratory gas burner.

Hydrogen (both elemental and that in CH_4 and NH_3) represented the thermodynamic energy needed to create a variety of different carbon-based molecules using CO_2 (i.e., the energy needed to create end products with higher energy than the starting ones). Hydrogen is by no means the strongest reducing element (or "reducing agent," in chemists' language): that accolade goes to the so-called alkali metals in Group 1 of the Periodic Table—essentially the first ("lightest") elements that can be regarded as true metals: lithium, sodium, potassium, rubidium, cesium, and francium, the last being the strongest reducing agent (reactivity increases as one goes down Group 1 to the heavier elements).

Now, hydrogen is also a Group 1 element, so is it also an alkali metal? Yes and no: hydrogen's typical properties at most temperatures and pressures are not particularly metallic, but liquid hydrogen, at very high pressure and low temperature, does behave like a metal: for example, it conducts electricity. The point about reactivity (reducing potential) is nicely made by two further schoolroom experiments: take a piece of potassium out of its jar of protective oil, cut it with a knife (alkali metals are very soft), and watch as the exposed face tarnishes very quickly. A large piece, if heated in air, will burn with a lilac flame. Put into a dish of water, a piece of potassium will race around producing a lilac flame as it reacts with the water: being a stronger reducing agent than hydrogen, it essentially

> **Box 1.3 More on the Miller-Urey experiment: simulating Earth's early organic chemistry**
>
> On the other "side" of the circuit from the reaction chamber (containing CH_4, NH_3, and H_2) was a "condensing arm" cooled by circulating cold water. Beneath that was a section of piping with an outlet valve, where the reaction products could be drawn off and analyzed at intervals. Now for the crucial part that caused chemical changes in the reaction chamber: electrical discharges—simulating, on a very small scale, the electrical discharges (essentially lightning) in the foggy and hot atmosphere of early Earth. The two researchers had done their homework on the gaseous composition of primordial Earth's atmosphere that was hypothesized at the time, and their set-up roughly maintained that ratio of gases. The first result was the formation of amino acids (the building blocks of proteins), and it caused quite a stir when revealed in 1953, in the journal *Science* (of the American Association for the Advancement of Science [AAAS]).[9] Miller published that one under his sole name. In 1959, Urey and Miller jointly published a paper, again in *Science*, that caused no less excitement.[10] Urey, Miller, and other researchers subsequently identified nucleotide bases (the building blocks of the DNA and RNA code) in the reaction mixture. The fundamental chemical principle under investigation, as explicitly stated in the 1959 paper, was whether such anorganic conditions (i.e., lacking organic molecules and living organisms) could produce organic substances with a *reduced chemical state*—that is, as opposed to oxidized. Oparin strongly suspected that they could, given the absence of oxygen in the atmosphere at that time on Earth and the chemical legacy, so to speak, of hydrogen in the universe at large. As we know, hydrogen is the most abundant element in the universe (being the first to be formed by cosmological physical processes), and it has a "reducing potential": it can, under the right conditions, react with substances in a more "oxidized" state and create ones that are more "reduced." The classic schoolroom example is the explosion of an oxygen-hydrogen mixture when lit (with a flame or a spark), producing water vapor: water is a compound of reduced oxygen.

usurps hydrogen and reacts with oxygen to produce first potassium peroxide, which then reacts with water to produce potassium hydroxide—an alkaline substance (hence the term "alkali metal"). The reaction is so *exothermic* (heat-producing) that the displaced hydrogen burns—hence the flame. So hydrogen isn't a very strong reducing agent, but it's certainly much more abundant in the universe than any of the alkali metals, and so it is expected to be the predominant partner in reduced forms of carbon and other compounds.

That is what Oparin had conjectured, too. However, on early Earth, as today, there is very little hydrogen in the atmosphere, most of it having escaped Earth's gravitation because it is so "light" (low in density). Bound hydrogen, on the other hand, is very abundant on Earth, as on other planets (Neptune, for example, has a mantle of water ice [H_2O], solid ammonia [NH_3], and methane ices [CH_4]). The starting gases in the Miller-Urey experiment were, therefore, logically chosen, and, by adjusting the relative concentrations of these three gases and adding a little hydrogen, the experimenters created conditions that would favor new equilibria of chemical compounds. This is another example of re-establishing the so-called *thermodynamic equilibrium* (long-term state at which the system is at rest and doesn't "want to go anywhere"). The respective reaction doesn't necessarily happen spontaneously because of the now familiar activation energy barrier (explained earlier) that can only be overcome by delivering activation energy: that is, the constituent substances need a "kick" to get them over that barrier to reactivity, after which they will make the products that are thermodynamically favored (long-term stability) under those circumstances (temperature, pressure, substrate/product concentrations).

The "kick" in the Miller-Urey experiment was an electric discharge—mini-lightning. The reduced equilibrium products that Miller and Urey expected to emerge were higher-order organics, where part of the reducing equivalents from the CH_4, H_2O, NH_3, and H_2 were manifest in C-C chains and C-H bonds. Naturally occurring amino acids, other organic acids, purines, and pyrimidines (the ring-shaped bases of DNA), as well as sugars, were clearly identified—and not only by Miller and Urey because many other excited researchers reproduced their experiments. An important principle is that life needs a reducing environment, and carbohydrates had long been known to have reducing potential: hence the schoolroom experiment "Benedict's test" in which copper II ions are reduced to copper I by simple sugars, thus producing a color change from blue, through green and yellow, to red, depending on strength and concentration of the reducing sugar. As we will see next, the Miller-Urey experiment was not an appropriate physicochemical scenario for the origin of life; however, it certainly proved that many organic chemicals typical of life could be produced in complete absence of living organisms.

A better idea, based on different chemistry and energy flux as the driver

Some things Miller and Urey could not have known, but others they could have

The principle of reducing equivalents holds up until today, but two major obstacles to Miller and Urey's experiments mimicking the early beginnings of life have since cast a dark shadow over the whole concept. The first one is chemical: the starting substrates. Miller and Urey could not benefit from the insights into Earth's primordial atmosphere that we have today. Already in the 1970s doubts were being voiced about the—albeit perfectly researched—assumptions that the two chemists had made. To cut a long story short, most recent research suggests that early Earth had a Venus-like atmosphere, containing predominantly CO_2 and N_2.[11] That atmosphere persisted well into the geological epoch when life is thought already to have been present. Reducing gases (NH_3 or CH_4) may briefly have made an appearance from vaporization events caused by meteorite impact, but only for a few million years at a time. Earth's unique, and rather constant, distance from the Sun ultimately led to the accumulation of liquid water, thought to have arrived via innumerable impacts from ice-containing comets. Gradually the oceans formed. It is in this aqueous, relatively temperate context that life is suggested to have evolved, according to latest theories.

The second problem with the Miller-Urey theory is physical: the state of the mixture. It is no coincidence that I've used the word "equilibrium" quite a lot up until now. The Miller-Urey experiment (and its derivatives) uses a given amount of reducing equivalent—or chemical energy, if you like—to produce an equilibrium mixture that gets no further than relatively simple organic molecules. Why? Because it is at equilibrium, and that means it's going nowhere. If the primordial soup on early Earth was also an equilibrium mixture—and even if it had formed in the presence of the "right" gases—then it was also going nowhere: it needed something else. Life is the antithesis of equilibrium and is now widely described as a far-from-equilibrium open physical system. And so to the theory that is currently attracting so much attention: that "something else" that life needs is a chemical gradient, a type of concentration gradient with important chemical energy implications, so to speak. Early life could only have started with some constant source of energy—a flux of energy that, over geological time scales, did not reach equilibrium. Furthermore, life is almost certain to have begun in a watery (aqueous) environment, and before the advent of photosynthesis it must have relied only on reducing equivalents from early Earth for its chemistry. It probably used minerals from the Earth's crust, given the latest insights into Earth's early atmosphere.

Origin in oceanic abysses: A new theory based on alkaline water and electrochemistry

This is where the theory of alkaline thermal vents on the ocean floor[12] comes into its own: it posits that the chemical gradient between relatively alkaline water rising out of the Earth's crust and close-to-neutral surrounding seawater created a kind of primitive electrical battery or capacitor. These particular vents have temperatures of 80–200°C—much cooler than the black smokers (around which life swarms today—interestingly also dependent only on chemical energy from the Earth because of the enormous depths and hence absence of light for photosynthesis). But how can alkaline water seeping out of the sea floor constitute a battery or capacitor, you ask?

Now we have to look at the structure of vents because these are not just holes in the ground; rather they are mineral deposits (caused by the dissolved contents of the warm vent water meeting the seawater and precipitating): the result is irregular formations that grow a modest few meters high (small, compared with the biggest black smokers). Their fine structure, however, is their crucial feature: tiny, micrometer-scale tubules and pores from which the solutes emerge. Under these conditions, which are also rich in CO_2, chemistry that is very similar to a branch of industrial synthesis chemistry (Fischer-Tropsch, which is used to make synthetic fuels) is strongly suggested to occur. The reactions have been simulated in laboratory experiments as early as the mid-1990s.[13]

Such reactions produce lipids (biologically relevant types of oil and fat) ranging from 2-carbon up to 35-carbon atom chains: right about in the range of biogenic lipids of cell membranes from bacteria to human cells. Why are lipids important? Because they form layers—sheets in which other molecules can be embedded (in a cell, lipid membranes house protein channels and machinery of respiration, for example): surfaces on which biogenic reactions can take place. We don't yet know what these lipid-based structures were, but it is possible that they underwent a type of chemical evolution, slightly changing composition/gaining extra components that made them more robust, for example. They likely began as small patches on very thin parts of the walls of the micropore tubes in the mineral deposit—at the very extremities where the water on the "outside" was plain seawater (Figure 1.10).

Next, some of them might have bridged the outer faces of pores—thus separating the alkaline fluid from the seawater beyond without a mineral wall. Finally, one can conceive them "budding" off and rounding up into the first proto-cells, ancestors of all cells (sometimes referred to as the *last universal common ancestor* [LUCA]). Proto-cells, which can also be made in the laboratory completely abiogenically, are likely to have been tiny micrometer-scale bubbles (vesicles) with a double-membrane (lipid bilayer). Experiments published in 2019[14] demonstrate that, under the conditions prevailing in and around alkaline vents, a few different types of lipid are all you need to get spontaneous vesicle formation—vesicles with bilayers very similar to modern cells (Figure 1.10).

Chemistry on ancient membranes: H^+ gradients may have worked with "green rust" as the crucial catalyst

But let us dig a bit deeper into the chemistry that is proposed to have happened on those first membranes before they became proto-cells. In its basics, this likely

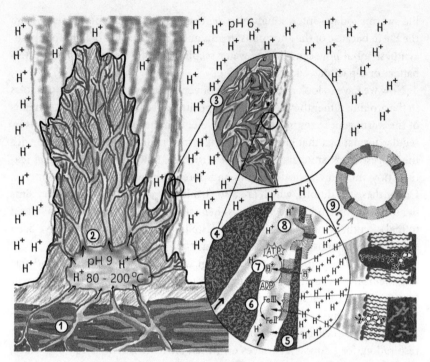

Figure 1.10 A potential origin of life in alkaline thermal vents on the sea bed, 2–5 km deep, inspired by the theory published by W. Martin (2011)[15] with minor modification and simplification. (1) "serpentinization" in the water-rich upper crust produces alkaline exudate at pH 9, laden with dissolved minerals; (2) on contact with seawater, minerals precipitate and accrete a growing structure (the vent) that is riddled with channels conducting the alkaline water; (3) at the outer surface, microscopic channels and pores exist; (4) micrometer-scale mineral walls separate the inner water from the seawater, punctuated with tiny pores; (5) here the pH gradient (inside pH 9, outside pH 6) and the constant movement of fluid cause net H^+ flux into the microchannels; (6) it is proposed that lipids (water-loving heads, with straggly C-C water-hating chains as tails) formed patches with membrane-like properties, in which a redox co-factor became embedded, thus allowing simple redox reactions with an iron compound; (7) membranes evolved to become bi-layers and incorporated primitive proteins containing a redox co-factor, enabling the H^+ flux to be harnessed to generate ATP, the universal biological energy-carrier; (8) membranes eventually bridged pores separating inside from outside; and may then (9) have budded off to form the first proto-cells (whose membranes by then had also acquired other proteins).

consisted of cyclical reduction/oxidation reactions (*redox chemistry*) driven by the pH gradient—difference in alkalinity—across the thin wall of the outermost micropore tubules in the vent. The basis of the pH scale is the concentration of hydrogen ions (H^+) in an aqueous solution, and vent water has a lower concentration of H^+ than seawater: seawater on early Earth was recently proposed to have had a pH of between 6 and 7.5[16] (7 being neutral), and deep ocean water is, as today, likely to have had the lowest pH. If we take 6 as the pH of parts of the ancient ocean floor and 9 as the pH of vent water (much more alkaline), that translates into an enormous H^+ gradient. The reason is that pH is a logarithmic scale, so each unit represents a 10-fold difference in concentration of H^+ ions (or protons, as chemists often call them, because a hydrogen atom that has lost its only electron is, essentially, a proton). That means that the three pH units separating the seawater from the vent water translate into a 1,000-fold H^+ concentration difference across the boundary between vent water and ocean.

The more highly concentrated protons in seawater therefore have a rather strong desire to equilibrate with the liquid inside the vent, and that is what is known as the *chemiosmotic principle*—a kind of force that is present at the heart of energy production in all cells, ancient and modern. We need to return at this point to the principle of oxidation/reduction driven by this force (referring to the redox chemistry mentioned earlier): first, the H^+ gradient can't do any "work" (anything useful) by simply existing; rather it must "move." That is proposed to happen via H^+ ions traveling through the separator between vent water and seawater. The separator was initially likely to have been merely the thin and porous wall of inorganic minerals naturally deposited as the vent grows, but it might also have been coated by lipids from the beginning: we may never know. Iron plays a role in redox reactions in living organisms today, and it is hypothesized that it was the crucial substance that allowed proto-life to harness the H^+ gradient between vent water and seawater. *Green rust*, a compound of two oxidation states of iron together with hydroxide (FeII/FeIII oxyhydroxide), is proposed to have been the "organizing" principle that enabled the quantum jump from primitive chemistry to complex energy pathways and material structures.[17] But without the connectivity of carbon expressed as naturally occurring lipids (i.e., the long chains of C-C bonds with hydrogens sticking out), fledgling life would not have gotten very far: membranes, formed by lipids, are the quintessence of all cells. Next I discuss how the first membrane-bound cells might have arisen.

How did life graduate from rusty membranes to the first membrane-bound cells?

I must now cut a very long story short and race through the rest of the hypothesis: lipids known to be formed abiogenically in alkaline vents coated the mineral "membrane" with their own bi-layer membrane; primitive proteins then condensed from the abiogenically present amino acids in the vent water and inserted themselves into the lipid bi-layer spontaneously, a process that happens in present-day living cells. Some of these proteins associated—again purely by happenstance and chemical affinity, as it were—with green rust centers; the proteins hence became tiny machines capable of transducing the energy from the redox reactions driven by H^+ influx (Figure 1.10). The energy was transduced by the proteins into bond formation between two phosphate ions (PO_3^-, phosphate also being a natural constituent of vent water) to create the higher-energy form of phosphate, *pyrophosphate* (PPi). This was the first step toward the universal energy carrier of cells, adenosine triphosphate (ATP).

A primitive type of metabolism probably evolved based on the conversion of inorganic reducing equivalents to organic molecules: we will look deeper into the organic synthesis on primitive vent-associated membranes later. Some of these organic molecules may well have acted as temporary storage substances and facilitated membrane growth via conversion to new lipids. Perhaps spontaneous division of these proto-cells occurred when they had reached a size at which the shearing forces of the surrounding fluid imposed a physical limit. The mere physical chemistry of lipid bi-layers is enough to remake vesicles after disintegration: it can be done in the lab very easily. Note, although I have mentioned DNA and RNA, neither of these two genetic materials plays a role in the chemical evolution of LUCA (the ancient ancestor of all cells) up to this point: that is to say, it is becoming increasingly likely that primitive life did not "need" a genetic material, and that, rather than replication of any genetic code, it was energy metabolism that represented the essence of life. There are even ancient chemical traces of primordial energy metabolism embedded in the metabolism of all extant organisms: the universal ability of cells to use inorganic phosphate (i.e., containing no carbon compound: PPi) as an energy source, and the universal presence of the two organic energy-carrier derivatives of phosphate: ATP and guanosine triphosphate (GTP). Further discussion of that concept is beyond the scope of this book; suffice it to add that most recent research (notably in the area of epigenetics) is revealing a very strong arrow of causality from metabolism to DNA, in addition to the reverse: that is, there is increasing evidence that metabolism modulates the expression of the genetic code and even its evolution.

As life evolved, it became more and more carbon-dependent

A genetic material certainly evolved at some point, probably close to the alkaline vents. Adenosine is a derivative of the DNA and RNA base adenine (a purine), which is believed to form spontaneously in conjunction with mineral catalysts and redox centers (e.g., metal sulfides) and CO in the alkaline vent milieu. This chemical step probably draws on single-carbon compounds such as formate (HCOO$^-$: easily formed from CO and methane present in the vent exudate). Here green rust—powered by the H$^+$ gradient between vent water and seawater— is hypothesized to have catalyzed the formation of formate (one-carbon acid), acetate (two-carbon acid), and pyruvate (three-carbon acid) via a ping-pong oxido-reduction (involving H$^+$, OH$^-$, and their uncharged radicals).[18] There was plenty of carbon in the ancient (*Hadean*) sea, in the form of dissolved CO$_2$ (i.e., hydrogencarbonate), which is what made it acidic. The small carbon compounds that formed are the building blocks of amino acids, and chains of amino acids (peptides) are known to form spontaneously under vent-like conditions; hence it was only a small step until, by chance, longer peptide chains folded into proteins with particular functionality. The organization of different chemical processes into different microregions on the ancient membranes was crucial[18] for this chemical evolution.

Formate in general acts as a very suitable building block for larger organic molecules. In addition to amino acids, purines (two-ringed) and pyrimidines (three-ringed)—as found in DNA and RNA—likely formed as well; however, their first function was probably not in a genetic material but in the biochemistry of energy-transduction (redox reactions). Sugars have also been shown to form under conditions mimicking alkaline vents. And so—likely after the emergence of the first proto-cells (Figure 1.10)—the scene was set for the emergence of RNA or DNA: these polynucleic acids are composites of a sugar-phosphate component bound to an organic base (purine or pyrimidine) and polymerized into long chains. The energy for polymerization comes from breaking a high-energy bond in the chemically evolved organic form of phosphate (ATP, where the "A" contains the purine base adenine).

Though Miller and Urey were probably wrong about the origins of life, their curiosity certainly has its present-day counterparts. Experimental setups abound—but this time to recapitulate chemistry in alkaline vents. In 2014, researchers from University College London and the University of Leicester in the United Kingdom reported the building of an "origin-of-life-reactor."[19] The apparatus was based on the concept of a microporous mineral structure exuding alkaline liquid containing minerals and rich in CO$_2$ at a temperature of 75°C: it initially produced small amounts of formate and formaldehyde (single-carbon organics), followed by sugars, including ribose and deoxyribose (the sugar moieties found, respectively, in RNA and DNA). They were produced by catalytic reactions promoted by iron-nickel-sulfur (Fe(Ni)S) clusters that formed on the walls of the micrometer-scale tubules from the mineral-rich water that was flowing through them. Life never broke from using metals and inorganic chemicals for specific functions, but they clearly were not to be the mainstay of energy production, storage, or material economies: instead, sunlight and carbon compounds took center stage.

Photosynthesis: The beginning of the shift away from mineral energy

Unfortunately there is no fossil evidence of the proto-life that we contemplated earlier; the first generally recognized fossils (called *stromatolites*) are of bacterial colonies. Most recently, even older fossils of archaea (cousins of bacteria) have been discovered at 3.4 billion years of age.[20] At around this time, it is believed that filamentous microorganisms were already photosynthesizing. Between 3.6 and 3.2 billion years ago (bya), cyanobacteria (sometimes called *blue-green algae*, though they are not algae at all) started to emerge as the principal photosynthesizers.[21] Yet later, as one of the two historic mergers between kingdoms of life, *eukaryotic* algae (cells with a nucleus and chloroplasts) were born. From here developed all photosynthesizing plants—organisms that form the base of 99% of all food-chains on Earth. Photosynthesis was the beginning of the evolution of most of life on Earth away from energy dependency on mineral- and Earth-bound energy sources (ion concentration gradients and inorganic reducing chemicals) to the energy of the Sun. It was the point of no return for carbon compounds as the almost universal currency for energy storage, distribution, and utilization: a massive global recycling operation.

How photosynthesizing plants convert the Sun's energy into chemically reduced compounds

Green plants, perceived by many as the apotheosis of environmental sustainability, convert CO_2 entering through their leaves and water through their roots into carbohydrates using the Sun's energy: photosynthesis. The first carbohydrate produced in photosynthesis by plants is *glucose*, a six-carbon simple sugar (monosaccharide) that is, for all its simplicity, quite fascinating (Figure 1.11).

This happens in the Calvin cycle, named after Melvin Calvin (1911–1997). From here, plants make various forms of starch (a complex carbohydrate made of many subunits of individual sugars) for energy storage, cellulose for structural purposes, and all the other constituents of their cells (proteins, lipids, DNA, RNA, etc.). Most of the energy that plants channel from sunlight into their central anabolism (the reactions that build structures and lay down stores) is used to break bonds in CO_2 and H_2O. Simply put, CO_2 is "reduced" to CO, and H_2O is split into oxygen and hydrogen. The oxygen is given off as a "waste" product,

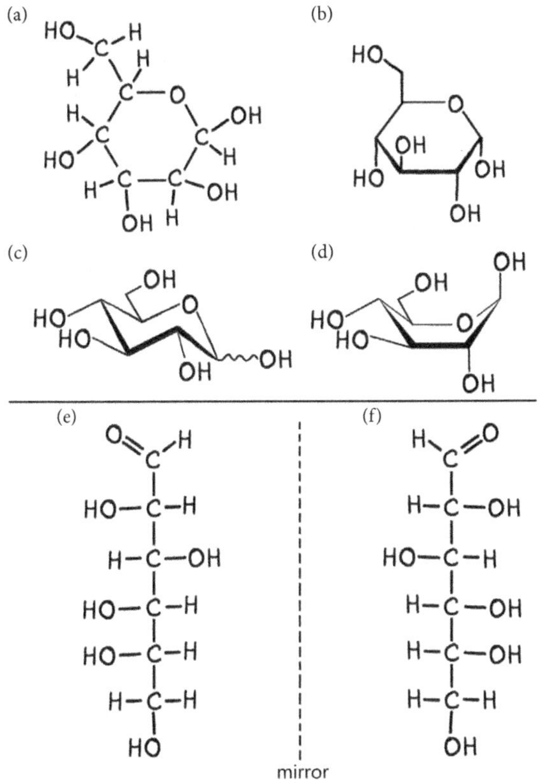

Figure 1.11 Chemical structure of glucose. In aqueous solution, glucose exists almost entirely in its cyclic form. (**a**) A Fischer projection showing all the atoms; (**b**) a projection showing only the so-called functional groups (i.e., the OH [hydroxyl] groups and not the carbon or hydrogen atoms). In such a projection, each kink is the point at which a carbon atom exists, and if there is only a single line on either side, two hydrogen atoms are attached to this carbon atom. Cyclic glucose exists in two geometrical forms that are constantly interconverting (i.e., in dynamic equilibrium with each other); (**c**) the "chair" form, or alpha-form; (**d**) the "boat" form, or beta-form, which is the more stable of the two, and hence the predominant species in ordinary solution. The alpha and beta forms are called "anomers." All forms of glucose can also be either right- or left-handed, depending on the exact geometry of the OH groups. This is most easily shown via the straight-chain enantiomers (also "stereoisomers"), which are exact mirror images of each other and are not in equilibrium (i.e., they cannot interconvert under normal circumstances); (**e**) L-glucose, which rotates polarized light in a left-handed (counter-clockwise) direction; and (**f**) D-glucose, which rotates polarized light in a right-handed (clockwise) direction, hence "D" for "dexter" (Latin "right"). D-glucose is, hence, also known as "dextrose." All forms of glucose have exactly the same chemical properties, but they differ in reactivity in biological systems because enzymes can only bind the D-form. (Note that, with amino acids, the building blocks of proteins, the opposite is true: the L-form is the more biologically relevant, being the exclusive constituent of proteins).

and the hydrogen reacts with CO as the first step in carbohydrate formation. I used the word "simply" because this process (1) doesn't happen in one step and (2) doesn't entirely depend on light. Briefly, photosynthesis is divided into the "dark" and the "light" reactions. The light reactions—a bit like a photoelectric cell in a solar panel—excite electrons and set them free to do useful work. In a plant, that work is done along a "molecular wire" of proteins that successively take a bit of energy from a transiting electron and catalyze a reaction that requires energy. Further down the photosynthesis pathway, this energy goes into chemical bonds (carbon atoms from CO_2 are joined to each other to form multicarbon sugars). For that to happen, the CO_2 is chemically reduced by a compound that the plant makes, called *nicotinamide adenine dinucleotide phosphate* (abbreviated NADPH). That H at the end does, indeed, stand for "hydrogen" (a ubiquitous reducing agent, as we now know). Free hydrogen is useless to a plant, so it makes it into NADPH. It does this by using the electrons that are released when the Sun's energy excites an organic pigment called chlorophyll (and so-called *antenna pigments*) and splits water into hydrogen and oxygen (basically, H_2O plus sunlight plus chlorophyll, plus protein molecular wire, plus $NADP^+$ produce oxygen and NDAPH) (Box 1.4).

It's a bit like the electrolysis of water that most people did at school, except that the energy for the water-splitting comes initially from sunlight instead of a battery, and the hydrogen is not released but captured to make a compound that the plant can use to synthesize materials. This light-dependent reaction of photosynthesis is also called the *Hill reaction*, after its discoverer, Robin Hill (1899–1991), and it must happen before the "dark reaction." In the dark reaction, the NADPH made via the light-dependent reactions of photosynthesis is used to enzymatically reduce CO_2, making C-C and C-H bonds in sugar. Though this reaction also occurs in the day, it mostly happens at night, and so we return to the beginning of the story because the dark reaction is none other than the Calvin cycle. The dark reaction is the true "synthesis" part of photosynthesis, and it creates the carbon-based intermediates that plants need for short-term and very long-term energy storage: think of a plant seed—packed with fats, oils, and protein—that can wait many years to germinate.

Photosynthesis is intimately integrated with material and energy metabolism

What was the point of my brief overview of photosynthesis? To make apparent that plants take CO_2 from the air and re-energize the carbon into a higher energy form by turning it, initially, into carbohydrates (basically, they reduce it with hydrogen). Most plants are true *autotrophs*, meaning that they generate all

Box 1.4 How green plants make the substrates of everything they need via photosynthesis

Chloroplasts, the photosynthesis factories (known as *organelles*) of green plant cells are derived from an "endosymbiotic event" in evolution, where a cyanobacterium (photosynthesizing bacterium) was engulfed by a cell that had already entered into symbiosis with an alpha-proteobacterium to become a eukaryotic cell (i.e., one with a nucleus and mitochondria derived from the ancient alpha-proteobacterium). This produced the first eukaryotic photosynthesizing organisms—single-celled algae—believed to have occurred around 1.9 billion years ago or even earlier.[22] Later, multicellular plants evolved, starting in watery environments and then colonizing land between 500 and 700 million years ago (mya).[23] All photosynthesizing plants have the same essential machinery that enables them to produce their own nutrition from nothing more than CO_2, water, and minerals. Many steps in photosynthesis are the exact opposites of biological energy production by oxidative burning of carbohydrates and fats (known as *cell respiration*) (Box 1.5). One process is, however, very similar: both photosynthesis and cell respiration in mitochondria generate H^+ gradients via electrons passing through a chain of protein complexes embedded in a membrane. To enter these so-called *electron transport chains* (ETCs), the electrons need an unusual level of energy. In photosynthesis, this energy is provided by sunlight in a narrow wavelength band. The resulting H^+ gradient is used to generate ATP, the universal cellular energy-carrier (see figure). ATP is essentially an organic compound bound to three inorganic phosphate groups (PO_3^-), and it is made by the addition of one phosphate group to adenosine *di*phosphate (ADP).

ATP consists of three parts: an organic base, adenine, which is a component of DNA and RNA; a five-carbon sugar, ribose; and three phosphate groups joined together. The high-energy bond is between the far-left phosphate group and the one next to it.

In the plant, this ATP is part of the chemical energy that the plant needs to fix CO_2 by reducing it and turning it into glucose, the first sugar of plant metabolism.

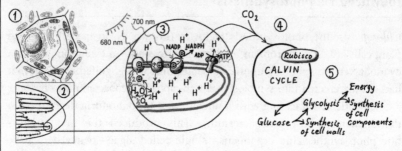

(1) Photosynthesizing plant cells contain chloroplasts situated near the cell wall, where they can capture most light. (2) Within each chloroplast is a complicated substructure of membrane-bound compartments called *thylakoids*, which house the protein complexes that perform the "light" reactions of photosynthesis (e.g., the Hill reaction—splitting of water), electron transport, and production of energy and reducing equivalents. (3) Two protein complexes absorb light via chlorophyll and antenna pigments at slightly different wavelengths (because these are largely in the red/yellow region of the visible spectrum, photosynthesizing plants look green). The first of these complexes is historically named *photosystem II* (PSII), and its core proteins had already evolved in cyanobacteria around 2.4 bya. PSII basically splits water into electrons, H^+ ions, and oxygen (this is, indeed, the oxygen that plants give off); the electrons pass through membrane-bound complexes (mainly proteins) and are given another kick of energy by photosystem I, thus enabling them, together with H^+ ions, to reduce NADP to NADPH (which is a strong biological reducing agent). Part of the electrons' energy is used to pump H^+ ions across the membrane, thereby creating a chemiosmotic gradient. H^+ ions flow back (trying to equilibrate the gradient) through a protein machine called the F_1F_0-*ATP synthase*, which produces ATP. (4) NADPH, ATP, and CO_2 from the air are the substrates for the dark reactions of photosynthesis, the Calvin cycle, where CO_2 is fixed (using energy from ATP and reducing power from NADPH) into glucose. The most important enzyme in this process is rubisco. (5) Glucose is then used for energy generation (via glycolysis and oxidative phosphorylation), production of cellulose (structural material), and starch for energy storage; glucose and metabolically related sugars are used for central metabolism, which generates the precursors of all other cellular constituents (amino acids, lipids, DNA, RNA, etc.).

of their food and all of the structural components of their organisms themselves, from scratch. Carbohydrates are used variously for construction purposes and "burning" by the plant's cells in a process known as *respiration*, to liberate the energy for useful work (growing and reproducing). The burning of carbohydrate-derived sugars for energy is obviously not a flame or explosion that we typically associate with combustion. All the same, it does consume oxygen, and the chemical end-point of the reaction is the same as burning: the carbon in the carbohydrates reacts with oxygen to produce CO_2; the hydrogen reacts with oxygen to produce water. This really is highly controlled combustion because it is enabled and finely regulated by enzymes and so-called redox protein complexes. The purely chemical part is done by individual enzymes, either with or without oxygen; releases a relatively small amount of energy; and is known as *glycolysis* (splitting of sugar). The part that has an electrical component, so to speak, takes place (similarly to photosynthesis) via another molecular wire made of protein complexes that successively takes energy from electrons that flow through it. These electrons come from high-energy compounds that are made in the mitochondrion from products of glycolysis. What happens in the mitochondrion is called *oxidative phosphorylation* or "oxphos" because it absolutely requires oxygen (Box 1.5). Both plant and animal cells do this "biological burning," and the product is the universal energy carrier, ATP.

How this quite elaborate process came into being is explained by evolution over billions of years. Equally important, it didn't happen in isolation, but was part of a global process of the evolution of Earth itself—not only the chemical composition of the atmosphere, but the nature of sediments on land and sea that became rock and organic deposits. It is no coincidence that plants use the Sun's energy to produce high-energy carbon compounds that both they and others consume, and, in return, plants give off oxygen that they and others use to "burn" these carbon compounds: it is a chemically stable system that cycles carbon molecules as energy carriers and materials.

Photosynthesis indirectly caused life on Earth to shift gear and explode in complexity

Before photosynthesis, Earth's atmosphere and oceans were 99% oxygen-free. The "great oxygenation event," initiated around 2.4 to 2.1 bya by oxygen emitted by photosynthesizing organisms (first the cyanobacteria), literally polluted the atmosphere with something that was toxic to most life at the time. Sometimes this is referred to as the "oxygen catastrophe" or "oxygen crisis" because anaerobic (living without oxygen) organisms (that was 99% of life on Earth at that time) had no way of dealing with oxygen: it literally poisoned (by oxidation)

Box 1.5 How plant and animal cells biologically burn carbohydrates: from glycolysis to oxphos

A small proportion of most cells' ATP is made in the cell's cytoplasm (gel-like fluid surrounding the nucleus and other structures) via the enzymes of glycolysis. When oxygen is very scarce or absent, most cells can still produce energy via glycolysis—though much less than with oxygen—in *anaerobic glycolysis*. Here, lactic acid (instead of pyruvate) is ultimately produced (a type of anaerobic fermentation, as happens in yogurt production): this lactic acid gives us muscle cramps during extended periods of anaerobic exercise (e.g., long runs), when insufficient oxygen reaches our muscles. Under conditions of plentiful oxygen, by contrast, glycolysis performs a type of burning to produce a partly oxidized carbohydrate derivative called *pyruvate*. This then diffuses into the cell's mitochondria, where a further key part of enzymatic metabolism happens: the *citric acid cycle* (also known as the *tricarboxylic acid* [TCA] *cycle* or *Krebs cycle*), where CO_2 is formed as one of the "combustion" products, and chemical energy from pyruvate is partly passed on to another temporary carrier by transfer of hydrogen: "NAD^+," becomes chemically reduced to NADH (where "H" is, indeed, hydrogen). NADH then bumps into the protein complexes in the millions of "molecular wires" inside the mitochondria (see figure). These, similarly to photosynthesis, are an ETC: they conduct high-energy electrons derived from NADH. As a formula: $NADH \rightarrow NAD + H^+$ e^- (one electron) + pumping of H^+ ions. At each stage of an electron's passage along the ETC, the electron's energy is, simplistically put, used to pump H^+ ions (also referred to as protons)—which exist naturally in all water-based solutions—from the inside of the mitochondrial double membrane into the space between the two membranes. This generates a measurable charge (around 150 millivolts for most mitochondria), a so-called *electrochemical gradient*. This is exactly the same gradient as the one believed to have initiated life at the alkaline hydrothermal vents discussed earlier, and storage of energy in this way is very similar to that in an electrical capacitor: note, it is *not* comparable with a battery, because in a battery the electrical energy is stored in a pair of substances that are chemically out of redox equilibrium with each other (one being much more reduced than it would prefer and the other much more oxidized). They are only prevented from reacting to reach redox equilibrium by an impermeable membrane: as soon as the membrane is bypassed by connecting something across the positive and negative poles of the battery, the chemical reaction starts to happen, converting the respective substances in the positive and negative "sides" of the battery into the forms in which they are at redox equilibrium with each other. All batteries, including Li-ion ones, work like that. Now, what happens to the electrons when they reach the end of the molecular wire (ETC)? Here, $O_2 + H^+ + e^-$ (electron released at end of ETC) → H_2O. This is known as a *respiratory pathway*, where oxygen is the "terminal electron acceptor," and, in the evolution of the eukaryotic cell, it was the symbiotic alpha-proteobacterium that had the enzymatic machinery (basically the ETC) to do the job with oxygen. In so doing, it prevented poisoning of the host (the archaeon) by the rising atmospheric oxygen. You will note that this reaction is the opposite of the splitting of water into electrons, H^+ ions, and O_2 in the light reaction of photosynthesis. Now back to our biological capacitor, the mitochondrion (which is the evolutionary derivative of the original alpha-proteobacterium): it builds up a very high imbalance of H^+ ions across its membranes, and, indeed, it must keep pumping to keep up the imbalance because (1) there is always a small amount of leakage—as there is in all capacitors (and batteries); but (2)—and more importantly—the H^+ ions are constantly flowing back across the membrane, namely through a massive enzyme complex (known as the F_1F_0-ATP synthase) that uses their flux to generate ATP.

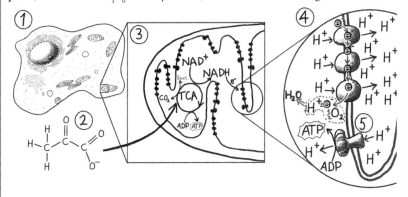

From pyruvate to oxidative phosphorylation in the mitochondrion. (**1**) Glycolysis in the cytoplasm of the cell produces (among other things) pyruvate from glucose; (**2**) Pyruvate diffuses into mitochondria, where it goes into the tricarboxylic acid cycle (TCA cycle). (**3**) In the TCA cycle some of the mitochondrion's budget of ATP is produced, and some of the reducing power of pyruvate is "burned" by oxygen in the process, making CO_2. Some of pyruvate's reducing power is used to make NADH, a strong biological reducing agent. (**4**) NADH undergoes a reaction at complex I of the mitochondrion's ETC, releasing a high-energy electron, which then passes along the "molecular wire" of membrane proteins (ETC), finally emerging from complex IV to be combined with oxygen and H^+ ions to make water—the other combustion product of carbohydrates and fats. During the electron's passage, H^+ ions are pumped across the membrane by each of the three protein complexes shown (I, III, and IV), thus establishing an electrochemical gradient. (**5**) H^+ ions flow back across the gradient through an enzyme complex, the F_1F_0-ATP synthase, making ATP. The role of oxygen as the oxidant, and the addition of a phosphate group to ADP in the TCA cycle and via the F_1F_0-ATP synthase, give the term "oxidative phosphorylation" to the combination of these two processes in the mitochondrion. For simplicity, only one "arm" of the ETC is shown (i.e., protein complexes I, III, and IV, also omitting coenzyme Q and cytochrome C). The other arm consist of four protein complexes (I, II, III, and IV) and derives its electrons from another product of the TCA cycle, namely succinate. Essentially the mitochondrion converts, by controlled burning with oxygen, organic hydrogen carriers to energy in the form of a phospho-organic compound (ATP). A crucial fact to mention at this point is that oils and fats are, ultimately, also "burned" in the same way via the ETC in mitochondria. The process of "aerobic" respiration (depending on oxygen as the combustant gas in glycolysis to produce pyruvate and in the mitochondrion as the partial combustant in the TCA cycle and the terminal electron acceptor of the ETC) is the powerhouse of cells that make up all plants and animals. This is true from the simplest single-celled ones through to multicellular organisms such as us. In fact, it is the major energy machinery of 99.9% of all eukaryotic life (cells with a true nucleus) on Earth. Its evolution gave life the energetic "push" that enabled it to graduate from unicellular early microbes to multicellular organisms of unimaginable diversity, behaviors, and life histories.

their finely tuned metabolism based on reducing agents (mainly redox cycling between iron II and hydrogen).[24] However, it also likely drove the evolution of the eukaryotic cell (cell with a nucleus, which bacteria and archaea don't have), the greatest evolutionary advance of all. This started with the merger of two individual cells from the two cousin kingdoms, archaea and bacteria: only the co-habitation of an alpha-proteobacterium inside an ancient archaeal cell could protect the archaeal cell from the toxicity of oxygen radicals. The alpha-proteobacterium was a relatively rare organism at the time because it could cope with oxygen by virtue of having a particular enzyme (terminal oxidase) that can transfer the electrons in its "molecular wire" (electron transport chain [ETC]) to molecular oxygen (O_2).

So when the chance cellular merger took place, it was fixed by evolution because it enabled the resultant hybrid organisms to survive higher oxygen concentrations much better than their predecessors. Some researchers believe that the alpha-proteobacterium literally burrowed into the archaeal cell parasitically (as some bacteria do with eukaryotic cells to this day). Equally intriguingly, there is evidence that it might well have been the stimulus for the evolution of sex, defined in its simplest form as the mixing of genomes upon reproduction (which in eukaryotic cells means meiosis followed by fusion of gametes). All of this, which included an ensuing explosion of diversity and mass (in tonnes) of life on Earth, was linked to being able to tap into an almost limitless source of energy: the Sun. We may never know exactly what happened, but it is clear that the ability to generate energy and survive the negative chemical repercussions was a major evolutionary driving force—one that shaped metabolism and nutrition right up to humans.

The evolution of photosynthesis caused the most striking shift in biological energy: away from minerals and simple chemicals from the Earth's crust as energy sources (reducing equivalents) and toward energy economies based on turning CO_2 into carbon-based energy carriers using the Sun's energy. This reaction can be done anywhere on Earth where there is liquid water (with minimal concentrations of certain elements), normal air, and light. It is a highly flexible, circular, and general energy economy based on ubiquitous ingredients. Hence it started the expansion of life to occupy all imaginable niches on Earth where the aforementioned basic components exist. Perhaps there is a parallel here from which human industry can learn.

Conclusion: The creation of carbon and its special chemistry—A toolbox of interconvertible compounds waiting for life to use

Carbon was created by the first stars to form in the universe, and it started making a bewildering diversity of compounds with itself, oxygen and hydrogen almost immediately—a consequence of its unique configuration of electrons. Organic chemistry (the chemistry of compounds of carbon with one or more C-C bond) clearly existed long before life on Earth; it also exists in diverse parts of the universe where no life has been detected, and none (as we know it) is likely to arise in the foreseeable future. If latest hypotheses on the origin of life on Earth gain further support, it seems ever less likely that information-coding polymers (a kind of primitive genetic material) were a prerequisite for life. The "secret" of life is not its replicable code, but its channeling of energy fluxes to store energy and organize matter. It is extremely hard to imagine it doing so with any other element than carbon.

Theories that life arose via an accumulation or seeding of organic molecules that are typical of living organisms today are unlikely to be correct because they don't enlighten us about the physicochemical conditions for life's inception: life "manages" energy fluxes in very special ways[25] to create intricate order in a process that plays with the Second Law of Thermodynamics. The rest of the universe—despite temporary order such as planetary systems, evolving galaxies, etc.—slowly succumbs to the Second Law; but life uses any energy fluxes (or gradients) into which it can tap to "deflect" the Second Law away from itself and out into its surroundings (where life's order is compensated by an equal creation of disorder!).

As hypothesized, proto-life started doing this with an inorganic chemical gradient, and the chemistry surrounding it was favorable for building flexible structures that could evolve: these were, and still are, absolutely dependent on carbon. The spontaneous appearance of the "right" organic chemicals in the right physicochemical environment (e.g., lipids in the micrometer fabric of alkaline thermal vents in the ocean floor) coupled with the presence of a chemical gradient (energy source) seems to be what sparked life in the form of the first cells. Life almost "chose" to work with carbon compounds because they were the best things "lying around" to produce structures that separated living matter from simple chemistry and embodied greater complexity than the surrounding inanimate milieu. This enabled proto-life to produce a diversity of structures that were either more or less successful in continuing their existence and replicating: the inception of evolution via natural selection. The process was merely physicochemical at the very beginning, but it bears the same hallmarks as biological selection. There is quite a lot of inorganic chemistry involved in the theory described earlier: protons (H^+), proton gradients, inorganic redox chemicals containing metal ions (FeII and FeIII), and inorganic phosphate (Pi). To this day, iron-sulfur clusters are, for example, indispensable in the ETCs of mitochondria and chloroplasts in eukaryotic cells. Inorganic chemistry, largely in the form of catalysts, teamed up with carbon compounds in the form of material and energy metabolism.

Organic substances clearly existed before organisms but in the absence of a usable energy gradient; thus it is important to emphasize the divide between the spontaneous emergence of organic compounds and the evolution of life. In so doing, and via a knowledge of the order of chemical events at life's origin, we may be able better to identify the conditions under which life probably arose. Organic chemistry was certainly necessary for life as we know it to emerge and evolve, but it was not enough. On the other hand, life clearly arose by using the enormous flexibility of carbon chemistry, something that it could not have done with any other element, even silicon, carbon's Group 14 cousin in the Periodic Table.

Photosynthesis and carbon compounds made a perfect pair. The food chains that evolved from photosynthesis, through primary consumers (herbivores) right up to majestic carnivores such as the great white shark (*Carcharodon carcharias*) and the somewhat less majestic *Homo sapiens*, are essentially dynamic recycling equilibria that have produced and sustained enormously diverse ecosystems for billions of years on Earth. Humans harnessed photosynthesis via crop plants during the Agricultural Revolution (starting around 14,000 years ago). Today, the livestock sector of agriculture the world over is amongst the fastest growing, as food intake shifts in unhealthy and environmentally unsustainable ways. We explore these aspects in the next chapter, where I discuss nutrition and biological energy economies. There we will see why carbohydrates and hydrocarbons were the compounds that life chose for energy storage, transport, rapid mobilization, and flexible response to an unpredictable environment.

2
The carbon economy of nutrition and food production
Getting out of control in most respects

In 2022, the world population reached 8 billion: the challenge of feeding humanity has never been greater. As if that weren't enough, the increase in meat consumption accompanying higher per capita earnings continues unabated. At one extreme, this is a standard-of-living quandary because wealth brings luxury and excess; at the other extreme, it is a sad reflection of the desire to break free from malnutrition and suffering. For the latter, we really don't need much livestock. Growing greenhouse gas (GHG) emissions from livestock are the most serious immediate problem, and one that can be addressed without any true decline in quality of life: wealthy countries should drastically reduce animal product consumption, mainly meat. Studying what humans need in their diet and the ways in which the various components are metabolized leads to insights that are relevant not only to reducing GHG emissions and environmental damage, but also to lessening disease and increasing healthy life span. There are also striking parallels between fuel considerations for human technology and the energy metabolism of food.

Human nutrition: What we consume and how our bodies use it

Arguably, the "natural" nutritional balance for humans is evolutionarily strongly tilted toward fats, oils, and, to a lesser degree, carbohydrates. In fact, heart muscle derives more than 70% of its energy from fatty acids (FAs, the main constituent of plant and animal fat). This enabled us to be active for long periods and cover prodigious distances because we can store relatively large amounts of these high-energy molecules. However, in the 20th century, nation after nation started to skew this balance as we were driven by the taste and status value of meat to consume more and more animal protein. This has led an already burgeoning industry of agriculture into ever greater production of animal products, which are, mass-for-mass, very inferior sources of energy for us compared with carbohydrates and fats. At the same time, because of convenience and cheapness, more people are eating more *refined* carbohydrate than before, a trend without evolutionary precedent.

What are we made of, and what do we need?

The interesting thing about recent human nutritional evolution is that it has increasingly less connection with our biological needs and biological evolution. At a very reductionist level, we can consider our chemical composition and what we need to eat to maintain it. We return now to carbon in the context of nutrition and the basic chemistry of metabolism. Life concentrates three chemical elements—carbon, oxygen, and nitrogen—more than anywhere else in the universe. Because the majority of our bodies is water (H_2O)—around 60% by mass—it is no surprise that oxygen (65%) and hydrogen (3.2%) together constitute most of our "wet" mass; not all of that is water, of course, because most organic molecules also contain oxygen and hydrogen. Carbon, at 18% by mass (wet), however, is the greatest contributor to our "organic" substance (i.e., the building blocks that make up cells, connective tissue, and our biochemistry of metabolism). All major classes of food, except one, contain compounds of carbon: in fact, their very essence *is* "carbon connectivity," as one can see from the schematics of the principal constituents of living organisms (Figure 2.1). Carbohydrates are, by mass, around 40% carbon; proteins are approximately 50% carbon; fats, oils, and lipids (the building blocks of biological membranes and coatings) are around 65% carbon. This is true of all life forms, whether animal, plant, bacterial, archaeal, or fungal. High-level structural elements such as skeletons and exoskeletons sometimes contain a lot of calcium and phosphorus (e.g., vertebrate bone and the calcified exoskeletons of crustaceans such as crabs and lobsters). But even in these cases, the largest single contributor to the dry weight of life is carbon. This is essentially because of the enormous flexibility of carbon to bond with itself and other atoms to make complicated molecules with myriad properties and—in life—evolved functions.

Together, proteins, and fats/oils (lipids), make up 29% (wet mass) of an average human body, divided equally by mass. Minerals (mainly calcium and phosphorus in bone and teeth) account for 5% and are the only category of food devoid of carbon; nucleic acids (DNA and RNA, which, respectively, encode

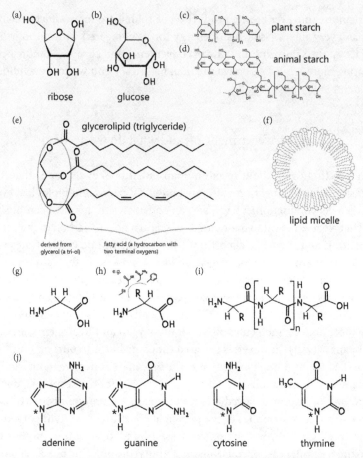

Figure 2.1 The principal constituents of living organisms. Carbohydrates: **(a)** the five-carbon sugar, ribose; and **(b)** the six-carbon sugar, glucose; **(c)** plant starch consists of unbranched chains of many molecules of glucose joined by a glycosidic bond (C—O—C); **(d)** animal starch (glycogen) consists of branched chains of many molecules of glucose. Fats and oils (lipids): **(e)** one type of lipid, a "glycerolipid" or "triglyceride," so named because three fatty acids are joined to glycerol (a triol, i.e., having three –OH groups); **(f)** a micelle of lipids, which forms spontaneously in aqueous solution, because the water-hating (hydrophobic) hydrocarbon chains associate with each other, leaving the more water-loving (hydrophilic) heads (glycerol moieties) on the outside. Amino acids: **(g)** the simplest amino acid, glycine, in which both positions at the crucial carbon atom are occupied by H; **(h)** to make the remaining 19 of the 20 amino acids encoded by the standard genetic code; the position marked "R" is replaced by different chemical groups—four are shown here as examples—thus making building blocks of proteins that have a large variety of chemical and physical properties; **(i)** a chain of amino acids (a peptide) folds into complicated structures that ultimately produce a three-dimensional protein with one or more specific functions (e.g., an enzyme or a structural protein). DNA/RNA bases: **(j)** the four bases of DNA (AGCT), showing the position at which deoxyribose (a variant of the sugar in **(a)**) is attached in DNA. RNA contains a slightly different fourth base instead of T, called "uracil" (not shown here).

and translate the cell's structure and machinery) constitute approximately 5%, most of which is RNA; finally, carbohydrates come in at a mere 1% (but ahead of vitamins, which are present in only minute amounts). If carbohydrates are such an important energy source, why are they so scant in our body? The answer is precisely *because* they are such a ready energy source: they can quickly be broken down to release their energy, and that is what typically happens inside us. In fact, that breakdown (known as *hydrolysis,* from Greek "splitting with water") starts immediately in our mouth in the case of starches, which are long chains of individual sugar units.

Carbohydrates: Most efficient use of land, easy to store, mobilize, and interconvert

In the pyramid of nutritional categories, carbohydrate usually appears at the base—representing the greatest mass per day of recommended intake for active people, though not in the *refined* form. Just 15 of the tens of thousands of edible plants on Earth provide 90% of global human energy intake; two-thirds of that 90% come from wheat, rice, and maize (corn).[1] Carbohydrate staple crops are the most efficient agricultural source of energy in human nutrition in terms of the energy required to make them, their GHG emissions per kilogram, and the area of land used per kilogram of crop. However, that said, they vary quite a lot in their contributions to global carbon dioxide equivalent ($CO_{2\,eq.}$) emissions (i.e., including non-CO_2 greenhouse gases), methane being the largest variable. Collation of wide sources of data[2] indicates that, when normalized for yield, rice agriculture generates four times as much $CO_{2eq.}$ as wheat or maize. Methane rising out of rice paddy fields where microbial processes are very rapid is the main culprit, and rice agriculture is estimated to contribute 12% of annual worldwide methane emissions. This is a recognized problem, but we must get it into perspective with one of greater magnitude that is growing faster as more people become able to afford animal products (or as more animal products *become affordable*). Compared with animal products, even rice is fairly harmless, having a $CO_{2eq.}$ footprint of around one-eighth (12.4%) that of meat per kilogram[3] (0.77 $CO_{2eq.}$/kg compared with 6.21 for meat on average).

Our bodies get to work on carbohydrates immediately. Our saliva contains an enzyme called *alpha-amylase* (from Latin "amylum" meaning "starch"), which breaks the long chains of starch (such as in bread) into shorter chains (e.g., dextrin) and disaccharides (two-component sugars) that we recognize as sweet-tasting (e.g., if we chew a piece of bread and keep it in our mouth for a few minutes). Amylases have a special place in the history of biochemistry because the amylase "diastase" was the first enzyme to be discovered—by French Chemist, Anselm Payen, in 1833. Many other mammals—mainly herbivores and omnivores—also have salivary amylase. An animal without energy is a dead animal because it can't move to find more food or water, and that is why evolution resulted in this very rapid means for us to gain energy from the first bite of carbohydrate-containing food that we take. On the other hand, consuming too much carbohydrate with a high glycemic index (GI; a measure of how quickly it is broken down and goes into our blood as sugar) eventually exhausts our pancreas, which secretes large quantities of insulin in response to large rises in blood sugar. Refined carbohydrates (e.g., white wheat flour) usually have GIs above 70; legumes, and whole grains such as oats, are typically less than 55.

Some complex carbohydrates (e.g., cellulose, the material of plant cell walls, which was also discovered by Payen) and celluloses combined with lignin (so-called *lignocelluloses*, from woody plants) resist our enzymatic attempts at extracting their energy: they pass through to our large intestine largely unaltered and constitute the last food group: roughage. But cellulose can be broken down by certain anaerobic microbes (e.g., certain *archaea*, cousins of bacteria, mentioned in Chapter 1). Ruminants (e.g., cattle, sheep, and deer) live in symbiosis with these microbes in their stomachs, and many of them are so-called *methanogens* (i.e., they produce methane by reacting H_2 with CO_2).[4] This is, of course, why cows "burp" methane. During a long process of chewing (to break up the plant cell walls) and oxygen-free fermentation by the methanogens, the cellulose is broken down into sugars. Ruminants work hard for their energy!

The amount of carbohydrate that animals can store in their body is typically very small: moderate deposits occur in muscle and liver only. The quantity of sugar (mainly glucose) circulating in the blood for immediate use is controlled very carefully. This is because sugars significantly change the osmotic potential of aqueous (water-based) solutions, making them draw water from their surroundings. In the case of blood, that can be very damaging in the long term, and it is the root cause of much of the vascular damage seen in diabetes. Professional endurance sportspeople will often eat high-carbohydrate meals (typically pasta) the day before an event. Much of that load will be converted to a type of storage carbohydrate found in animals, fungi, and bacteria (but not plants), termed "glycogen." Stored glycogen can be converted to glucose very quickly on demand. Regularly eating a lot of carbohydrate-rich foods (particularly containing highly refined carbohydrates) obviously outstrips the body's capacity to convert it into glycogen for storage in muscle (and liver): most of the excess (which also can't be burned) is turned into fat. Simultaneously, diabetes risk increases greatly.

A bi-fuel metabolic machine: What happens when carbohydrates run low

The problem with fat is that, once laid down, it takes a lot to get rid of it. Most humans (endurance athletes excepted) do not regularly reach the end of their ingested and stored carbohydrate reserves, and it is only at that point that humans start to burn fat reserves. One can literally smell this process, known as *ketosis* or *ketogenesis*[5] because it produces a mixture of ketones (organic compounds containing a C=O bond within the C–C chain—see Figure 2.2) that smell sweet or pungent—sometimes offensively so. Acetone (used as nail varnish remover and as an industrial solvent), the simplest ketone, is, indeed, in the breath of people in ketosis. So-called *ketone bodies* (collectively acetone, acetoacetate, and β-hydroxybutyric acid; also known as 2-hydroxybutyric acid [Figure 2.2] can also be found in their urine. This sounds like quite a serious condition if it produces solvents used by the chemical industry, but ketosis is, in fact, a perfectly normal metabolic response to carbohydrate depletion. It only becomes dangerous in combination with extreme hypoglycemia (sudden massive drop in blood sugar), leading to *ketoacidosis*—a type of acidification of the blood.

Ketogenic diets (ones that produce ketones in us) have become quite popular of late, but their origins probably go back to Hippocrates in Ancient Greece, who realized that fasting changed the body's chemistry in such a way that certain diseases or abnormal conditions became milder (e.g., epilepsy)—but don't radically change your diet without consulting your physician! In 1910, a French study investigated the effects of dietary ketosis on epilepsy[6]—as described in a 2004 historical review[7]—apparently with some success.[8] The term "ketogenic diet" was coined by Russell Morse Wilder from the Mayo Clinic, following from Rollin Turner Woodyatt's work on reviewing diabetes literature and identifying the three ketone bodies. This diet is relatively high in fat, moderately high in

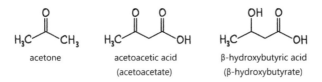

Figure 2.2 Ketones circulating in humans undergoing ketosis because of hypoglycemia. In fact, the third one, despite arguably being the most important, is not really a ketone, because the C=O bond is attached to one of the *end* carbons of the molecule, rather than within the carbon chain (it is a hydroxy-carboxylic acid).

protein, and low in carbohydrates. Hence, it does not cause the large spikes of insulin secretion that lead to obesity and diabetes. Instead, it engages enzymes in the mitochondria that perform beta-oxidation of FAs (long-chain molecules derived from fat break-down).[9]

The resulting ketones can be burned by our brain, heart, and skeletal muscles instead of the usual sugars. Certain amino acids (building blocks of proteins) can also be converted into ketone bodies (via beta-oxidation). In humans and other animals undergoing starvation, where fat reserves are depleted, protein starts to be broken down to provide energy. This is the last resort because proteins, and tissue with high protein content (e.g., muscle and connective tissue), are very costly for an organism to make. However, at lesser extremes of ketosis, medically beneficial effects are continually being revealed.[10] These can involve "epigenetic modifications" that modulate gene function: the most abundant circulating ketone body (D-beta-hydroxybutyric acid/D-beta-hydroxybutyrate [DBHB]) has been shown to readjust gene expression by changing methylation patterns on DNA and possibly also acetylation of DNA-binding proteins (histones). The explanation for ketones' positive effect on epilepsy seems to rest largely on these epigenetic mechanisms.[11] Basically, it seems that, beyond their role in energy economy, certain ketone bodies have important "tuning" functions for genes involved in disease and aging.

Ketones can also be produced from protein, but dietary protein intake is not "abused" by organisms to make energy when carbohydrates or fat are present, nor to make more protein-based structures and structural repairs than "necessary." Excess protein is simply excreted, mainly in the urine. Necessary protein intake for most healthy adults under 65 years is around 0.8 g dry weight per kilogram body weight per day.[12] For a 70 kg person, that's 56 g dry protein mass, which would be contained in 210 "wet" grams of lean steak, or 220 g of Cheddar cheese, or seven average chicken eggs, or 430 g of dried soy beans, for example.[13] By contrast, 225–325 dry grams of carbohydrate are recommended by the Mayo Clinic[14] for an average adult per day (e.g., roughly 375 g of rice, before cooking). But there's more to it than that because researchers have also shown that protein requirement overrules carbohydrate intake: humans eat until they have satisfied their *protein* need. Within reason, the more that protein is diluted in a meal with carbohydrate, the more of the meal people will eat. This is what happens on a fast-food diet, which contains a high ratio of carbohydrate (cheap) to more nutritionally valuable food types. To understand why carbohydrates provide us with so much readily mobilizable energy, and why they are so distinct from fats and oils, we need to look more at the bonds between their atoms.

Energetic specifics of carbohydrates and fats: It's all in the chemical bonds

The carbon economy of our body is influenced very strongly by the types and strengths of bonds that carbon makes in carbohydrates and fats in contrast to proteins. Basically, carbohydrates and fats contain relatively high-energy bonds.[15] How much energy are we talking about? Glucose is one of the most important biological sugars in terms of energy. Combustion of glucose (Formula 1) releases mass-for-mass around two-fifths (0.4 times) as much energy as combustion of diesel (2,870 kJ/mol, which translates into 15.94 megajoules per kg [MJ/kg]).

$$\text{Formula 1: } C_6H_{12}O_6 + 6O_2 \rightarrow 6CO_2 + 6H_2O$$

Fats and oils are even more concentrated in energy: typically, olive oil has between 38 and 39 MJ/kg—a little less than diesel. There are more bonds between carbon and hydrogen in fats and carbohydrates than there are in proteins. Of the three, fats are the most "hydrogenated," so to speak, then come carbohydrates and, last, proteins. Hydrogen is carried by carbon, and though it does not, on its own, represent the energy in carbohydrates and fats, there is a correlation: typically, the less hydrogen that carbon compounds contain (via C-H bonds), the less is their energy density.

What are biological fats and oils in terms of chemistry?

Dietary fats and oils are, in chemical terms, "triglycerides" (see the section "What are we made of, and what do we need?" Figure 2.1e), differing essentially in melting point (oils are liquid at ambient temperatures; fats solid). They are hydrophobic (water-hating) and therefore don't dissolve in water. Some, as in milk, may be emulsified in water in combination with protein. Their hydrophobicity gives them a great physical advantage over carbohydrates: they are stored in our body free from water, they don't affect our water balance, and they are dehydration-resistant. Fats in our body shuttle between different forms and locations, and, in a form known as "lipids," they are basically the most important component of the external and internal membranes of all living cells on Earth. The archetypal structure of a fat (or oil) is presented in Figure 2.1e. Such fats/oils are built on a backbone (the three-carbon "beam" at far left of the molecule in Figure 2.1e) that came from glycerol, a so-called *triol* (because it contains three OH "-ol" groups [as in alcoh*ol*]). Each of the fronds attached to the glycerol was originally a "fatty acid"—about as close to a hydrocarbon as you can get—and became joined to ("esterified" with) the glycerol by reacting with one of glycerol's OH groups.

The FAs are the main energy carrier in triglycerides, and they are cleaved off the glycerol during digestion and may then be stored or broken down directly in cells (in a process called *beta-oxidation*, which produces four molecules of adenosine triphosphate [ATP]—the universal bio-energy currency—per cycle). The exact energies that change hands in the cell under physiological conditions are extremely hard to measure, but the number of ATP molecules produced is a very convenient "proxy" for them.

Why fats and oils are denser in energy than sugars

Glucose, with its six carbon atoms, produces 38 ATPs in the process of glycolysis, followed by oxidative phosphorylation in the mitochondrion. One FA of 16 carbons' length—that is, *palmitic acid* (completely saturated)—produces 130 ATPs if similarly fully catabolized in the cell.[15] Hence, the ratio of biological energy (ATP) produced by glucose compared with palmitic acid is 1:3.4 (i.e., 38:130). Compensating for the molecular size difference between glucose and palmitic acid, we get a ratio of 1:1.28—that is, the FA produces 28% more ATP energy per carbon atom than glucose. The fact that, in triglycerides, there is also the glycerol molecule to consider (and which is also catabolized to make ATP) does not make a great difference to the ratio (see Box 2.1). From here on, we'll therefore only consider the FA part of fats/oils. An important note at this point: though the terms are deceptively similar, the distinction between *hydrocarbons* (to which, crudely speaking, fats and oils belong) and *carbohydrates* (to which sugars belong) is very important.

The reason that fat produces more ATP than carbohydrate is simple: energy in aerobic (oxygen-respiring) cells (that's most of our cells, most of the time) comes from the oxidation of "fuel" containing hydrogen and carbon—a highly controlled form of burning, but chemically indistinguishable. The "more reduced" and "less oxidized" that the fuel is to start with, the more energy it can release upon oxidation. Focusing on the carbons that are bound to a (reducing) hydrogen, in contrast to an (oxidizing) oxygen, this concept becomes very clear (Figure 2.3).

If we inspect the "oxidation state" of the carbons in our two energy molecules, we see that glucose has five carbon atoms bonded with one oxygen each, and one carbon bonded with two oxygens. This is a significantly higher ratio of carbon to oxygen than in a fat or oil, and hence glucose is basically closer to complete oxidation (burned) than a fat or oil (for more details, see Box 2.1). *Mass for mass,*

Box 2.1 Triglycerides produce large ratios of ATP per unit mass by containing FAs where all available carbons are "tanked up" with hydrogen

The ratio of carbons to ATP molecules produced is not constant for FAs present in triglycerides because the first beta oxidation requires one ATP to be *spent*; so, the longer the FA, the greater its energy per carbon atom. The longer the chain of the FA, the more "efficient" it is in this respect. What effect does the glycerol part have on the ATP ratio? If we consider a whole triglyceride containing three molecules of palmitic acid, we must multiply by three and add the ATPs produced by oxidative metabolism of the glycerol, which is 20.[16] In total, therefore, this triglyceride would produce (130 × 3) + 20 = 410 ATPs. Adjusting for number of carbon atoms again (the triglyceride has three, hence making the total (3 × 16) + 3 = 51) gives a ratio of 1:1.27, glucose to triglyceride, in terms of ATP per carbon atom. Thus the FA, and not the glycerol, is clearly the distinguishing chemical unit between sugars and fats. The glycerol molecule is very similar to glucose in terms of ATP yield because it has a very similar oxidation state to glucose (ratio of carbons to carbon-bound oxygens).

Any given carbon atom can make four bonds with neighboring atoms. In the case of glucose, the remaining bonds (not occupied by oxygen) are made with hydrogen. That is important to know because hydrogen is a "reducing agent" (i.e., it is the opposite of oxygen, so to speak). In terms of the oxidative energy metabolism of living cells, the more hydrogens that a carbon-based fuel carries bonded to each of its carbon atoms, the more "reduced" it is, and hence the more energy it can give out when it is oxidized by oxygen. If it is already partly oxidized (as is glucose, and to a much greater extent than the FA), the consequence for its energy content is obvious. In our 16-carbon FA from earlier (palmitic acid), one of the carbons has very little reducing potential because it is bonded to two oxygens and its neighboring carbon; the remaining 15 carbons have a total of 31 hydrogens, and the ratio of hydrogens per carbon (roughly two per carbon) is much higher than in glucose (glucose is 1.33 hydrogens per carbon). Unfortunately one can't see this as a simple ratio calculation where energy in the form of ATP relates to this ratio of hydrogen to carbon because there is quite a lot of complicated biochemistry and non-equivalences in between. However, the basic principle is sound: the more highly reduced (with hydrogen) the carbons in a fuel are, the more energy it will liberate per carbon atom upon oxidation. A small qualification must be included here: for simplicity, I have presented an FA with an even number of carbons; the figures (and one step of the metabolism) are slightly different for odd-number FAs, but the details go outside the scope of this book. The extra

ATP energy produced per hydrogen-carrying carbon atom in palmitic acid (compared with glucose) is not in proportion to the hydrogen ratio (see caption of Figure 2.3): that is, palmitic acid would be expected to produce even more ATP per carbon than glucose does. This appears to be a biochemical tradeoff for the physical advantages of fat.

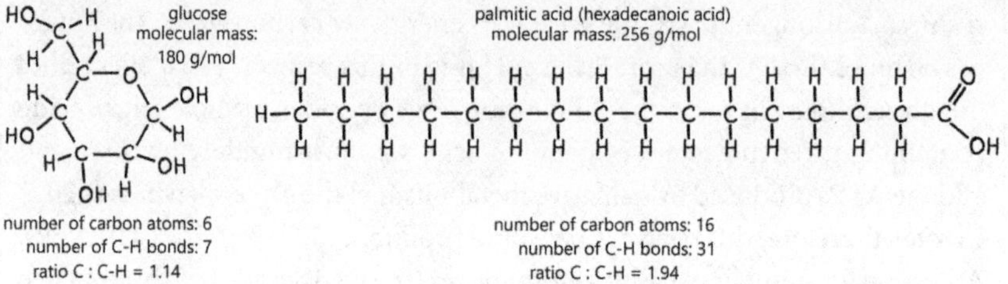

glucose
molecular mass: 180 g/mol
number of carbon atoms: 6
number of C-H bonds: 7
ratio C : C-H = 1.14

palmitic acid (hexadecanoic acid)
molecular mass: 256 g/mol
number of carbon atoms: 16
number of C-H bonds: 31
ratio C : C-H = 1.94

Figure 2.3 Glucose and palmitic acid (as a linear Fischer projection) presented for the purpose of comparing their potential for oxidation (burning) to produce biological energy (adenosine triphosphate; ATP). The emphasis is on the proportion of C to C-H bonds in the respective molecules. In glucose, the ratio is 1.14; in palmitic acid it is 1.94. That means that in terms of C-H equivalences, palmitic acid has 70% more than glucose. However, in terms of ATP produced per carbon atom, the "superiority" of palmitic acid is much lower: glucose produces 6.33 ATP per carbon, and palmitic acid 8.13, which translates to 28% more ATP per carbon for palmitic acid. Palmitic acid is a fully saturated FA (i.e., all available H-bonding positions are occupied by H); other FAs are partly unsaturated, meaning that between one or more pairs of C atoms there is a C=C double bond. Note that if we replace the final COOH group of palmitic acid with CH_3, we have hexadecane, a principal alkane in diesel.

palmitic acid produces significantly more than double the quantity of ATP compared with glucose (glucose: 180 g/mol divided by 38 ATPs = **0.21** mass/ATP ratio; palmitic acid: 256 g/mol divided by 130 ATPs = **0.51** mass/ATP ratio). Glucose that has been polymerized into glycogen for short-term storage produces a little more ATP upon oxidation than plain glucose, but it also takes a little more energy to be polymerized into glycogen in the first place. There is an intriguing parallel here between biology and the energy carriers for human technologies:

- Biology: Relatively oxygen-rich glucose, easy to make into glycogen, store, and mobilize at short notice, but of comparatively low energy density; relatively oxygen-poor fat/oil, more laborious to make, but of much higher energy density.

- Technology: Relatively oxygen-rich alcohols (mainly methanol and ethanol), easy to make, but of comparatively low energy density; relatively oxygen-poor hydrocarbons (e.g., gasoline, diesel, kerosene), more laborious to make, but of much higher energy density.

We will see more of this parallel in Chapters 4 and 5.

From its mere chemistry, we would expect palmitic acid to produce even more ATP than it does, but evolution seems to have engineered a biological tradeoff to reach a compromise between high energy density and suitability as a storage substance. In addition to having a very high energy-to-weight ratio (Joules per kg), fat is water-insoluble and buoyant. Glucose, by contrast, is 56% *denser* than water, and glycogen (the animal storage carbohydrate) is 83% denser than water. In animals, fat is an essential component of long-term energy storage, and it also has other functions (e.g., insulation, buoyancy). Animals in normal physiology (i.e., excluding obesity) can accumulate prodigious quantities of fat, but the same is not true of glucose and glucose-derived animal starch (glycogen). Per carbon atom, fat produces more ATP than glucose, but its mass-to-mass superiority over glucose or glycogen as an energy store, and its additional functions, distinguish it strikingly from carbohydrate. Animals, regardless of which evolutionary class, cannot carry enough glycogen around with them: it's too heavy and bulky. The parallel with different energy sources for human transportation is obvious.

On this note, there is something even more interesting about the bond energies in FAs, and the thought might well have occurred to you: What about *unsaturated* FAs (and, hence, unsaturated fats/oils)? These are important in terms of diet, because many—in particular, omega-3 polyunsaturated FAs—are known to be healthier for us than saturated fats. As it happens, there are omega-3 unsaturated FAs (e.g., hexadecatrienoic acid [HTA]) that also have 16 carbon atoms. However, because of the three unsaturated C-C bonds, they have six C-H bonds fewer than palmitic acid (Figure 2.4). Are they less energy-dense? Yes, but only slightly.

Biochemistry tells us that the unsaturated HTA (Figure 2.4) makes six ATPs fewer than palmitic acid (i.e., 4.6% less than the saturated palmitic acid).[9] Unsaturated FAs provide slightly less energy per molecule than saturated ones, but mammals absorb unsaturated FAs more efficiently.[17] We consume a mixture of saturated and unsaturated FAs in our diet, and the more animal products we eat, the more the balance shifts toward saturated FAs (an exception is oily fish, which contains the classical omega-3 unsaturated FAs). The so-called cis-forms of unsaturated FAs tend to be more healthy for us than the trans-forms (Figure 2.5), which are mainly found in meat and dairy products of ruminant animals.

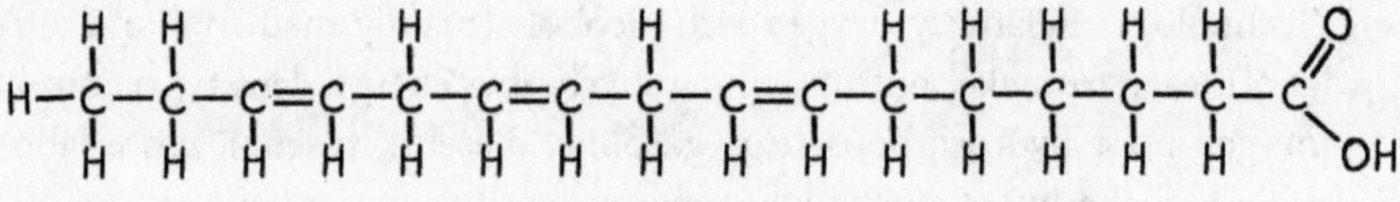

$$H-\overset{H}{\underset{H}{C}}-\overset{H}{\underset{H}{C}}-C=C-\overset{H}{\underset{H}{C}}-C=C-\overset{H}{\underset{H}{C}}-C=C-\overset{H}{\underset{H}{C}}-\overset{H}{\underset{H}{C}}-\overset{H}{\underset{H}{C}}-\overset{H}{\underset{H}{C}}-\overset{H}{\underset{H}{C}}-\overset{O}{\underset{OH}{C}}$$

Figure 2.4 Linear Fischer projection of one form of hexadecatrienoic acid (HTA), an omega-3 unsaturated FA, which is a tri-unsaturated cousin of palmitic acid. Relatively straightforward calculations can show that, molecule for molecule (mol for mol), the "sacrificing" of six C-H bonds for the formation of three new C-C bonds in the form of the three C=C double bonds does not significantly reduce the plain chemical energy value when burned with oxygen (i.e., the higher heating value). However, it slightly reduces the biological energy value (number of ATPs produced) because of the intricacies of beta oxidation.[9]

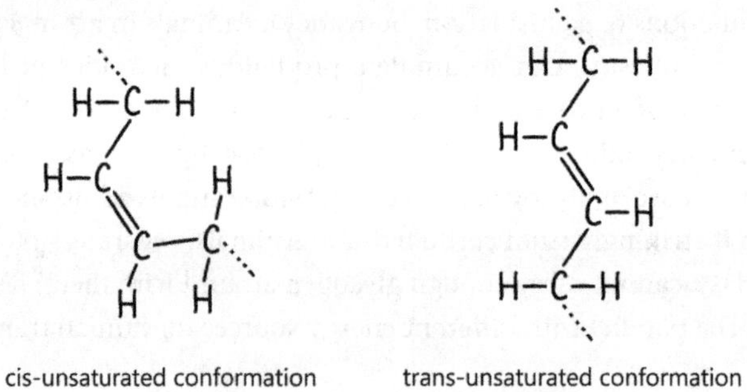

cis-unsaturated conformation   trans-unsaturated conformation

Figure 2.5 How the chemistry of trans-unsaturated fatty acids (FAs) makes them potentially unhealthy. If two carbon atoms in an FA chain are unsaturated with hydrogen they have an extra bond free, and therefore exist as C=C. Because of physical inflexibility of a double-bond, the remaining groups attached to these carbons can be in one of two configurations, but cannot interconvert. Unsaturated FAs from plants are overwhelmingly of the cis-form (i.e., C-shaped), whereas those in animal fat are overwhelmingly of the trans-form (i.e., Z-shaped). All C=C bonds in FAs are more reactive than C-C bonds, but the trans-configuration is greatly more reactive than the cis-configuration. This gives trans-unsaturated FAs greater propensity for forming reactive oxygen species, hence reacting more with other biological molecules and with themselves (the last of which produces cholesterol, for example). The "electronics" of the trans-form are more "direct" than those of the cis-form, and hence electrons can pass through the C=C carbon atoms to other molecules more easily. Interestingly, industrial trans-fatty acids (e.g., those that used to go into margarine to turn it from a liquid into a semi-solid) are much more reactive than naturally occurring trans-fatty acids.[18] Oxidized fat is a principal cause of atherosclerosis, and a variety of trans-FA oxidation/reaction products are involved in many diseases, including inflammatory conditions and cancer.

C-C and C=C are more "oxidized" than C-H, but all are greatly less oxidized than the products of their burning with oxygen (i.e., H_2O and CO_2). We delve into this topic more in Chapter 4, "Fuels and Efficiency," but already in Chapter 3, "The Carbon Cycle," we meet a substance that burns to produce a respectable amount of heat per kilograms and consists mainly of carbon-carbon bonds and relatively little C-H: coal. Essentially, energy is stored in C-C/C=C bonds as well as C-H.

Metabolic economies and health: Why biological hydrocarbons have the edge over carbohydrates

Glucose from carbohydrates was long thought to be an absolute and constant requirement for long-term human survival, and fat simply a high-energy reserve. The state of ketosis was believed to be abnormal, and fasting a quaint practice related to religious fervor. When insulin is low, when we're not eating carbohydrate, and we are simultaneously either enduring fasting or heavy physical exertion, something rather interesting happens in our fat breakdown (catabolism). Under "normal" circumstances, we would do beta-oxidation and send those two-carbon chunks of FA to the mitochondria and into the *Krebs cycle* (in the form of Acetyl-CoA, a biological carrier molecule used in degrading energy molecules). Named after one of the two biochemists who discovered it in 1937— Hans Adolf Krebs and William Arthur Johnson, at the University of Sheffield— the Krebs cycle (or tricarboxylic acid cycle [TCA] cycle) is where glucose is also burned. However, how do we define normal circumstances?

What is "normal" human fuel intake?
"Normal" is only what we define as "normal" based on our "fuel" intake and resultant metabolism today. It's almost certain that humans (in general) throughout most of their evolution had no "constant, guaranteed" intake of any of the major food groups until around 12,000 BC, when the agricultural revolution started; rather they had, for example, fruit one day, nuts the next, freshly caught antelope the next, root vegetables the next, nothing the next . . . and so on. On the "nothing day"—or maybe it extended to two days—they wouldn't have had much carbohydrate left, and yet they would have had to keep moving to find food. Now stored fat would be mobilized in large amounts: instead of entering the Krebs cycle, the two-carbon chunks of the FAs would be converted into ketone bodies in the liver (mainly acetonacetate and beta-hydroxybutyrate, mentioned earlier). These ketone bodies circulate and enter cells, where they are directly burned in the mitochondria. The resulting smelly breath wouldn't have bothered anyone, by the way, because everyone except suckling babies would have been in ketosis.

The biological reason for ketone metabolism takes us into some of the most fascinating questions of biology today: What determines aging and life span, and can we extend healthy life span? The story starts with the mystery of why the body does something seemingly futile: interconverting fat to acetyl-CoA, subsequently to ketones, and then back into acetyl-CoA. The reason started to be unraveled back in 1994 and 1995, when a research group in Japan did some reading 40 years backward in the literature[19] and studied the effect of adding the ketone body beta-hydroxybutyrate (also known as beta-hydroxybutyric acid [BHB]) to isolated, beating rat heart. The result was striking: a significant increase in cardiac output, but at lower oxygen consumption than usual.[20] Basically, the heart became more energy-efficient. The upshot of this bench research and reading was the discovery that the BHB ketones formed in this seemingly futile cycle are not identical: they are mirror images of each other (one right-handed, the other left-handed); one acts more as a fuel, whereas the other acts more as a "signal" to our metabolism to change gear and become more efficient. That greater efficiency has important survival benefits for an organism at times of food shortage, and it seems to increase life span, too.

Metabolism is highly sensitive to the "handedness" of molecules

In Chapter 1 we briefly touched upon glucose's "right-handed" and "left-handed" forms, defined by the direction in which they rotate in polarized light. Carbon excels at this mirror imagery, because it so easily forms four bonds to different chemical groups (tetrahedral bond geometry; see Chapter 1, "Carbon's 'unsatisfying' electronic configuration is its greatest virtue"). Such molecules are said to have a *chiral center* (from Ancient Greek *kheir*, "hand"). This is the case with the second (or beta) carbon atom of the ketone body BHB: it holds a CH_3, OH, H, and CH_2COOH group. Two arrangements of these are possible, which are mirror images of each other in exactly the same way as D- and L-glucose (Chapter 1, Figure 1.11).

Enzymes are almost always highly specific for one or other chiral form, and most amino acids are L, whereas most sugars are D. Most molecules exist only in one or other form in organisms, and that is what makes the discovery of D- and L-BHB acid in mammals particularly interesting. It seems as if evolution has created a situation-dependent switch, tuning metabolism to a different level. This is what happens when burning (beta-oxidation) of FAs exceeds a certain level, as we run out of glucose. The result is an "overflow" of acetyl-CoA, and a signaling to the body that it is going into starvation and needs to convert its remaining energy stores into power for muscle via an unusually energy-dense substance: fat. At this point, the excess acetyl-CoA is converted by mitochondria in the liver to *D*-beta-hydroxybutyrate (the mirror image) and released into the circulation. In the mitochondria of destination cells, it is recognized by an enzyme that only binds to this D-form, and that sets off a chain of reactions that produces acetyl-CoA in unusually high concentrations in those mitochondria. (This doesn't happen in the liver mitochondria, because they lack that crucial enzyme in the chain.)

The L-form of beta-hydroxybutyrate can certainly be used by mitochondria as a moderately efficient energy source; but it is the D-form that really turns mitochondria on. The burst of acetyl-CoA produced in this particular way in the mitochondria greatly increases their capacity for producing ATP, hence giving heart, skeletal muscle, and brain an energy boost. The body's energy economy is tuned to a higher "gear"—ketotic metabolism—for increased food-foraging activity. Recent research indicates that a ketogenic diet can even improve cognitive function from normal baseline,[21,22] and can ameliorate cognitive symptoms of certain neurodegenerative diseases.[23,24] Could mitochondria raising their game when burning fat in ketosis be considered a parallel to high-performance technology applications that exploit the high energy density of hydrocarbons (e.g., jet aircraft, cement manufacture)? Just a thought.

Signs that ketogenesis really can extend life span as well
What evolved millions of years ago in mammals had already evolved billions of years ago in single-celled organisms from bacteria through to protozoa, many of which use a polymer of DBHB as an energy storage substrate. Supposing that this pathway would make mitochondria in brain more efficient, some researchers wondered whether the defect that causes Parkinson's disease (PD) (death of cells in the substantia nigra of the brain due to failing mitochondria) could be treated with DBHB. In 2003, a promising proof-of-principle study (using mice) was published,[23] and, at the time of writing, researchers are investigating the feasibility of the ketogenic diet and ketone dietary supplementation for adjunct treatment of PD.[25]

There is also evidence that ketogenesis can help against non-alcoholic fatty liver disease (NAFLD).[26] The reason is obvious: fat already in the liver—which, in excess, causes liver disease—is a prime material for the liver's job of making DBHB. Clear out that fat via ketogenesis (production of the three ketones, including DBHB), and your liver should become more healthy. But could it be that a ketogenic lifestyle might also prevent general features of aging? Indeed, it might, and, to find out why, researchers are looking ever more closely at what's going on in the mitochondria when they are burning DBHB.

So why don't we do ketosis all the time?
So why, if it's so great, isn't this ketone-based energy economy the norm? The reasons are several, but in evolutionary terms they probably boil down to the following:

1. It consumes more energy per unit mass to make fat stores—as opposed to glycogen—from carbohydrate intake, so this is not an economy that the body wishes to put through short-term cycles at short intervals.
2. There is a risk of reducing muscle mass and becoming weaker: on a ketogenic diet one essentially removes one source of dietary energy (almost) completely—that is, carbohydrates. Proteins can, as we know, be converted to energy, though that is a very inefficient use of them. Still, if, for a period, we only have access to fat and protein, and one day we eat too little of each, we might start catabolizing our own muscle and connective tissue.

The crux of the matter is that fat is a longer-term storage substance in our bodies than carbohydrate. It's a crucial reserve for times when carbohydrates, and food in general, are scant. Evolution has equipped organisms with numerous fail-safes, redundancies, parallel systems, bootstraps, etc. Ketogenesis might be an interesting and relevant topic for potential interventions in pathologies; it's also good to know that it's a normal part of our metabolism and can have beneficial effects.

A degree of ketosis may well improve the efficiency of our mitochondria and might even extend life span. However, long-term obligatory ketosis is probably not what we evolved for—otherwise we wouldn't have much of a pancreas to secrete insulin. On the other hand, overuse of the pancreas causes diabetes, the scourge of modern culture and a major contributor to aging. Latest insights into the mechanism of ketosis indicate that it probably works to extend life span by decreasing signaling through the insulin/insulin-growth factor receptor, which leads to a cascade of changes (too complicated to describe here). This ultimately causes cells to produce more antioxidant enzymes (superoxide dismutase 2, catalase, glutathione peroxidase).[27] The expression of hundreds of other genes is also affected, but what is basically going on here is that the cell is increasing its defense against the toxic oxygen radicals produced by normal energy metabolism. By doing so, it increases the life span of all of its components. Reduce the rate of aging of individual cells, and the rate of aging of the whole organism is reduced.

We certainly don't have enough experience of people performing long-term ketogenesis, at one extreme, but we do have plenty of evidence of the bad effects of a high-carb diet at the other extreme. So reducing carbohydrates—often radically—is certainly not going to harm us: rather the reverse. The easiest route to occasional ketosis is simply to eat much less, and much less often, than usual—that is, to practice the caloric restriction that has been shown to increase life span in many model organisms, from tiny worms to mice. Furthermore, by reducing oxidative damage, the increased life span is also less plagued by the nastier degenerative conditions of aging (dementia being one of the most distressing).

The place of protein in the larger picture of metabolism and environmental impact

Linkages between energy metabolism and protein metabolism

The carbohydrate and protein metabolisms of cells are closely linked: central metabolism of glucose produces a key intermediate called *alpha-ketoglutarate*. Depending on the needs of the cell, the alpha-ketoglutarate can either continue in the Krebs cycle, making ATP (energy), or it can be drawn off by an enzyme that reacts it with ammonia to produce an amino acid: glutamate (or glutamic acid). This is the simplest of the amino acids and the precursor for most others. Plants do this by using the Sun's energy to make glucose and taking nitrate from the soil to make ammonia (proteins rely on nitrogen for the bonds between the amino acids). Joined together as a "polypeptide chain," amino acids are the building blocks of all proteins (Figure 2.1 g,h,i), structural or enzymatic. If a cell wants to grow and divide fast, it will tend toward this side of the equilibrium, which also produces the compounds that it needs to replicate its DNA (growth requires new cell nuclei with new DNA). Cancer cells, for example, invariably shift toward this kind of metabolism.

What comes from glucose can also return to glucose because all chemical and biochemical reactions represent an equilibrium: at the right temperature/pressure, concentration of starting substrates, and final products—and often with the right "catalyst" (in biology, an enzyme)—a reaction can be made to go backward as it were. To spare you further details, suffice it to say that many amino acids can be converted to glucose, others can be converted to glucose or ketones, and two amino acids out of the 20 that our cells use can *only* be converted to ketones.[*28] As long as the body still has ample fat to burn, only around 1% of the ketones circulating during ketosis come from protein degradation.

In a diet containing plenty of fat and carbohydrate (a typical "First-World" diet), excess protein is not used for energy production (ketogenesis); rather,

Ketogenic amino acids	leucine, lysine
Glucogenic and ketogenic amino acids	phenylalanine, isoleucine, threonine, tryptophan, tyrosine
Glucogenic amino acids	alanine, arginine, asparagine, aspartic acid, cysteine, glutamic acid, glutamine, glycine, histidine, methionine, proline, serine, valine

it is simply excreted in our urine as urea (made in our body by the reaction of toxic ammonia with carbon dioxide) and some creatinine. What an unforgivable waste, particularly when we consider the GHG emissions from meat production.

Some important health-related qualifications

The details of general nutrition and vegetarian diets are beyond the scope of this book, but there are a few qualifications that I should make at this point. The first is about the vitamin B complex (eight vitamins that are essential for metabolism and red blood cell production): the only substantial natural source is animal products; in fact, vitamin B_{12} is not found in plants, except certain algae, and vegans must take it in pill form. Second, it is undeniable that meat contains a higher concentration of protein than vegetable matter per unit mass. People recovering from substantial weight loss will find it easier to do so via moderate meat consumption; however, these days, concentrated forms of vegetable protein with similar nutritional qualities are also available. Even body-building is possible on a vegetarian, or even vegan, diet. Last, iron: we need this element for our blood's hemoglobin. Inorganic iron (as in rusty iron nails, for example) is very hard for the body to absorb. The form that the body is "used" to absorbing is *heme*, the organic iron-containing molecule that makes vertebrate animals' blood red and that carries oxygen.

Doctors continue to advise anemics and women who are still menstruating to think hard before giving up red meat; hospitals, wishing patients who have lost much blood (as a result of an operation or accident) to recover as quickly as possible, will include a moderate amount of red meat in food as a routine measure. Animals, and *only* animals, possess heme. This leads us to an unavoidable reality: humans have evolved biologically to be omnivores. There is nothing that we can do to change that, and we should not be ashamed to be occasional meat-eaters. What we *should* view critically is the source and quantity of the meat that we eat against the background of global population growth. Essentially, by raising animals for human consumption we are converting energy-rich carbohydrate- and hydrocarbon-containing vegetable feeds (with some plant protein) into a different class of food (animal protein) that provides us with no energy except in extreme situations. This protein is very energy-intensive to make compared with plant protein, because large creatures must be kept alive and fed for years before slaughter. If a normal healthy person eats too much meat, the body simply excretes the excess. That is a high price to pay for the taste and texture.

We evolved biologically and culturally as mere opportunistic meat-eaters

It is, moreover, obvious that we did not evolve to eat much meat, and one doesn't even need to go back to ancient history to witness that truth: 1960 will suffice, a year when, conveniently, the world population was almost bang-on 3 billion, and total meat consumption was around 70 million tonnes per year.[29] That equals roughly 23 kg per head per year. Already in 2010, with a population of almost 7 billion, the human world was consuming 300 million tonnes of meat per year, or roughly 43 kg of meat per person. China exceeded Europe in total meat consumption in 2000 (though, of course, not in per capita consumption), making China the fastest growing meat consumer globally. Just as remarkable is the observation that India, where vegetarianism is a widespread cultural attribute, barely increased its total meat consumption between 1960 and 2010, let alone per capita. I won't go into details of the proven health benefits of a low-meat diet (particularly low red meat). Briefly, they include lower incidence of coronary heart disease and bowel cancer. However, one health-related incident produced an interesting temporary downward turn in European meat consumption, letting China past, as it were: it occurred around 1990, amid increasing public concern about beef from cattle with the bovine form of scrapie (a disease of sheep)—bovine spongiform encephalopathy (BSE). After the practice of feeding minced-up sheep remains to beef cattle was stopped, and public trust in the meat industry had improved, the trend of increasing meat consumption in Europe picked up again.

But we are continuing to eat more and more meat per capita
A 2018 study presented a model showing that removing animal-raising from agriculture in the United States would reduce GHG emissions from agriculture by 28%.[30] The UN Food and Agriculture Organization (FAO)'s Food Outlook (a biannual report on global food markets) for 2018[31] notes that, from 2000 to 2014, global production of meat increased by a staggering 38% as some nations' gross domestic product (GDP) grew and more people could afford more meat; but also—disappointingly—as meat production in disgraceful conditions became ever more efficient, and the price fell. The trend in meat consumption until 2030 is predicted to be quite similar. Water shortages also threaten: 1 kg of beef for human consumption is estimated to take 15,400 L of water (15.4 tonnes) to produce.[32] In the process, it creates around 51 kg of $CO_{2eq.}$ just on the farm, and a further 9.4 kg from the many processes involved with feed, waste management, slaughter, and transport to the consumer[33]: a total of 60.4 kg $CO_{2eq.}$ per kg beef. The energy requirement to produce that 1 kg of beef is around 251 MJ.[34]

Here's a thought experiment based on being able to produce 1 L of renewable diesel (e-fuel) for a price of €2 (not an outrageous prediction). The diesel equivalent of the 251 MJ per kilogram beef is 6.7 L, and, if it's synthetic diesel at a cost of €2 per liter, then we're in the ballpark of around 13 Euros per kilogram—basic cost—if we want to produce that beef with renewable hydrocarbon fuel. I surveyed the price of 1 kg of supermarket beef in Germany in January 2022, and it typically ranged between €5 and €11 Euros. In some supermarkets it can even be cheaper than fruit and vegetables per kilogram. There are some expensive cuts, ranging up to €45 (e.g., for prime tenderloin), but that is a small proportion of the whole animal. When one takes into account minced beef (which can sell for as little as €3 per kg), we're talking about a highly discounted animal—discounted at the cost of the environment. Furthermore, its environmental impact during rearing until slaughter is enormously disproportionate to its contribution to human nutrition (Box 2.2).

How on Earth can everyone in the chain from farm to fork make any money out of this meat economy? Easy: many don't. The farmer makes very little money but is compensated by large farming subsidies (from the European Commission or, in the United States, from the Department of Agriculture); the freight lorry driver barely earns enough to feed himself; the people stacking the shelves in the supermarket are on an absolute minimum wage; the supermarket itself may well

---

Box 2.2 The carnivore's quandary: growing a whole animal over several years

Consumption of meat from ruminants (mainly beef and lamb) has the greatest impact in terms of GHG emissions—CO_2 mainly from the energy needed to raise and keep the animals until slaughter and a substantial component of methane produced by the animals' digestive system. It was estimated in a 2019 publication that US beef has a GHG footprint of 21 kg CO_2 equivalents (i.e., including methane) per kilogram carcass weight.[35] By the time it reaches the butcher or supermarket, the meat (which is around 65% by mass of the carcass[36]) has a footprint of at least 32 kg $CO_{2eq.}$ per kilogram meat: that is, without factoring in butchering, transport, and packaging. To put that into perspective, an average American (and Europeans are not far behind) currently eats upward of 120 kg meat per year[37]: 25 kg of that are beef, and a further 23 kg are other red meat,[38,39] making a total of 48 kg of boneless trimmed meat. Taking just the beef now: 25 kg of meat at the counter infers 38.5 kg carcass, and that equates to 809 kg $CO_{2eq.}$, which is the same $CO_{2eq.}$ as produced by combusting 355 L of gasoline. That would equate to driving an average modern gasoline car more than 5,000 km. Multiply that by four for a four-person family, and one gets 20,000 km driving equivalents (12,400 miles) just from beef; add in the rest of the red meat (another 23 kg per head per year), which has a markedly lower $CO_{2eq.}$/kg, and the figure is probably around 7,500 km driving equivalents per person, or 30,000 km (18,600 miles) per four-person family. That's 30% *more* than a full year's driving (based on US averages) for one mid-size modern gasoline car. Next we could add milk products. . . .

make the largest "earned" margin. The only reason why this "works," mathematically at least, is that the energy used to produce that 1 kg of beef is not remotely renewable, and it (the fossil fuel, that is) is traded at a very low price—basically a tradeoff against environmental damage. That this situation must change is undeniable, but the trend is the opposite.

Energy inefficiency and GHG footprints of animal products: The cases of milk and meat

It's so accepted in the West that hardly anyone questions its justification. However, averaged across the globe, 1 L of conventional milk is estimated to make 2.4 kg of $CO_{2eq.}$ in its production and metamorphosis into food or drink[40]—roughly the same amount as released by burning 1 L of gasoline. Many people don't think twice about eating animal products several times per week. Might they, if it were compared with car driving?

A sobering analysis of milk consumption in fossil-fuel equivalence

Question: How many liters of milk are consumed each year compared with liters of gasoline/diesel burned in road transport? First, I'll define milk consumption as the total volume of milk used to make all dairy products consumed by people: in this context, averages of milk consumption in liters per person per year range between around 250 (USA) and 300 (Europe).[41–43] Considering the United States, if all that milk is produced via "conventional methods" as opposed to more sustainable agriculture and animal-husbandry, it equates to the burning of 250 L of gasoline per person per year in terms of consumer-side $CO_{2eq.}$ emissions. If we now imagine an average family of four people, that's 1,000 L of milk = 1,000 L of gasoline in CO_2-equivalence terms. If that family has an average mid-sized car, with a gasoline consumption of roughly 7 L per 100 km (33 US miles/gallon; 40 UK miles/gallon), that equates to around 14,300 km driven in terms of pure consumer-side CO_2 emissions (a compensation for well-to-tank [WTT] emissions is made in Figure 2.6).

This is an upper limit figure, because some milk products are based on more environmentally sustainable milk production,[44] but various comparisons of the GHG emissions from conventional compared with ecologically sensitive farming show the difference to be relatively small (e.g., the study by Flysjö[45]). More importantly, it shows the scale of the problem and how much impact overconsumption can have. On a European level, differences in method produce a range of 1.77 to 2.4 kg $CO_{2eq.}$ per liter milk, with a mean of 2.05.[44] Even at the lower figure, milk consumption by an average four-person family is comparable

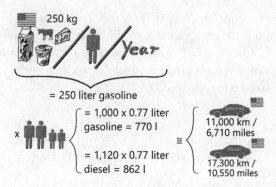

Figure 2.6 Average milk product consumption in the United States produces $CO_{2eq.}$ emissions in the same ballpark as conventional car-driving. Assuming an equivalent of 250 L of milk (as milk and milk products) consumed per year per person in a variety of dairy products in the United States (see earlier text references), and a CO_2 emission equivalent of burning 1 L of gasoline per liter milk, we have 250 L of gasoline per person per year; and in auto diesel CO_2-equivalence, that works out at 280 L of diesel. Applying a well-to-tank energy adjustment (fuel production "costs" roughly 20% of the final fuel energy) these volumes amount to roughly 192.5 and 215.6 L, respectively, in the tank. For a four-person family, that translates to 770 L gasoline and 862 L diesel. Assuming modern medium-sized car real-world consumption figures of 7 L gasoline and 5 L diesel per 100 km, these volumes of fuel produce around 11,000 km/6,170 miles for the gasoline car and 17,300 km/10,550 miles for the diesel car. Translated into months driving, at an average US value of 20,500 km (12,800 miles) per year,[46] we arrive at 6.4 months driving for the gasoline car and 10 months for the diesel car.

in magnitude of GHG emissions to other sectors of fossil fuel use averaged across the population. It's certainly on a par with car driving (Figure 2.6).

How can milk product consumption equate to so much car driving? Are we roughly in the right ballpark, or has something gone horribly wrong in the calculations? Here festers one of the largest problems in the accounting of GHG emissions. Figures of between 13% and 20% of total anthropogenic GHG emissions arising from agriculture are given in a wide variety of literature on the subject, but they overwhelmingly leave out the following: fossil fuel consumption by farm machinery, feed transportation, and fertilizer production (which are usually reported, respectively, under energy, transport, and industry); even more worrying, emissions due to land-use changes are usually reported under forestry-related emissions instead of agriculture.[45] It has been known for more than a decade that if one classes all of these as agriculture-dependent—which seems the justifiable thing to do—the GHG emissions from agriculture are between 30% and 40% of the anthropogenic total (e.g., see Foley[50]). And more recent full-picture analyses are even more worrying: a 2020 report by Greenpeace concludes that, in 2018, the total GHG emissions ($CO_{2eq.}$) resulting from the industry of just livestock-keeping in the European Union was significantly more than the CO_2 emissions from all private cars and commercial vans in those countries (704 million tonnes of $CO_{2eq.}$ compared with 656 million tonnes for cars/vans).[51] Many other high estimates for the GHG impact of agriculture exist (e.g., see WorldWatch[52]), but one also has to be critical when reading them, check their justifications for including various sources, and be sure that they are talking about $CO_{2eq.}$. A 2021 reanalysis of 2010 data concludes that more than 40% of global $CO_{2eq.}$ comes from food production, and more than half of that arises from animal-based food and feed for animals.[53] Next, I look at livestock in general, and beef in particular.

Livestock competes with transportation in terms of GHG emissions
Let's consider $CO_{2eq.}$ emissions from agriculture in the United States and in Europe. Recent figures from the European Union's statistics department[54] claim that agriculture accounts for about 10%, in $CO_{2eq.}$ terms, of total EU GHG emissions. However, that report only takes us up to 2015. The next scheduled report was due in 2020, but, at the time of this writing, it had not yet appeared. What *did* appear was the Greenpeace report mentioned earlier,[51] basically stating that animal farming in the European Union produces more $CO_{2eq.}$ emissions than all cars and vans combined: 17% compared with 14.5% (vans produce an extra 3.7 on top of the 10.8 from cars). How can that be, given that, according to the European Commission only five years earlier, the *whole of agriculture* was emitting only 10%? One explanation could be that the most recent analysis was done by Greenpeace rather than a governmental organization, and included more components of the larger picture. Analyses of farming are no less complicated than analyses of other sectors of human activity, so there could well be reasonably large errors. However, when one puts the afore-mentioned figures together with the worst-case scenario for $CO_{2eq.}$ emissions caused by milk products, it all "fits" quite well. At least they don't seem to be out by a factor of two: more likely within the range of normal data collection errors. Looking at energy consumption is another way of assessing the scale of agricultural impact. According to a 2018 study, animal farming consumes a little over 50% of the total energy that goes into farming in general,[55] a figure that is essentially supported from other perspectives (e.g., mass of feed consumed).

As reviewed recently,[56] already at the turn of the millennium 60% of biomass entering agriculture went into livestock. Again, figures from different sources seem to "ring true." Agriculture is an extremely large energetic, environmental, and climatic burden. It is also a very sensitive matter, being linked to culture and the economy—features that probably make it very difficult for the European Union (or any other geographic area, for that matter) to make major inroads in terms of energy balance and CO_{2eq} emissions. The European Union, for example, is shackled to farming subsidies and increasing mechanization of meat production, driven by consumer habits. The challenge of reducing the environmental impact of farming is every bit as great as that of reducing the environmental impact of transport: in fact, I believe it's greater. We may hear much more about the transport-related aspects of the European Green Deal because, with respect, it's simpler to appear to be achieving transport-related goals. It is much easier to externalize (transfer out of the European Union) environmental impacts of car production than it is to externalize environmental impacts of agriculture.

One thing is certain: by reducing our consumption of animal products, we can achieve significant reductions in CO_{2eq} emissions from agriculture—reductions that are, on the basis of the figures mentioned earlier, also significant in terms of total CO_{2eq} emissions. Already in 2019, an international group of scientists, drawing on reliable data that are updated annually, had come to the following conclusion:

> Profoundly troubling signs from human activities include sustained increases in both human and ruminant livestock populations, per capita meat production.... Eating mostly plant-based foods while reducing the global consumption of animal products (figure 1c–d), especially ruminant livestock (Ripple et al. 2014), can improve human health and significantly lower GHG emissions (including methane in the "Short-lived pollutants" step). Moreover, this will free up croplands for growing much-needed human plant food instead of livestock feed, while releasing some grazing land to support natural climate solutions (see "Nature" section).[57]

How much, in concrete terms? Using figures as of 2016, we see that beef production is by far the largest single emitter of CO_{2eq} per kilogram in animal farming: in terms of efficiency of conversion of feed calories into product, beef comes in at 1.9%, whole milk at 24%.[58] One kg of beef produces upward of 60 kg of CO_{2eq}.[59] in its rearing. Other studies claim even higher values (99.48 kg per kg beef product; in second place being lamb and mutton at 37.92 kg[60]). Current beef consumption in Europe is between 10 and 11 kg per person per year.[61] That equates to roughly 600 kg of CO_2, or the equivalent of 2,900 km driving in a medium-sized modern gasoline car, or 3,500 km in a medium-sized modern diesel car (including correction for well-to-tank emissions). For a four-person family, therefore, simply eating an average amount of beef produces CO_2 emission-equivalents of comparable impact to a year's car driving (Figure 2.7). A recent literature review concludes that, in comparison with an average omnivorous diet, switching to a vegetarian diet reduces GHG emissions by 24–56%, and becoming vegan can reduce GHG emissions by 21–70%.[62] The large variability in these figures can be explained by, for example, certain vegetarian food

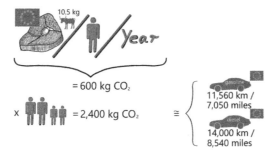

Figure 2.7 Average beef-eating in Europe produces $CO_{2eq.}$ emissions comparable with average car-driving. 2,400 kg of CO_2 equate to 2,400/2.28 L burned gasoline, which = 1,050 L gasoline; and 2,400/2.64 L burned diesel, which = 909 L diesel. Applying a percentage of well-to-wheels emissions of 23%, produces figures of 809 L gasoline and 700 L diesel. Assuming modern medium-sized car real-world consumption figures of 7 L gasoline and 5 L diesel per 100 km, these volumes of fuel produce 11,560 km/7,050 miles for the gasoline car and 14,000 km/8,540 miles for the diesel car. Translated into months driving, at an average European value of 12,000 km per year, we arrive at 11.5 months driving for the gasoline car and 14 months for the diesel car.

sources that are not environmentally sustainable (e.g., soya products from soy grown on land stolen from rain forests).

Methane and nitrous oxide are the most worrying components

Attempts to be comprehensive when calculating GHG emissions from agriculture lead to figures between 23% and 28% of total global GHG emissions.[63] However, a major question remains about whether reported figures include land-use changes. Including them certainly seems justifiable, and, if one does that, figures as high as 35% can emerge as percentages of total anthropogenic $CO_{2eq.}$ emissions (figure calculated from Our World in Data[63]). Methane (CH_4) and nitrous oxide (N_2O) emissions are particularly problematic agricultural emissions because they contribute many-fold more than CO_2 to the global warming potential of agriculture over 100 years (GWP_{100}) per unit mass. Furthermore, together they are released in larger quantities from agriculture than CO_2: as a percentage of total agriculture-related GHG emissions, CH_4 constitutes 36%, and N_2O 32% (estimated from data in Our World in Data[63]). This gives global agriculture a role in global warming over 100 years that surprises many: Could it, as speculated, be responsible for 50% of current increases in global average temperature?

The largest percentage of methane in the mix comes from enteric fermentation in cattle: latest figures from Europe reveal cattle methane as 36% of the total agricultural $CO_{2eq.}$ emissions including fossil fuel uses.[64] Furthermore, some researchers believe that the GWP_{100} of methane has been significantly underestimated and that its mere $CO_{2eq.}$ values do not present the true picture.[65,66] Maybe we need a new framework for assessing the currently very uncertain impacts of emissions of $CO_{2eq.}$ In addition, land-use changes—which are prominent emitters of GHG—are rarely counted in agricultural impact, and recent research suggests that these changes are four times greater than previously estimated.[67]

The largest single CO_2 emitter (as opposed to $CO_{2eq.}$) in the agricultural chain is production of ammonium nitrate fertilizer from ammonia made via the Haber-Bosch process[68] (reacting nitrogen with hydrogen at high temperature and pressure with a metal catalyst). That CO_2 is relatively small in quantity: each tonne of ammonium nitrate creates 3 tonnes of CO_2 in its production, and 50.6 million tonnes of ammonium nitrate were produced in 2020,[69] equating to 0.42% of CO_2 emissions from fossil fuel consumption. However, it leads to a much larger problem: ammonium nitrate that soaks into soil and is not taken up by crop plants is largely converted by microbes to N_2O, which has a GWP_{100} of 273 times that of CO_2.[70] Land-use changes are a major contributor to GHG emissions, and it is invariably reported that taking land from forests and turning it into agricultural land results in a net release of GHG. Using as little land as possible for the greatest yield is, therefore, important.

Concerned scientists emphasize the urgent need for more vegetarian diets
Summarizing the worst of humans' part in climate change, a report supported by 11,000 scientists appeared in the journal *BioScience* in late 2019 (publication date 2020).[57] Drawing on primary literature publications extending over a period of almost 40 years, from the 1980s onward, the report's most alarming graphic shows plots describing trends in human civilization, from population growth, through livestock production and loss of rain forest, to GHG emissions and energy trends. All crucial measures are going in the "wrong" direction: further and further away from sustainability. Interestingly—and probably not by coincidence—the rates at which ruminant livestock production, human meat consumption, and energy consumption from fossil oil are increasing are very similar. In 2021, Ripple et al. published a depressing update.[71] Already since 2014, it has been known that the raising of ruminant animals for meat and dairy production is the largest single sector of agriculture contributing to anthropogenic GHG emissions.[72]

One obvious question is "Do we need to each so much meat?" The answer (known for centuries, if not millennia) is "no." And if we want scientific support for it, how about 11,000 scientists' call for humans to adopt more plant-based diets (as published in the 2019 report[57]). Seen across human civilization, many cultures have always been almost, or completely, vegetarian. If many of us don't need to eat as much meat as we do at present, what are the consequences for our bodies if we do, never the less? Do we benefit? The answer to that one is also "no," but there are some exceptions, as I discussed in "Some important health-related qualifications." On average, diets high in meat—particularly red meat—are correlated with higher incidences of cardiovascular disease, colorectal cancer, diabetes, and stroke than diets poor in red meat and rich in vegetables, as reviewed in Clark et al.[73] Notable is the significant difference between the health risks of a high intake of processed red meat (e.g., hamburgers) compared with unprocessed red meat: a 500 g steak is "better" for you than a 500 g burger. Considering all indicators of health risk and environmental damage, fruit, vegetables, and plant-derived foodstuffs emerge as significantly better than meat.[73] Fish might be said to be somewhere in between and have an environmental impact depending strongly on its origin (organic fish farm or caught in the open sea). The likely mechanisms of increased risk of certain diseases in a meat-heavy diet are beyond the scope of this book; interested readers are directed to Oteng and Kersten[18] and Seiwert et al.[74] and for an overview of the health-benefits of plant-based diets to Cena and Calder.[75] One interesting by-product of occasional meat consumption during human evolution is the use of fire and cooking, as I'll explore before closing this chapter.

How wasting food exacerbates an already terrifying trend
Wasting animal products is as bad as it gets, but wasting anything must be avoided. How much are we wasting? For Europe it's 173 kg of foodstuffs per person per year,[76] compared with total consumption of around 540 kg per person per year.[77] That's 32% by mass! In the United States, we're talking about annual wastage of around 40.7 million tonnes total per year[78] or around 123 kg per person. Again, there are complications in estimating such figures, but in Europe and the United States, somewhere in the region of 25–30%-plus mass of food is wasted.

What are the figures for poor parts of the globe? Unfortunately they don't look good, but for different reasons. Whereas in Europe and the United States much food is wasted by consumers or simply not sold by stores because of formalities such as sell-by dates, in Africa an enormous amount of produce goes bad before it even reaches the shops because of inefficient harvesting, storage, and distribution: up to 25% of grain and 50% of fruit.[79] What remains is most certainly not wasted by consumers in anything like the same quantities as in richer countries! The figures globally are probably in the region of 30% by mass of food wasted. And the $CO_{2eq.}$ of that wastage was recently estimated at 6% of total anthropogenic GHG emissions.[80] In comparison with the reporting of the need to reduce transport-related CO_2 emissions, this hardly gets a mention, despite being twice the $CO_{2eq.}$ of global air travel.

Food in flames: How humans' long relationship with fire formed tolerance to many carbon compounds

If we imagine that, once in a while—and much more rarely than our present-day visits to the supermarket to buy a hunk of super-cheap meat—our paleolithic ancestors caught an animal, how would they have eaten it? We know that humans had fire by then (paleo-archaeological finds indicate the taming of fire around

2.7 million years ago—the lower paleolithic). By around 200,000 years ago, our ancestors were building simple hearths with stones. Though not sedentary at that point in history, humans stayed in one place long enough to make a central point of meeting, socializing, and eating: back then, as now, the kitchen was the center of the clan. Even our modern word "focus" (the center of a human activity) is derived from the Latin for "hearth," *focus*, from which the Italian *fuoco*, meaning "fire," stems. This paleolithic hearth served as a source of warmth, deterrent against predatory animals, and as a place to cook meat: literally the first barbecue. What carbon compounds in the meat and wood did during the paleo-BBQ is quite revealing about what the livers of paleo-humans had to start dealing with (Figure 2.8).

Figure 2.8 The Paleolithic barbecue: Principal components of wood smoke and charred meat, showing only generic compounds to represent larger classes. Methane is included for completeness, despite not being classically harmful. Soot particles have a distribution mode (most frequent size) of between 0.1 and 0.2 micrometers,[81] similar to cigarette smoke. The polyaromatic hydrocarbons (PAHs) can further be subdivided into anthrenes, anthenes, and anthracenes. In meat, both PAHs and heterocyclic amines (HCAs) are produced at high temperature. The HCA shown is formed by reaction of L-phenylalanine (an amino acid) with creatine (an energy storage amino acid derivative that is particularly concentrated in muscle).

Cooking directly with fire put many "unusual" carbon compounds into our diet (and into our lungs)

Our closest modern-day equivalent to paleolithic meat cooking is, indeed, the barbecue, and studies show that cooking meat in this way produces a variety of compounds that are potential or actual mutagens: they can mutate the DNA in our cells, and mutations in our DNA have the potential to lead to cancer. Carcinogenic compounds have been shown to be present in barbecued meat (i.e., meat cooked at relatively high temperatures of 180–250°C). The two classes that are most relevant to every-day life are:

- *Heterocyclic amines* (HCAs): These are formed via the *Maillard reaction*, named after the French physician and chemist, Louis Camille Maillard who, in 1912, published a paper describing how amino acids and carbohydrates react to produce cyclic compounds containing nitrogen. Hence, during searing or grilling, the brown color on meat is produced by the Maillard reaction of amino acids from the meat proteins, and creatine, with glycogen (animal starch) and glucose contained naturally in the muscle tissue of the animal. These HCAs taste great to humans, and that's why we tend to prefer frying and grilling meat to boiling it. Paleolithic humans didn't have that choice because free water was often unavailable, and they hadn't invented the cooking pot yet. More pragmatically, direct fire was probably simply an expedient way of getting meat up to the kinds of temperatures where it becomes more tender, is more digestible, and tastes better.
- *Polyaromatic hydrocarbons* (PAHs): These are produced mainly by cooking involving smoke, and hence they also occur in smoked foods, such as smoked fish, smoked sausages, etc. PAHs are produced to a certain extent when we fry meat in a pan, particularly if the temperature is high (e.g., when searing beef) or when we have taken our eye off the pan for too long, and the surface of the meat/meat fat and frying oil start to burn. But the superlative domestic PAH production facility is the good old barbecue. Here, charcoal burns, producing smoke, often invisibly (the particles are too small to see); meat fat drips into the glowing embers, ignites and burns inefficiently, producing ample smoke: basically, we get charring and smoking at the same time, and the smoke contains PAHs identical to the ones produced in stars, interstellar clouds, and upon fossil fuel combustion. Fullerenes (or Buckminster fullerenes, the geodesic bucky balls) are not, however, present, because we don't reach that kind of stellar energy level on our patio, back garden, or terrace. Still, the PAHs are interesting enough because they are also DNA mutagens.

Some are mutagenic, but the link with human cancer is not simple

How do we know that these carbon compounds in cooked and smoked meat are mutagenic? It's certainly not by studying people who eat a lot of barbecued food or studying people at all. For this, we have to go back to the second half of the 20th century and a New York-born biochemist and geneticist by the name of Bruce Ames (1928–). In the late 1960s, Ames was developing a test for mutagenicity of compounds with which humans come into contact—a test that would be simple and sparing of animals. Animal welfare was starting to gain momentum with the publication of the three Rs of reduction, refinement, and replacement in research animal use by Russel and Burch in 1959. *Salmonella typhimurium*, a close relative of the salmonella bacterium that causes typhoid fever, was Ames's object of study. Ames genetically engineered the *S. typhimurium* to contain three types of mutation: one in the enzyme pathway that makes the amino acid histidine, one in part of the bacterium's DNA damage repair machinery, and a third in an enzyme involved in building and maintaining the bacterial cell wall. The details of the mutants need not concern us here, but the idea was clever: with these mutations in its DNA, the bacterium cannot grow and multiply in a medium that lacks the amino acid histidine. Expose it to chemicals that mutate its DNA, and, sooner or later, mutations will occur in the spots that Ames engineered to cripple the bacterium and return them to their original (or equivalent) identity: a strain will emerge that is able to grow in the absence of histidine and/or repair its DNA and/or multiply because it can build new cell walls. The sooner/later bit gave an indication of the potency of the mutagen: the more mutagenic the chemical in question, the sooner a so-called rescuing mutation would occur, and the sooner one would see colonies of bacteria expanding from a discrete spot on the otherwise barren agar petri dish. The test was officially implemented in 1973.

What happens in our liver makes a big difference

What does it mean if we say that compounds identified as mutagenic are also carcinogenic, inasmuch as they increase the risk of cancer in humans and animals? Here we enter a topic that increasingly frustrated the Ames test and contributed significantly to a major debate on the usefulness of model systems to mimic human conditions. What the Ames test could not do was to mimic the human liver or that of any other animal. And why should it need to do that? Because many organic compounds that we ingest in our diet—either as constituents or contaminants—are altered by the liver in so-called *first-pass metabolism*: the blood from the lining of our intestines that absorbs nutrients basically goes through the liver before it goes anywhere else. Here enzymes (principally the P450 cytochromes) detoxify noxious chemicals that we don't want to hang around inside us, hence making them more soluble for excretion via our kidneys in our urine. P450 cytochromes are also present in other cells—notably those lining the lung, where they deal with noxious substances that we breathe in.

The problem that evolution only partly "saw" (as we will see later) was that in "dealing with" toxins, P450 enzymes convert certain organics known as *pro-carcinogens* into carcinogens: they are made less acutely harmful to us, alright, but the tradeoff is that they contribute in the longer term to an increased risk of cancer. This is exactly what happens to heterocyclic nitrosamines from food: as such they are not carcinogenic, but, after P450 processing in the liver, they can attack DNA and chemically modify it. So, probably to the horror of Russell and Burch of the three Rs, researchers started mashing up rat liver and adding it to the Ames test to mimic the first-pass metabolism that created the true carcinogen from the precursor. Rats were chosen because they are omnivores that live around humans and consume human food waste. Hence they were assumed to have quite similar liver metabolism to humans; pigs are too large and ethically perhaps more questionable. The rat liver is, indeed, a pretty good research substitute for the human liver. However, the whole rat is not a good substitute for the whole human, and the metabolism of PAHs and its relation to cancer is complicated.[82] Here is the rub.

Why modeling humans with rats is not ideal in cancer research
Rats live for a maximum of two years in the wild; in captivity perhaps two-and-a-half. As rat owners know, most animals die from one or other type of cancer. Over their short and very fecund life spans, rats were left by the wayside in evolutionary terms with respect to resistance to cancer: by the time they die, they will have been able to reproduce many times, passing their genes on to many surviving offspring. Humans have a different life-history strategy, being able to endure conditions of famine, during which they become infertile, and saving resources in order to reproduce in small numbers but over a much larger time span. Put crudely, humans, on average, are more resistant to cancer than rats: evolution has endowed us with better cellular antioxidant-based protection and better DNA repair efficiency. Furthermore, and of relevance to carbon compounds from our paleolithic barbecue, humans have evolved consuming potentially carcinogenic products of high-temperature, smokey meat cooking, and smoke in general (much more so than rats).

Because of our nutrition and obsession with fire and meat-carbonization, we have been eating and breathing in all manner of aromatic and polyaromatic carbon compounds on quite a regular basis for hundreds of thousands of years. Natural selection has done its invisible work on us. Humans and rats have very similar liver metabolism response to *immediate* toxicity because, regardless of longer-term impacts, highly acutely toxic substances are a game-stopper for all organisms, whether long- or short-lived. Humans are, however, much more resistant to many carcinogens than are rats. Interestingly, Ames himself published a paper in 1990 that concludes "at the low doses of most human exposures, where cell killing does not occur, the hazards to humans of rodent carcinogens may be much lower than is commonly assumed."[83] A further related observation is known as "Peto's Paradox"; that is, although one would expect cancer incidence to correlate with the number of cells in an organism, larger organisms (having more cells than smaller ones) overwhelmingly have *lower* cancer incidence than do smaller ones. The explanation might well be related to life span: large animals usually live longer than smaller ones, and their reproductive strategy has evolved together with life span to maintain longer fitness: this they do by having more effective protection against reactive oxygen species (ROS; shorthand for "strong oxidants"), which are one product of the reactions of PAHs and HCAs that enter us. There are a few notable exceptions (e.g., the small, naked mole-rat), but these organisms also have exceptional biochemistry and/or biology in general.

As well as oxidizing the fabric of cells, ROS cause carcinogenic mutations in DNA. Studies on rats have shown, however, that even when an environmental toxicant causes changes in DNA (so-called *DNA adducts*), these are not universally linked with a mutation in the DNA sequence; furthermore, cancer does not always arise at the location (tissue) of highest DNA adduct formation in an animal. Cancer is a very complicated topic, and studying and comparing cancer between rats and humans is even more complicated. As concluded by a 2008 literature analysis, "in several mammals a highly species-specific metabolism of PAH could be observed allowing a critical view to the extrapolation from animal experiments to the human situation."[84] Moreover, even within the group of rodents, susceptibility to toxicity from a variety of environmental pollutants can vary greatly, posing the question "Which rodent should we use to model human susceptibility?"

What does all of this mean for substances in our environment? Interestingly, there was a move to classify styrene (a fully unsaturated cyclic hydrocarbon and well-known as its polymer "polystyrene" and "expanded polystyrene," as well as a solvent for plastic resins) as a human carcinogen, largely based on studies in mice. A 2013 research paper, by contrast, shows that the only potential carcinogenic mechanism of styrene (one that, in this case, didn't mutate DNA) was an irritant effect specific to the mouse lung and not relevant in humans: styrene was recommended not to be included in the list of known and likely carcinogens.[85] Clearly, we can't become frightened of everything that surrounds us in everyday life, so it's important to investigate claims and become concerned about the right things. Much more research is needed. While we are in the lung, I want next to complete the picture of the carbon compounds that started entering humans in much larger quantities as of their taming of fire.

The smoking that came before smoking: Fumes from the ancient barbecue
Let's start with modern smoking. There is no doubt that it is—among other things—certain products of incomplete combustion of tobacco leaf that are the main causes of lung cancer in smokers. But which? Tobacco smoke certainly contains PAHs and HCAs; furthermore, numerous studies of these compounds in whole rodents or in suitable rodent cells in culture have associated them with both chemical derivatives of DNA (the so-called DNA adducts, caused by reaction of PAHs and HCAs with DNA bases) and carcinogenic alteration of rodent cells (so-called transformation to the cancer phenotype). However, there has been increasing skepticism about whether these DNA adducts in *human* lung cells are truly caused by PAHs and HCAs,[86] in particular suggesting that it could well be non-aromatic compounds in tobacco smoke that are the major DNA mutagens. It's important to note that the authors of the cited study are all from independent academic institutions (University of Louisville, USA; King Faisal Specialist Hospital and Research Center, Saudi Arabia; University of Kentucky, USA; Fox Chase Cancer Center, USA; National Cancer Institute, USA). These are not the only authors who suspect that something less intuitive is going on in tobacco-smoke-induced lung cancer: in 2018, an unrelated group of researchers from similarly independent institutions published a paper claiming that aldehydes, and not PAHs or HCAs, are the dominant carcinogens in tobacco smoke.[87] Other research supports this notion.[88] Even the purely gaseous phase of tobacco smoke can stimulate lung cancer in rodent models; but are rodents representative of the situation in humans? Again, a complicated matter.

Probably produced selection pressure for "genetic tolerance"
There are clearly "modifier" genes that can greatly influence the risk of lung cancer in humans: naturally, smoking is unhealthy per se, and it increases the

risk of lung cancer in everyone who smokes, not to mention chronic obstructive pulmonary disease (COPD). However, the majority of smokers (85–90%) don't develop lung cancer,[89] an observation that might be explained by nutritional components such as vitamin B intake and different individuals' slightly different single-carbon metabolism.[90] As an aside, between 1992 and 2021, National Institutes of Health (NIH) figures show lung and bronchus cancer incidence to have declined by 35% in the United States.[89] Together with lifestyle and health in general, genetic variability contributes importantly to the difference in cancer susceptibility across the population. Though progress is being made, the exact mechanism whereby PAHs and HCAs cause cancer in *humans* is not unequivocally known. We essentially conclude that they are carcinogens (or pro-carcinogens) from the following observations: they are present in tobacco smoke; they are converted from pro-carcinogens by P450 enzymes into chemicals that can attack DNA; there is a positive correlation between tobacco smoking and lung cancer risk; and mouse (and other rodent) studies, where they are intentionally administered in dose-controlled ways, indicate carcinogenic properties.

Humans are unlikely to have evolved smoking cigarettes, and mice or other rodents certainly haven't. But humans *have* evolved for between 200,000 and 2.7 million years with relatively regular exposure to wood smoke and oil smoke constituents (both inhaled and ingested). If we take a median value, that translates to at least 48,000 generations: more than enough reproductive cycles for natural selection to take place. Is there evidence that some kind of *genetic adaptation to combustion products* took place? We cannot reliably reconstruct our paleolithic past, but there is growing evidence that such adaptation may well have happened. The observations that only 10–15% of smokers develop cancer and that some can live to a ripe old age prompted two researchers at medical schools in California to look for a genetic cause. In 2015, they published a paper revealing results of genetic analyses of elderly smokers compared with non-smokers.[91] Their conclusion: a particular genetic make-up (to cut a long story short) was associated with a 22% increased probability of reaching the age of 90–99 compared with individuals not possessing that genetic network.

It seems that humans have higher tolerance than other animals to organic combustion products
Could this network be part of a larger genetic make-up that protects all of us against combustion products, and how strong would its protection be when we have greatly outlived our short-lived cave-dwelling ancestors? Mounting evidence suggests "yes, it probably is," and the selection pressure would be relatively strong, because—contrary to most infants today—infants would have been inhaling and eating combustion products from birth. We must now consider

biological age (i.e., how degraded an organism is, regardless of years [chronological age]). Neolithic people aged much more quickly than most present-day people, and so the biological age at which resistance to smoke-related carcinogens played a role would be equivalent to a much older present-day person in terms of years. All of the antioxidant defense and DNA repair mechanisms degrade much more slowly in present-day humans, hence probably extending the anti-cancer genetic benefit of our forebears to much later points in our life.

This scenario is relevant to all humans. Fire, and hence smoke, is not some kind of luxury that only well-to-do humans could enjoy; rather, it was an important selective advantage—certainly at the time that *Homo sapiens* radiated from Africa. If you couldn't make and use fire correctly, you were at a severe disadvantage. It protected, warmed, and improved nutrition. This smoke contains anthracenes, pyrenes, and derivatives of benzene,[92] as does exhaust gas from kerosene, diesel, and gasoline combustion.[93] And humans who didn't succumb easily to wood smoke, either via acute toxicity or longer-term cancer, were bound to reproduce more on average. Those who were more sensitive likely reproduced less, and hence we can assume that extant humans have a higher tolerance to the constituents of wood smoke than do humans who lived before the taming of fire. Incidentally, wood smoke contains the highest concentration of PAHs of any combustible organic matter per mass burned.

It then follows almost automatically that modern humans would have a higher tolerance to such compounds than mice and rats. The case of tobacco smoke—which comes from a different combustible material and is inhaled in a very different set of conditions (temperature, concentration, gaseous composition)—might well have contributed to clouding the issue of the human carcinogenicity of PAHs in smoke. This complicated picture will take many more years of research to resolve, and we should bear this in mind when developing cultural stances and policy toward combustion of hydrocarbons and weigh up advantages and disadvantages in a technology-neutral way. Clearly, there are significant health impacts from breathing in combustion products of any sort (and I am absolutely not excusing extreme situations such as smog- and smoke-laden cities), but there are plenty of "natural" health risks that we don't often think about avoiding; and we have evolved with them.

Parallels between human technological evolution and biological evolution: Perfection is relative

I would like to suggest a further interesting parallel before concluding this chapter. It's between biological metabolism and human sources of fuel for energy and the means of extracting that useful energy. Not only does biological evolution equip organisms with the ability to use a variety of substrates for energy, but it also equips them to use those substrates in different ways and with different efficiencies. Sometimes efficiency is traded off against recyclability: for example, carbohydrates can be laid down (as glycogen in muscle) and reliberated in the form of glucose very easily, but they are a much less efficient source of energy compared with fats. Fats, on the other hand, are very high in energy density but more complicated to lay down and reconvert to available energy when needed. Furthermore, as we have seen earlier, one and the same energy substrate can be used in different ways with different efficiencies: fats are broken down into FAs, which may then be used directly in metabolism, being broken down further (beta-oxidation) into two-unit carbon compounds (acetyl groups that are then converted to Acetyl-CoA) that enter the Krebs cycle and start to produce energy in the form of ATP right there. When things get really tough, however, these two-carbon units are increasingly redirected to the liver, where they are converted into ketone bodies. These then circulate in the body, enter muscle and brain cells, and produce the most energy per unit of carbon of any biological carbon-based fuel that our body has at its disposal. Why such diversity? To enable us to be flexible and survive changing circumstances, both regarding the available source of nutrition and with respect to changing environments.

This is certainly not about an all-out maximization of efficiency in and concentration on a single concept or energy source. Evolution doesn't work that way; rather, it results in the "optimization" of organisms to be *sufficiently successful* in the environment in which they must survive to reproduce and raise offspring to independence (in the case of mammals, for example). There are many examples of evolution not reaching (or "caring about" reaching) perfection as we humans would define it with reference to an isolated biological system. Richard Dawkins wonderfully illustrates this in his book "The Blind Watchmaker."[94] A biological organism does not exist in isolation, and it does not suddenly come into existence without a past that has "conditioned" its evolution. If we wish to speak of biological perfection, we need to use the word not in absolute terms, but in relative terms—terms that are conditioned by the realities of evolutionary history, including tradeoffs, resulting in benefits of flexibility. Most biological systems embody this in the existence of parallel metabolic pathways and redundancy, thereby making organisms robust to unpredictably changing environments.

The same principles apply to human technology. We risk staking the success of energy production and use on an insufficiently large range of technologies. We would be very unwise to throw away all technology and infrastructure of the past and think that we can replace them with "more sustainable solutions" in one go because biology tells us that sustainability means working efficiently with what one has—both in terms of modifying it and not overusing it. We may strive to perfect a given novel technology and become completely reliant on it;

and we might do so without sufficiently acknowledging the potential—or even quite likely—pitfalls of such a strategy because it will take longer than the developmental phase for them to become pressingly apparent.

I believe that that is exactly what we would do if we concentrated on a single type of "propulsion solution" for mobility, or a particular means of energy storage for peak energy production (e.g., batteries for electricity storage). Fortunately, we are also considering hydrogen, which has been a source of energy for microbes since the earliest forms appeared on Earth. Bacterial and archaeal hydrogenases use hydrogen to reduce other substances, and we know that on early Earth there was hardly any oxygen for the hydrogen to reduce (to water). Hence a major route of reaction was with CO_2, which was abundant on primitive Earth: hydrogen was (and still is to this day by many microorganisms, e.g., the gut microbes in ruminant animals) enzymatically reacted with CO_2 to form methane. This is the simplest and earliest manifestation of using carbon to carry hydrogen as a fuel. But as we saw in the section "Why fats and oils are denser in energy than sugars," it is the unique *combo* that carbon and hydrogen form that makes C-H compounds so useful and sustainable: both the C-H and the C-C bonds carry the energy. In summary, there are innumerable parallels between biological metabolism and the energy questions facing us at present, and we will investigate these in later chapters.

Conclusion: Countering further increases in agriculture's enormous GHG impact, getting human nutrition back on track, and learning from biology

It is clear that large parts of agriculture—notably livestock-keeping for animal products—are environmentally unsustainable. Humans' unintentional destructiveness is both legendary and repetitive, but we must now learn urgently from the more recent history of our environmental abuse. One area in which I believe that we must learn is the biochemistry of carbon compounds, specifically the natural energy and material economies of living organisms: here there are many parallels to our uses of energy carriers in human technology use and development. I also hope that I have described human nutrition in ways that make clear how key dietary components work, how much of them we really need, and how we can reduce environmental impact by adjusting our diets in simple and healthy ways.

We have seen how cooking, and fire in general, exposed humans to a variety of organic combustion and meat-charring products that are mutagenic (and potentially carcinogenic) and that evolution likely endowed us with substantial resistance against them. Our paleolithic ancestors would, however, be quite astounded by how much meat many humans now eat. The burgeoning and ever-growing sector of animal agriculture is clearly unsustainable in many ways (from environmental to social and ethical): the energy, resources, and land use (to name but three factors) involved in this sector steal from the more important and nutritionally healthy sectors of plant agriculture and nutrition. They pump inordinately large amounts of $CO_{2eq.}$ into the atmosphere, methane being the fastest-growing and most concerning component (largely from microbial digestion in cattle). Dairy product and meat consumption in First-World countries produces a $CO_{2eq.}$ footprint that are comparable with that from car-driving. Eating more meat than necessary is a plain waste of resources and mainly harbors suffering, both at global and biochemical/molecular levels.

The most interesting and promising biochemistry is that of carbohydrates and biological hydrocarbons (my shorthand for oils and fats). Here we see fascinating insights emerging from biomedical research that may well significantly promote healthy aging (and for those who would vote for life span extension, that too). These insights further elucidate the reasons why carbohydrates need to be kept in check in our diet, and why fat is the biologically preferred energy storage material. These lead us into areas such as ketone metabolism and the epigenetics of disease, longevity, and health in general. They also nicely illustrate why it is the combination of C and H, rather than either on its own, that make C/H fuels so advantageous: the carbon backbone is so much more than a mere "carrier" of hydrogen-based energy. This leads to insights that I believe are important for technology—insights that we must understand and exploit better in the interests of energy sustainability and flexibility in uncertain times.

The learning that we can and must do from biology is enormous. I conclude with the obvious, but rarely addressed, parallels between the biological energy and material economies of carbon compounds. Here there are tradeoffs that biology has made for very good "reasons" (evolutionary selection pressures), meaning that the "most energy-efficient" solution to human eyes is not necessarily the one that biology chose; rather, it is the solution that led to greatest sustainability in the context of ecosystems, changing environments, and uncertainty. The use of fats and oils (simplistically put, "biological hydrocarbons") as energy storage substances is key because the energy and material flexibility are manifest in the C-H and C-C bonds. This enables many features (chemical and physical) to be combined, thus producing a perfect compromise.

3
Sources and sinks

Where carbon compounds accumulate on Earth, and what they do there

In 2019, according to the International Energy Agency (IEA), 33.4 gigatonnes (Gt) of carbon dioxide (CO_2) were released globally from the energy generation industry. A further 3.4 came from other industries. That represents a slight flattening of the emissions plot, as more regenerative energy is produced (e.g., solar, wind, and hydroelectric),[1] but it's still an all-time record. In 2020, the global COVID pandemic pushed emissions down by between 1.5 and 2.6 Gt,[2] the largest single-year decline on record. Energy-related (fossil fuels and industry) CO_2 emissions in 2021 returned to pre-COVID levels, amounting to 36.3 Gt,[3] a fraction less than their 2019 record of 36.8 Gt. These emissions are causing the tiny percentage of CO_2 in the atmosphere to grow faster than ever before. However, to understand the massively disproportionate effect of CO_2 in global warming properly, we must delve into much more science than the (inaccurate) cartoons of radiation/absorption/re-emission presented in everything from popular media to scientific papers. As we watch, horrified, the Earth's response to our emitted CO_2—and increasingly methane (CH_4)—we realize how lacking is our knowledge about some of the largest carbon fluxes: a massive organic–inorganic recycling that slowly produced enormous carbonate rock strata and deposits of oil, gas, and coal. By burning these so fast and degrading the environment in other ways, we are destroying key parts of the very biosphere that made them.

In the atmosphere: Anthropogenic CO_2 inputs to the carbon cycle continue to rise at around 2.5 ppm per year

The CO_2 emissions increase in 2021, at 6%, was the highest ever witnessed. An unprecedentedly high proportion of coal in the global mixture was the culprit. In China alone, electricity demand rose 10% in 2021. If one includes CO_2 emitted by changes in land use, the estimated global total for 2021 is just over 42.2 Gt,[4] but I will exclude land use from here on because it's not directly a consequence of fossil-fuel burning, is highly variable, and is harder to estimate. No matter, these are enormous figures, but what really counts is how much of that CO_2 was not reabsorbed by the Earth (via all routes, biological and physical).

From the parts per million (ppm) rise, we can work out the rate of CO_2 increase in net terms. The website "CO_2 Earth," tracks the rise in ppm CO_2 in the atmosphere almost in real time and whether "we are stabilizing."[5] The residual CO_2 is the figure that matters, and it's not stabilizing but creeping up at around 2.5 ppm per year at present (average from the Keeling Curve 2010–2020, covered later). If we take total fossil CO_2 emissions for 2021 as 36.3 Gt (energy production and industry), it's quite easy to calculate how much of it failed to be absorbed if we annually track the total mass of CO_2 in our atmosphere. That total (based on analysis of global carbon fluxes) is around 3,200 Gt atmospheric CO_2; that is, there are 875 Gt *carbon* in the form of atmospheric CO_2: that's 3.67 (the conversion factor for carbon → CO_2) × 875 = 3,211 Gt CO_2).[6] Knowing the increase of 2.5 ppm per year, we arrive at the widely quoted figure for net increases in CO_2 per year of 19 Gt (or 5.2 Gt elemental carbon)—that's 0.6%. Not much, right? A small percentage of the population can't believe that this 0.6% can make such a difference, but the arguments persuading them are deeply flawed, as we will see later.

Why the truth about CO_2-driven global warming is complicated

The unintuitive link between rising CO_2 concentrations and global warming was, for many years, a rich source of ammunition for CO_2- and climate-change deniers who regularly produced utterances such as: "How can a small increase in a gas that is such a small proportion of the atmosphere lead to such effects?" or "A 1°C increase in global temperature is tiny on the Kelvin scale, and can't possibly produce significant effects." Deniers still exist, but they are less numerous and less vociferous these days. The first important point about CO_2 is that it is *not* by soaking up energy and itself becoming warm that it creates the greenhouse effect: there is far too little CO_2 in the atmosphere to act as a kind of thermal repository; however, it is also not, physically speaking, by reflecting radiation either: there is a more complicated but much more scientifically credible explanation. Few readers are likely to have encountered it in the form that I will present.

I will do my best to navigate us engagingly through the mechanisms of anthropogenic climate change. There's no better place to start than the claim that small increases in an already small concentration of a "normal" atmospheric gas can cause such disproportionate effects: though this is an alien concept to many people, many have also heard of the "butterfly effect"; that is, that weather is an extremely complex system (mathematically speaking), embodying unpredictability that increases non-linearly with time, and that, at some point, a small turbulence caused by a butterfly's wings could be the "seed" of an Atlantic storm. We can predict the weather over a few days with reasonable accuracy,[7] but small initial measurement and modeling errors will follow a doubling rule with time. This is typical of a so-called *chaotic system*, as studied by the American mathematician Edward Norton Lorenz (1917–2008), starting in the 1960s. In fact, it was he who coined the expression "butterfly effect." In 1982, Lorenz, having studied years of meteorological records from the European Centre for Medium-Range Weather Forecasts, produced an estimate for the doubling time for small measurement errors: 2.4 days. A starting error of 0.5°C would, after two days, have grown to an uncertainty of 1°C, and, after 5 days, to +/–2 degrees. After 10 days, such a relatively small starting error would have become +/– 8 degrees. Lorenz's estimate is similar to others' and is consistent with today's practice of applying ever higher measures of uncertainty as weather forecasts approach the one-week mark.

Interestingly, a physical system doesn't even have to "appear" to be complex in order to display complex behavior: the "three-body problem" of physics is a classic example.[8] Here, three bodies of identical size, shape, and mass exist suspended in space. Like planets, under gravity, they attract each other, and, when released from their starting positions, they move according to gravitational laws. These laws are certainly reliable enough to predict individual movements, momentarily, and the *two-body* problem is mathematically quite easy to solve; however, it is absolutely impossible to predict the movements of the *three* bodies relative to each other over anything but a short period of time. Furthermore, infinitesimally small changes in the starting positions of the bodies relative to each other will quickly produce completely different trajectories: the epitome of chaotic behavior and one of the simplest systems that demonstrates it. It represents a type of "emergent behavior"—behavior that is not deducible merely by applying the simple rules of interaction of one part of the system with another, but rather "emerges" as a feature of the whole. Such systems, which are all highly non-linear, are found throughout the universe: life itself embodies countless examples. That a seemingly simple system can deceive us with its emergent complexity turns out to be a big impediment to "belief" in unexpected effects, even among trained scientists: we all tend to reason intuitively when faced with an unfamiliar problem.

The point is this: when things as astounding and fascinating as chaos theory and emergent effects break into the public consciousness, they are capable of generating genuine and accepting fascination. Unfortunately, if a nominally tiny increase in CO_2 concentration in our atmosphere has major implications for the way we live our lives, it is easy for that insight to be viewed with great skepticism if it's hard to explain. And, to be honest, it is. Indeed, some of the effects produced by CO_2 emissions in conjunction with the Earth's physics, chemistry, and biology are unintuitive.

> **Box 3.1 Short-wavelength radiation**
>
> Here photons (the particle form of radiation in the wave-particle duality) encounter electrons buzzing around the atom or molecule, and, if a given electron is at the right energy level (as defined by quantum mechanics), it will absorb a given photon and jump up one or more energy levels. In simplified terms, the electrons exist in a series of levels or tracks along which they move—the farther away from the atom's nucleus a level is, the higher the energy of the electrons whizzing around in it. Having been promoted a level or two by a photon of the right energy, an electron is, under ambient conditions, not allowed to enjoy its new-found superiority for long: the atom or molecule is "uncomfortable" with this imbalance of hierarchy, and two things can happen: (1) the electron can fall back to its original level, and in doing so, emit a photon of the same energy as the one that promoted the electron; (2) the atom or molecule can react and change into a different chemical entity: an atom might react with another element that it encounters at the same time; a molecule might do the same, undergo internal energy transitions (ultimately capable of producing thermal motion—i.e., heating up the bulk medium), or even split into two or more parts as a result of the "energy overload" (so-called *photolysis* [splitting by light]). These types of reaction are what happens, for example, when a substance is degraded by UV light: the high-energy electrons basically trigger chemical reactions.

CO_2 absorbs infrared radiation in particular wavelength bands

All atoms and molecules can absorb electromagnetic radiation (the type to which heat, light, ultraviolet, microwave, and X-ray radiation belong). Depending on the wavelength of the energy, it has particular properties and interactions with matter. Although overlapping to some extent, I've broken these down to three categories.

1. *Short wavelength radiation* (including visible light and ultraviolet) moves electrons around in atoms or molecules (Box 3.1) and is relatively high in energy. Its interaction with matter can lead to its energy being converted to longer wavelengths of re-emitted radiation (e.g., infrared). This is what happens to much of the sunlight that falls on Earth.

2. *"Heat" radiation and microwaves* (in order of increasing wavelength), which move whole atoms and molecules in what is known as "thermal motion"—a random motion that can be observed under a normal light microscope as tiny solid particles (e.g., smoke) are buffeted by air molecules: Brownian motion, after the Scottish botanist, Robert Brown (1773–1858). All objects absorb these wavelengths and, in the process, become "warmer" (see Box 3.2). This is known as "black body" radiation, and it refers essentially to heat radiated from the surface of perfectly absorbing object (i.e., one that reflects no radiation). In the process of radiation, if not fed with more energy, the object loses kinetic energy (energy of movement), which corresponds to its particles (atoms or molecules) jiggling around ever more slowly. In a solid, that motion is vibrational, around an average position; in a gas or liquid, the motion is largely unrestrained. The important concept here is that thermal energy is an average across all molecules in a body, or, in the case of a gas (e.g., air), across all different molecules in a composite of different molecules. In such a mixture, each type of molecule has variations in energy distributed in a bell-shaped pattern around a maximum: this is pictured as a curve of numbers of particles versus speed of particle (Box 3.2). The shape of that curve is identical for all types of particle (e.g., molecules), be they large ones or small ones. In our atmosphere, the distribution of kinetic energies in the nitrogen molecules is exactly the same as the distribution of kinetic energies in the CO_2 molecules: there's nothing special about CO_2 in that respect.

A solid, liquid, gas, or mixture of gases that is cooling uniformly radiates heat to its surroundings identically, regardless of whether the substance is a mixture of molecules (e.g., a mixture of nitrogen, oxygen, and carbon dioxide, as in our atmosphere). It is *not*, for example, the case that the CO_2 molecules in our atmosphere are "warmer" than the nitrogen molecules. However, as we will see later, they differ from most other molecules in the air in respect of the third type of radiation transfer:

3. *Infrared radiation*. Here, in what we perceive partly as "heat radiation," reside the wavelengths from 1 mm down to 0.00075 mm (0.75 µm) that can be absorbed by molecules to produce stretching/contracting and bending of their bonds. This is the realm of *infrared* (IR) *spectroscopy*, and within, and close to, the range of 7–17 µm, this kind of energy exchange definitely does depend very characteristically on the type of molecule, hence often being called "finger print" radiation (see Box 3.3). Here atmospheric physicists research the absorption/re-emission of energy that the Earth absorbs in the form of "sunlight" (ultraviolet [UV], visible, IR) and re-emits largely in the IR range.

Many molecules have a characteristic "fingerprint" absorption pattern in the IR part of the spectrum on account of having bonds between different types of atoms: an O-H bond in H_2O absorbs different frequencies than does a C=O double bond in CO_2; but an O-H bond in an alcohol also has a slightly different spectral fingerprint

Box 3.2 Black body radiation

For objects at identical temperatures, black body radiation has exactly the same range of wavelengths (same peaks and troughs in its emission spectrum), regardless of the material out of which it is made.[9] This is because black body radiation derives from the jiggling around of the particles as a whole (and *not* what is happening within the particles, e.g., bond vibrations). The hotter an object, the faster its molecules—or, for a pure element, its atoms—are jiggling around. This jiggling is random in direction and extent, but the average extent of jiggling gives the temperature that we can measure accurately or perceive qualitatively (cool, hot, very hot). This corresponds to the average of the energy distribution, which is a skewed bell-shaped curve known as the *Maxwell-Boltzmann curve* (after its discoverers, James Clerk Maxwell [1831–1879] and Ludwig Boltzmann [1844–1906]).

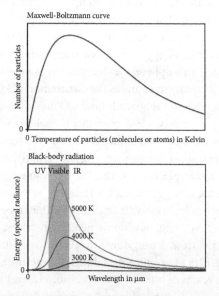

At room temperature, the black body radiation of an iron bar is entirely in the infrared: if we completely darken the room, we cannot see the bar. As we heat it up to past 700°C (973 K) we start to see it glowing faintly red, then orange, yellow, and finally "white" at 5,527°C (5,800 K). Visible wavelengths that fall on the Earth are converted (via numerous processes of absorption and intramolecular and intermolecular energy transformations) to lower wavelength energy manifest in the thermal energy of motion (jiggling around of molecules and atoms) in all substances on the Earth's surface and atmosphere: these then re-radiate in the IR part

> of the spectrum at ambient temperatures, and that radiation is the energy source for the greenhouse effect because it does not escape completely back into space: a proportion is hindered by components of the atmosphere, mainly H_2O, CO_2, and CH_4.

from an O-H in H_2O. Let's look at the IR absorption spectra of nitrogen, oxygen, and CO_2, the main components of air. The first two are easy: they have none. Molecular nitrogen (N_2) and molecular oxygen (O_2) absorb no IR. Why? The basic explanation is that they are very simple molecules of two atoms so their bonds don't have the ability to bend (change angle between each other), and, when they stretch and contract (at resonant frequencies, like a violin string), the overall charge distribution across the molecule doesn't change. Essentially, a bond will only absorb infrared radiation if, in the process of stretching/contracting/bending, the charge distribution of the molecule as a whole changes (i.e., the distribution of electrons between the two atoms in the bond changes toward one or the other atom). This is known as a change in "dipole moment." Because the atoms in N_2 and O_2 are, respectively, identical, there is no imbalance of charge between the two, and hence no dipole to change by stretching.

Molecules comprising dissimilar atoms are more interesting. All of these already contain bonds with unequal distribution of electrons between the atoms: there are so-called *electro*philic—electron-loving—atoms, and so-called *nucleo*philic atoms—the opposite of electron-loving. As it happens, oxygen is more electrophilic than carbon, and so, in any carbon-oxygen bond, the electrons tend to be drawn more to the oxygen, leaving the carbon, relatively speaking, positively charged, and the oxygen relatively negatively charged: a classical molecular dipole. As the bond vibrates in and out, so the strength of the dipole across the whole molecule changes, hence making it absorb, and emit, infrared radiation. It does so at characteristic wavelengths—the fingerprint mentioned earlier. These correspond—via quantum mechanics—to "vibrational modes" of the bond, similarly to the vibrational modes of a string on a musical instrument. On my violin I can play a particular note by pressing the string hard down onto the finger board. If I then lightly rest another finger between the hard-down one and the violin's bridge, I can encourage the string to vibrate in one of its other vibrational modes, a so-called harmonic. There are many such harmonics for a given "starting note."

Molecular bonds also have numerous harmonics. That is why a bond with dissimilar atoms on either side (or even the same atoms, but each having different electronic pulls [see Figure 3.1]) vibrates at different, very defined (quantized) frequencies corresponding to different energies. These energies correspond to the different wavelengths of IR radiation that the bond will absorb in transitioning between the different resonance modes. The relative "electronic pull" created by the partner atom of the bond—hence creating the initial dipole—partly determines the frequency of resonant vibration (and hence IR absorption wavelength). Two further variables contribute: (1) how "stiff" the bond is (force constant in N cm^{-1}) and (2) the mass of each atom (plus whatever else is attached to it). For example, a heavy and a light atom bonded to each other via a very stiff bond with a dipole will produce basic and additional resonant vibrations that are of much shorter wavelength (i.e., higher energy) than two heavy atoms bonded to each other via a very "stretchy" bond with a dipole. However, the exact absorption spectrum arises by properties that are very akin to emergent effects because they are so hard to calculate and hence predict (Box 3.3). Note also the analogy to the three-body problem in physics: CO_2 and H_2O are both tri-atomic molecules in which the three atoms are in motion relative to each other.

Atmospheric H_2O dominates IR absorption, but CO_2 cycles much more slowly and absorbs strongly at two crucial wavelengths

Now comes the beginning of the "story" that ultimately frames increasing CO_2 concentration in our atmosphere as the cause of most of current global warming. It starts with the question: How many other gas molecules with a dipole are there in our atmosphere, and what are their relative concentrations? First, we have the so-called *permanent gases* (Table 3.1), all of which are elements and exist either as individual atoms (e.g., the noble gases) or maximally as simple diatomic molecules, hence having no IR absorption.

Next we have the so-called *variable gases* (Table 3.2), which are cycling relatively quickly through the atmosphere, earth, and oceans; often having concentrated areas of production; and are able to change concentration relatively quickly (on geological timescales). All of these variable gases have a bond with a dipole *and* a bending mode. All can absorb and emit IR radiation via their bonds. CO_2 and N_2O vie for the longest half-life (CO_2 around 120 years; N_2O between 114 and 132 years, depending on the exact measurement method). The concentration of water, both locally and globally, in our atmosphere is part of a gigantic cycle of evaporation from oceans, land, freshwater bodies, and plants, roughly balanced by precipitation (rain, snow, hail) and crystallization directly from the air onto the land (as happens in the Arctic and Antarctic). Almost 13,000 km^3—or 1.3×10^{13} metric tons (13 teratonnes)—of water are suspended in our atmosphere at any one time; on average, this enormous mass of water cycles once every 10 days. Not only are glaciers in the Arctic and Antarctic melting into the sea as a consequence of climate change, but the world's oceans

Box 3.3 Long-wavelength radiation in the context of "fingerprint" radiation

Though it overlaps in terms of wavelength with black body radiation (and absorption), IR radiation of this type has a fundamentally different source: not the jiggling of a molecule, but rather the small movements in its bonds—stretching or bending.

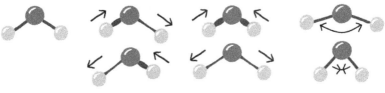

Ground state   Asymmetric stretching   Symmetric stretching   Bending

Vibration frequency, and hence energy, depend *largely* on (1) the mass of the atoms at either end of the bond (expressed as "reduced mass," which for *linear* molecules is $m = (M_1 \times M_2)/(M_1 + M_2)$ or for triatomics $(M_1 \times M_2 + M_2 \times M_3 + M_1 \times M_3)/(M_1 + M_2 + M_3)$ and for non-linear ones requires complicated corrections for bond angles); and (2) the stiffness of the bond (force constant, F, in Newtons/cm—see the ChemNet article[10]). Essentially, frequency is proportional to $\sqrt{(F/m)}$ and wavelength to $1/(\sqrt{(F/m)})$, so the "heavier" the composite of atoms, the longer the wavelength and hence lower the energy; but the "stiffer" the bonds, the *shorter* the wavelength and hence the *higher* the energy. Roughly speaking, H_2O has quite stiff bonds but a relatively low reduced mass (because hydrogen is "light"); CO_2 has very stiff bonds (C=O double-bonds) but a relatively high reduced mass. The two features somewhat compensate, leading H_2O and CO_2 both to have principal absorption peaks close to each other in the "fingerprint" part of the IR spectrum (from 4 to 25 μm), but not in identical positions.

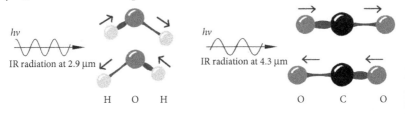

hν IR radiation at 2.9 μm   H O H

hν IR radiation at 4.3 μm   O C O

Even such primary spectra and their extent of overlap are hard to calculate with reliability; the shoulders of peaks, produced by first-, second-order, etc. harmonics, are even harder to calculate. *Reliable and detailed* spectra and assignment to vibrational modes can only be determined by observation (practical spectroscopy), and even that is not trivial.

Table 3.1 Permanent gases

Gas	Chemical formula	Percent volume of dry air/ppm (parts per million)
Nitrogen	N_2	78.08/780,800
Oxygen	O_2	20.95/209,500
Argon	Ar	0.93/9,300
Neon	Ne	0.0018/18
Helium	He	0.0005/5
Hydrogen	H_2	0.00005/0.5
Xenon	Xe	0.000009/0.09

Data reproduced with modification from University of Arizona website.[11]

are receiving 37 gigatonnes of water per year from land run-off and evaporation because of changes in plant cover: we are slowly drying out.

The point about water is this: there is a massive amount of it cycling between land, atmosphere, and oceans via natural, fast, and completely uncontrollable mechanisms (at least in the short-term). Water vapor is, as it turns out, the dominant greenhouse gas (GHG)—responsible for around half of the radiative forcing from atmospheric gases.[12] However, we are unlikely ever to be able to adjust its atmospheric concentration in ways that are large-scale and controllable enough to regulate global temperature; more worrying, global warming leads to around 7% more water vapor in the atmosphere per °C rise in temperature[13]: a positive feedback system (i.e., non-linear) that increases global warming yet further. Atmospheric water is what made Earth warm enough to support life. It is estimated that, without water vapor, the average temperature of the Earth's surface would be around −18°C.[14] The next most concentrated IR-absorbing molecule is CO_2. The four vibrational modes of its bonds (see Figure 3.1) give it two characteristic "fingerprint" absorption peaks (at 4.3 and 15 μm) because one of them doesn't absorb radiation (it doesn't change the overall dipole) and two of the others are geometrically identical. There are other "overtones," mainly 2.7 and 2 μm, but they are weaker and hence absorb less energy than the principal modes.

The shorter the wavelength of radiation, the more energy it packs. It's quite intuitive: at a short length of wave cycle (peak to peak) there are more oscillations (waves) per unit distance than at a long one. But both waves are traveling at the same speed (the speed of light), and so, because the oscillations

Table 3.2 Variable gases

Gas	Chemical formula	Percent volume/ppm (parts per million)
Water	H_2O	0–4/0–40,000
Carbon dioxide	CO_2	0.042/420
CH_4	CH_4	0.00017/1.7
Nitrous oxide	N_2O	0.00003/0.3
Ozone	O_3	0.000004/0.004
Chlorofluorocarbons	Cl / F / C	0.00000002/0.0002

Data reproduced with modification (e.g., actualization of CO_2 ppm) from University of Arizona website.[11]

are essentially electromagnetic energy, there is higher energy density in a ray of short-wavelength/high-frequency radiation (Figure 3.2).

Of the IR-absorbing modes of CO_2, the asymmetric stretching mode creates the greatest change in charge distribution with respect to the position of the central carbon. Hence it produces a shorter wavelength—higher energy—absorption than the bending mode (4.3 μm compared with 15 μm). The 15 μm band has historically been studied more in relation to the greenhouse effect of CO_2 (radiative forcing) because it was easier to identify—in fact, it was spotted more than 130 years ago by two physical chemists but in its "inverse" form (i.e., *emission* at 15 μm): Rubens and Aschkinass were studying the flame produced by a laboratory gas burner and trying to work out whether the energy emitted at 15 μm came from a chemical reaction during combustion or from bond vibrations in CO_2. They postulated the latter, but it was only confirmed in 1948, by Earle Plyler, a US physicist and pioneer of infrared molecular spectroscopy.[15] However, both the 15 and the 4.3 μm bands—the strongest absorption regions of CO_2 in the troposphere[16]—are absolutely crucial to understanding CO_2's surprising interaction with radiation that is reflected or reradiated from the Earth. To appreciate why, we must return to H_2O.

CO_2's main absorption bands "close windows" in the H_2O spectrum

Water has the same kinds of bond-vibrational and -bending modes as CO_2, but the atoms in H_2O have different masses from those in CO_2, and the bonds different "stiffnesses." Thus the main absorption peaks of H_2O are at different wavelengths from those of CO_2. We need not concern ourselves here with the exact wavelength values of the H_2O peaks, but rather with their positions relative to the CO_2 spectrum: as shown in Figure 3.3, the main peaks (and their "shoulders") from CO_2 occur in positions overlapping the two troughs in the H_2O spectrum. Climate scientists often refer to this as "carbon dioxide closing the windows in the water spectrum": the troughs in the H_2O spectrum are where H_2O absorbs very little IR (re-)radiated from the Earth, instead letting most of it pass by to be absorbed by something else, or to escape into space. The fact that CO_2's principal peaks overlap with these troughs means that it impedes the escape of these wavelengths of radiation from the Earth.

Now, you might be thinking, "but hang on a minute, H_2O can constitute up to 4% of the gas in the atmosphere by parts per million, but CO_2 is just 0.04%—roughly one-hundredth that of H_2O." However, CO_2 and H_2O as individual molecules don't absorb equally. It's all about absorption coefficients (i.e., how strongly a molecule absorbs certain wavelengths at a given concentration of the molecule in the air).[17] For example, between 4.3 and 3.9 μm CO_2 absorbs between 1,000 and 10 times as strongly as H_2O per mass of gas.[18] To make a crude approximation, in that spectral region, CO_2 absorbs, on average, 100 times as strongly as H_2O. That is why, at one-hundredth of the concentration of H_2O, it effectively blocks a large part of the H_2O window between 4.3 and 3.9 μm.

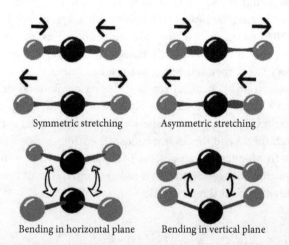

Figure 3.1 Stretching and bending modes of CO_2. The symmetric stretching mode does not absorb infrared (IR) radiation because it causes no net change in molecular dipole across the whole molecule; the asymmetric stretching mode absorbs at 4.26 μm (usually rounded up to 4.3); bending in the horizontal and vertical planes absorbs at 15.0 μm. The vertical and horizontal bending modes are geometrically identical (i.e., "degenerate"), hence producing only one absorption wavelength.

If we take just the CO_2 absorption band at 4.3 μm and a CO_2 concentration of 400 ppm at normal atmospheric pressure at sea level (ground level of Earth), we have an absorption coefficient of 0.44 m^{-1}. That basically means that IR radiation of wavelength 4.3 μm traveling through air containing 400 ppm CO_2 is reduced to 44% of its original value in the first meter of air, and then to 44% of that new value by the next meter of air, and so on: it's strongly non-linear, producing a curve called a *negative exponential*. Importantly, the IR absorption wavelengths of any molecule are also present in its IR (re-)emission peaks, so what an IR-active gas can absorb, it can also re-emit. However, in absorbing and re-emitting, it interrupts the straight-line passage (and escape) of IR energy through the atmosphere, partly bouncing the energy back down to Earth. Furthermore, if a molecule of CO_2 (or H_2O) that has been set vibrating in one of its modes bumps into another molecule (of whatever kind, nitrogen, oxygen, water, or even another CO_2) before it has re-emitted, it can *conduct* (rather than radiate) its absorbed energy into the thermal motion of that other molecule. In this way, it directly heats up its collision partners—not always, but frequently, depending on density. This happens much more in the lower strata of our atmosphere, where air is denser than higher up.

Condensing the details into basic messages: (1) CO_2 absorbs reflected/re-radiated radiation in two parts of the spectrum where water doesn't, hence having the potential to block two radiation escape windows in the water spectrum; (2) CO_2 has a much higher coefficient of IR absorption than does water, thus allowing CO_2 to absorb similar amounts of energy at a tiny concentration compared with water; (3) CO_2 is often presented as simply re-emitting IR (in the same wavelengths that it absorbs), but we need to explore how it *converts*

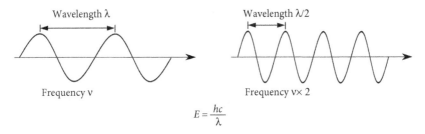

Figure 3.2 Short-wavelength radiation has more energy than long-wave radiation because energy is inversely proportional to wavelength, and directly proportional to frequency. Hence halving the wavelength (doubling the frequency) doubles the energy that the radiation carries. In the formula, E is energy in joules, h is Planck's constant (6.626×10^{-34} J s), c is the speed of light (3×10^8 m/s), and λ is the wavelength. For simplicity, only one of the waves of electromagnetic radiation is shown: the orthogonal one is omitted.

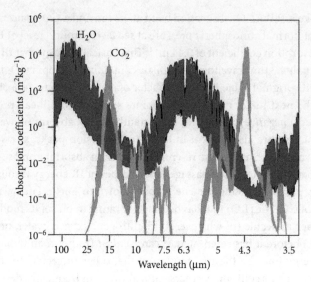

Figure 3.3 Superposition of infrared (IR) absorption spectra of H_2O (dark gray) and CO_2 (light gray). The principal absorption peaks of CO_2, at 15 and 4.3 μm, largely overlap with the respective troughs in the H_2O absorption spectrum. Though CO_2's principal peaks are considered to be "saturated" (i.e., no further blocking of the H_2O windows can occur) in 10 m of atmosphere at ground level, this is not so at higher levels of the atmosphere; furthermore, the shoulders of CO_2's principal peaks, and the nearby secondary peaks, will, as CO_2 concentration rises, "block" additional regions of the H_2O troughs, hence leading to significant further absorption of radiation that would otherwise escape. Figure redrawn with minor modifications from Zhong and Haigh.[16]

its bond-vibrational energy to *thermal energy* in collisions. H_2O does much less of that, as we will see later, hence distinguishing CO_2 from H_2O yet further as a GHG.

The climate-changing effects of CO_2 started to be researched around 150 years ago

Now for a historical interlude. Acute public concern about global warming is relatively recent, having started around the turn of the millennium. We might even suppose that not much was known about the trend and its causes until recently; how wrong we would be. We had more than 30 years to "wake up." In the late 1950s, sitting at what we would today regard as a primitive computer, surrounded by paper tape printouts in the Mauna Loa observatory on Hawaii,

American chemist Charles Keeling (1928–2005) was making a momentous discovery. He was the first to make an instrument capable of accurately measuring CO_2 concentrations in air samples. Now, in 1958, and with an academic grant to set up a CO_2 monitoring station, he was producing data that would reveal something very worrying. Three years later, in 1961, the data showed with very little doubt that the average concentration of CO_2 in our atmosphere was increasing steadily, and faster than ever before in documented history. This trend is now known as the *Keeling Curve* (Figure 3.4).

Others picked up on this research, or were commissioned by corporations to do related studies and investigate the link between CO_2 emissions and global warming. Scientists working for the Aeronutronic Ford Corp., Aeronutronics Division, California, conducted a study analyzing the absorption of different atmospheric gases at different altitudes. The results were published in 1976, and they state: "Absorption and emission by the well-known 15 μm bands of CO_2 forms the basis for experiments on the remote sensing of the atmospheric temperature profile from satellite-borne instruments."[21] Renamed the "Ford Aerospace & Communications Corporation," this division of the Ford Motor Corp. conducted research with military applications, but it clearly had much broader implications. Equally interesting—and certainly no coincidence—a year later, in 1977, Exxon, the global oil giant, had at its disposal heaps of related data and modeling that enabled it, and other oil companies, fairly accurately to predict the doubling time for CO_2 concentrations in the atmosphere given trends in fossil fuel burning.[22] They foretold an accompanying increase in mean global ground-level temperature of 2 to 3°C per doubling: how prescient!

But all of that is recent history compared with the very beginnings of the research into CO_2's role in global climate. For this we must go back to the second half of the 18th century. One of the great mysteries of the time was what had caused Earth's Ice Ages and, equally, what had caused their end. There was no lack of ideas, but the more convincing ones involved the Earth's atmosphere: Joseph Fourier (1768–1830), the French mathematician and physicist, reckoned that the atmosphere played a major role in the thermoregulation of the Earth, but he never identified a mechanism that could correlate changes in the atmosphere with long-term changes in climate. The science of radiation and heat transfer in the atmosphere developed in response to curiosity about how glaciers vary in extent under the influence of the sun's radiation. This was first studied methodically by the Irish natural scientist, measurement expert, and pioneering mountaineer John Tyndall (1820–1893). Tyndall's interest in variations in glaciers in the Alps prompted him to make heat-transfer measurements on gases—something that until then had only been done on solids and liquids. He showed that N_2 and O_2 are practically transparent to radiation, and that H_2O and CO_2 are the main constituents of the atmosphere that absorb and re-radiate/

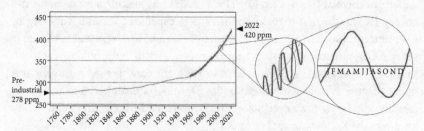

Figure 3.4 Atmospheric concentrations of CO_2 in ppm since pre-industrial times. The plot since 1958 is known as the "Keeling Curve." Variation in CO_2 concentration across the year in the Northern hemisphere is due to the cycle of photosynthesis: deciduous trees and other plants increase foliage from spring to summer, hence leading to a temporary decrease in CO_2 concentration from the average value; in autumn, values start to increase again as foliage is shed. Identical yearly variation and general trend are seen in data collected from the Zugspitze, Germany's highest peak. Values in the above plot: since the 18th century from Macfarling et al.[19] (ice-core data); values since 1958 from Keeling et al.[20] according to data collated by the Scripps CO_2 Program, University of California San Diego, CA (see https://keelingcurve.ucsd.edu/permissions-and-data-sources/).

transfer IR. He calculated the absorption coefficients in his laboratory using a copper cube as the heat source. He also emphasized the phenomenon of solar radiation absorption by the Earth, followed by re-radiation at longer wavelengths (IR)—wavelengths that could not leave the Earth so freely as they had entered. In a scientific transaction in 1859, he wrote "when the heat is absorbed by the planet, it is so changed in quality that the rays emanating from the planet cannot get with the same freedom back into space. Thus the atmosphere admits of the entrance of solar heat; but checks its exit, and the result is a tendency to accumulate heat at the surface of the planet." Hence Tyndall can truly be considered the father of atmospheric climate research, and he essentially discovered the greenhouse effect before it was so named.

The Swedish physical chemist (and first recipient of the Nobel Prize for Chemistry) Svante Arrhenius (1859–1927) was also a pioneer of atmospheric climate research. Arrhenius was sure that anthropogenic CO_2 was a principal determinant of global temperature, and he predicted a continuously rising global temperature as a result (publication in 1896).[23] In testing his theory that rising CO_2 levels had led to the end of the last great Ice Age, he developed a system for estimating global temperature increases according to increases in CO_2 in the atmosphere. Following the thinking of Tyndall and Arrhenius, the heat was on for CO_2. In 1892, drawing also on measurements made by two Austrian physicists, Ernst Lecher (1856–1926) and Josef Pernter (1848–1908), Swedish physicist Knut Ångström (pronounced "Ongstrome") (1857–1910) published a paper "On the Significance of Water Vapor and Carbon Dioxide in Absorption in Earth's Atmosphere." Incidentally, Knut was the son of the famous Anders Ångström, one of the founders of the science of spectroscopy, after whom the unit "Ångström" (one ten-billionth of a meter) is named. Ångström the younger's paper was published in German under the title "Ueber die Bedeutung des Wasserdampfes und der Kohlensäure bei der Absorption der Erdatmosphäre" (On the Importance of Water Vapor and Carbonic Acid in Absorption in Earth's Atmosphere) because it appeared in probably the most influential physics journal of the time, which happened to be in German: *Annalen der Physik* (*Annals of Physics*)—where Einstein would later publish his relativity theory (in 1916).

Ångström's paper is seen by some partly as an attempt to correct what he perceived as a measurement mistake that Arrhenius had made in reaching a finding with such negative economic repercussions: Arrhenius was one of the first (if not *the* first) to call publicly for a reduction in coal burning in the interests of the climate; Ångström basically reckoned that CO_2 would quickly reach absorptional saturation and cause no further effect. A scientific exchange followed in which not only Arrhenius inspected and updated his own results, but also discovered mistakes that Ångström had made! This coaxed Ångström into revisiting his thinking, and he thus improved his understanding of downward terrestrial radiation: he even designed and made a measuring instrument for the purpose (in 1905).

And the riddle of the Ice Ages? That was solved by the work of two men whose lives didn't even overlap: the Serbian mathematician, astronomer, and geophysicist Milutin Milanković (1879–1958) discovered the work of Swiss biologist and geologist Louis Agassiz (1807–1873). On this basis, he proposed that the cyclical variations in the "wobble" of the Earth's rotational axis (precession) caused ice ages: correct answer . . . almost, because the cooling driving the onset and the warming initiating the end of ice ages was very strongly dependent on falling and rising atmospheric CO_2 concentration, respectively. And that was mostly influenced by the depths of the Southern Ocean, as a net absorber or net emitter of CO_2, as precessional variation brought it infinitesimally further from or closer to the sun.

CO_2 exchanges energy with its environment in ways that produce a surprising effect

Heat capacity of gas is only a small part of the picture

A grasp of both fingerprint radiation and black body radiation (basically stemming from thermal movement of atoms) is fundamental to understanding the warming effect of atmospheric CO_2. The crux of the matter is the following: black body radiation, being independent of the composition of the solid, liquid, or gas in question is a relatively simple physical system—there is absorption of energy in the process of increasing the average speed of movement of the particles in the substance, and there is radiation of energy in the same proportion and distribution of wavelengths as the absorption at a given temperature. Basically, we can, in a laboratory setting, take a container of normal atmospheric air, heat it up with an IR source, measure the energy that it has absorbed, and then watch it re-radiate that energy to the surroundings.

Seen as a bulk substance, the air will absorb and emit the same wavelengths of radiation and absorb and emit the same amount of energy (by the simple and universal principle of energy conservation). If we change the concentration of CO_2 in the experimental air, all we do is change the amount of energy that we need to supply to it in order to increase its temperature by a given amount: the more CO_2, the more energy we will need to increase the air's temperature by 1°C, for example. If we didn't increase the intensity of the heat source, we would need more time to heat the air containing more CO_2 to the same temperature as normal air, and—conversely—the air with more CO_2 would radiate and conduct energy for a longer time before falling back to the starting temperature. This effect per se plays a small role in global warming, increasing the heat capacity of the atmosphere with increasing CO_2, but something much more important is happening, as we'll see later.

Why the saturation argument against CO_2-driven global warming doesn't work

Some skeptics of CO_2-driven global warming present the following argument: CO_2's effect of absorption of IR radiated from the Earth is "saturated." They mean that there is already more than enough CO_2 in the atmosphere to swallow up the small percentage of this radiation within a matter of 10 m of atmosphere. Hence, the argument continues, further increases in atmospheric CO_2 concentration will not lead to *more* heat being "trapped"; rather, the radiation will merely travel through a shorter distance of the air before all of it is trapped. What this explanation doesn't address is what happens *after* the energy is "trapped." Clearly, the CO_2 in the atmosphere can't indefinitely trap energy and never release it. However, that part of the process is conveniently neglected in this reasoning.

The point about saturation (at least for 4.3 and 15 μm absorptions) is true, but it misses a very important part of the larger system: the atmosphere is more than 10 m thick. What is absorbed by H_2O and CO_2 in the first 10 m is partly transferred via re-radiation and conduction into the next 10 m. Here it is instantaneously held, then transferred again (backward, sideways, upward), and so on. In fact, what is absorbed by *anything* in the first 10 m (by water vapor, other gases, dust) is re-radiated and conducted further, hence reaching higher strata where, at a certain height, CO_2 concentration is *not* (yet) saturating because of the lower density of air. Adding more CO_2 to the atmosphere essentially raises that threshold layer at which radiation escape into space can start to happen; in concert, the thickness of atmosphere experiencing the greatest CO_2-driven forcing increases, hence increasing the energy held in the atmosphere in total. Furthermore, larger portions of the H_2O spectrum "windows" are occupied by the "shoulders" of the principal CO_2 peaks (4.3 and 15 μm), and absorption on the shoulders is still far from saturation. Both these effects mean that CO_2 definitely continues to contribute greatly to global warming as its concentration in the atmosphere rises.

And air currents don't disrupt layers on fast enough timescales

Now imagine that we have layer upon layer of CO_2-containing atmosphere. We essentially compound the effect at each layer, and that is one reason why a relatively small nominal increase in CO_2 concentration can have such a large effect: CO_2 *appears* to act as a "thermal brake," slowing the diffusion of heat out of the atmosphere and into space. If we double the concentration of CO_2, reducing the saturation distance to 5 m, say, then in effect we double the number of insulating layers of atmosphere that slow heat escape to space. It's a bit like first having double glazing—which certainly does a lot to slow heat escape from your living room—and then halving the distance between the two glass panels by adding an extra layer of glass, making it into triple-glazing—which is an even better insulator; or, seen another way, it "brakes" the escape of radiated and conducted heat from your living room even better than double glazing because it has three thermal brakes rather than just two.

You might hear arguments about why this is "wrong." One example of this reasoning appears on the website https://nov79.com/gbwm/ntyg.html (accessed on February 21, 2020), where the opponents of anthropogenic CO_2-driven climate change state "Supposedly, the radiation will be re-emitted and re-absorbed more often, when distances are shorter. But they err in two ways. One is in not taking into account the convection which removes the relevance of short distances. The other is in assuming the direction is toward space." Let's consider convection. The "layers" of air that contain sufficient CO_2 to be saturating for the reflected and re-emitted radiation are only "notional layers." It's obvious that air moves via convection, wind, and updrafts caused by mountains, but those movements are very slow compared with the speed with which energy transfer processes occur. Radiated heat travels at the speed of light (because it's electromagnetic radiation), which is 3×10^8 meters per second in a vacuum (and minimally less in air). Even the updraft of a severe storm is a mere 50 m per second (180 km per hour). Molecules moving *through* air and *conducting* heat are also much faster than the bulk air.

An average updraft/downdraft may seem like an argument-breaker for humans trying to "see" atmospheric layers; however, for electromagnetic radiation, it's piffling. In the time taken for a ray of IR, unbroken, and in a straight line, to reach the outer edges of our atmosphere at around 50 km, a cubic centimeter, say, of "wind" moving at 50 m/s would have moved 8.5 mm. And because the CO_2-deniers are talking about the mixing of 10-m layers of air, if we concentrate on the scale of 20 m (two layers mixing), then a ray of IR radiation would cover the whole of that distance in around 0.00000007 s (70 nanoseconds). Within that tiny time interval, a small volume of air would—even at the speed of a storm-related updraft—move the grand distance of 0.0000035 m, or 0.0035 mm, or 3.5 microns: around 20 times thinner an average human hair.

In fact, absorption/re-emission/conduction is happening over much shorter distances within and between our notional 10-m layers. Hence the energy transfers of the whole system are playing out (over both small *and large* scales) at timescales that are many orders of magnitude smaller than those of even the strongest winds, let alone comparatively sluggish "normal" convection currents. For electromagnetic radiation completing cycles of absorption/re-emission many times between molecules (zig-zagging through large depths of atmosphere), the volumes of air through which it moves stand practically still.

More to the point, as soon as radiation has passed from one notional layer to the next, the clock is reset to zero for that particular ray of radiation. We hence must see this in a completely different frame of reference compared with slow mass movements of air: we're instead talking about extremely fast, individual, transfers of energy, the rapid physics of which happens in almost complete nonchalance to air currents. But there is something even more important going on because, whereas water vapor is doing most of the absorption/re-emission in this system, CO_2 appears to be doing absorption followed mainly by *conduction*.

Ultimately, thermal energy, not just radiation, is transferred, but how?
A temperature increase that we can measure with some kind of thermometer manifests not mere radiation, but an increase in the energy of atomic/molecular movement in the air, water, land—everything: this refers to a very particular class of movement: *translational motion*, the movement of a whole molecule with respect to its center of mass, also known as *thermal motion*. Bond-vibrational energy (stretching/bending in response to IR absorption) doesn't directly contribute to the classical *temperature* of a molecule. However, bond-vibrational energy can be transformed into translational motion, which *does* manifest as a rise in classical temperature (e.g., in °C). Thus, more important than the human construct of "temperature" (which quantifies *one facet of movement*) is the physical measure of *total internal energy* of molecules (the sum of translational, rotational, bond-vibrational, and bond-associated [potential] energy [spring energy]). The law of energy conservation and physical transitions channel IR

radiation via bond-vibrational modes into thermal motion (temperature), but only via collisions between molecules.

Transformation of a CO_2 molecule's bond-vibrational energy into thermal motion can only happen by this molecule bumping into another (*any* other). If the energy is transformed into translational movement of the other molecule, we can regard that as "energy *conduction*." A collision can also result in the redistribution of energy *within* a molecule, hence, amazingly, causing its bond-vibrational energy to mutate into translational energy: that is, the internal "buzzing" of the molecule is reduced, and its speed through space is increased (greater thermal motion)! Regardless, the key events in the energy transfers that result in greater thermal motion are molecular collisions.

In summary, beside (re-)radiation of IR in the atmosphere, there is also a great deal of *conduction and conversion* of absorbed energy between molecules going on: CO_2 molecules bumping into other molecules, and molecules of air bumping into land, water, and objects on Earth. These collisions, not radiation per se, are the crucial step of energy transfer in global warming. And we need to know how fast they are happening because that will model how CO_2 acts so disproportionately to increase ambient temperature and whether air currents disrupting notional layers might interfere with the model.

- *How fast are CO_2 molecules moving at 20ºC and one atmosphere pressure* (NIST standard temperature and pressure [STP]: 293 Kelvin/100 kPa)?
- *How long does it take for a CO_2 molecule to encounter another molecule of air (of any kind) whereupon its bond vibration has the opportunity to be converted into thermal motion*—so-called *translational energy* (as opposed to bond-vibrational energy)? This one needs more explanation: when IR radiation is absorbed (as a photon), it causes a higher-energy electronic state in the CO_2 molecule (a bond vibration). That state doesn't last forever because, as we know, CO_2 can also re-emit IR radiation. The state may either relax on its own to a "lower-energy" state and/or re-emit IR radiation, or it may relax upon collision with another molecule, thereby largely converting vibrational bond energy into translational thermal energy (potentially initially via rotational energy). The relative proportions of internal and external relaxation going on help us understand how much CO_2 re-radiates, and how much it converts energy into heat in the surrounding air. This balance in turn depends on the last point:
- *What is the probability at each collision that such an energy transition (bond vibration into thermal motion of the collision partner) takes place?*

The time taken, on average, for a CO_2 molecule at STP to collide enough times with (any) other molecules to transfer bond-vibrational energy is relatively easy to calculate (see Chapter 3, Calculation 1, online at https://doi.org/10.1093/oso/9780197664834.001.0001). It turns out to be 10 millionths of a second

(10 microseconds [μs]). So, if the CO_2 molecule had absorbed IR radiation, would it still have that energy 10 μs later (on average), or would it have re-emitted it? The information for answering this question was produced back in the late 1960s and 1970s as part of work to develop CO_2 lasers (and CO_2/N_2 lasers—a relevant gas mixture when we consider our atmosphere), and, separately, by researchers studying the atmosphere of Venus (which is more than 96% CO_2). More information on this part can be found in Calculation 1. The bottom line is that conversion of IR-induced bond-vibrational energy in CO_2 to thermal energy (tangible heat) via collision with another molecule appears to occur between 66,000 and 220 times faster than re-emission of that IR radiation (see Calculation 1 and Figure 3.5). CO_2 is not chemically changed by this collision, and it can immediately absorb more IR and convert it to thermal energy via collision. Largely going through cycles of IR absorption and thermal energy generation once every 10 microseconds (see Calculation 1), CO_2 acts as a thermal catalyst, a concept very different from the "thermal brake" that is generally communicated. If the collision (with whatever molecule) results in the bond-vibrational energy in the CO_2 molecule transforming into its own increased thermal motion, the larger picture doesn't change: a tiny part of the atmosphere has become warmer.

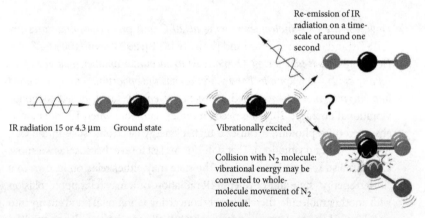

Figure 3.5 CO_2 absorbs infrared (IR) radiation, but most of the energy is converted to thermal motion before it can be re-emitted as IR radiation. In air at normal atmospheric pressure and 20 °C, the probability of one or other eventuality (signified by a question mark) is very strongly tilted toward the lower: i.e., a bond-vibrationally excited CO_2 molecule passing on its bond-vibrational energy to another molecule (in this case, it happens to be nitrogen (N_2)) (and potentially converting it to its own thermal motion too), via collision, before it has the "chance" to re-emit it (see Chapter 3, Calculation 1, online at https://doi.org/10.1093/oso/9780197664834.001.0001). This results in the IR energy (that initially caused the CO_2's bond vibrations) largely being converted to thermal motion energy (heat) in air molecules, including CO_2 itself. Nitrogen is statistically the most likely collision partner.

CO_2 as a thermal catalyst

Now, from another perspective, we see why CO_2 does something so disproportionate to its atmospheric concentration: CO_2 acts as a catalyst in thermal energy transitions. If CO_2 were to re-emit IR radiation in 0.1 *nano*second instead of 1 second, things would look very different: a large part of the radiative energy that CO_2 received would be re-radiated. Variously through a chain of absorption and re-radiation, and also directly, much would leave the atmosphere in a matter of milliseconds. CO_2 performs the energy conversions from bond vibrations to thermal energy (heat) in its surroundings so quickly that a small concentration of CO_2 can have a very large effect. Water, on the other hand, converts very little of its absorbed IR radiation to trapped heat because H_2O has an unusual property compared with other tri-atomic molecules: it re-emits IR relatively quickly and converts much, much less of it to thermal energy than does CO_2. This is why we can have 100 times as much water as CO_2 in our atmosphere and yet not boil to death.

The speed of the CO_2 energy transitions is also what finally destroys the argument that CO_2 cannot produce insulating layers rather like double or triple glazing because air currents cause the layers to mix. As mentioned earlier, a CO_2 molecule experiences an energy transformation-collision at STP in 10 µs on average. In that time, during an extreme updraft (50 m per second), the bulk air in which it finds itself would move no more than 50 micrometers. CO_2 absorbs reflected IR from the Earth and, in a few millionths of a second, forces it to go a circuitous route between bond vibrations and thermal interactions with all types of molecule in the atmosphere: what finally escapes into space is only a fraction of what is emitted from the Earth. The more CO_2 in the atmosphere, the smaller that escaping fraction.

Further compounding CO_2's catalytic behavior is its remarkable chemical stability: to form CO_2, carbon reacts with oxygen in a kind of perfect relationship: oxygen's and carbon's electrons mingle to create two so-called *double bonds* that are shorter (and hence stronger) than almost all other bonds that carbon makes with oxygen in other compounds, organic or inorganic. CO_2 is perfectly linear, O=C=O, giving it further "strength" against reactivity because the electrons in its two double bonds can swash around between them easily, reinforcing the bonds and making them so unusually short. Oxygen loves to pull electrons away from things, leaving them positively charged when it is bound to them and hence relatively reactive; however, in CO_2, the shortness of the C=O bond, and the dissociated nature of the bonding electrons, leaves the central carbon atom much less positively charged than in most other carbon–oxygen compounds. This carbon is, hence, rather unreactive. It takes a lot of energy, or quite some chemical cunning, to separate the oxygen and carbon in CO_2: once in the atmosphere, a molecule of CO_2 can last a very long time

(CO_2's half-life in the troposphere—the layer where weather phenomena occur—is 120 years).

Because CO_2 is so stable, a rapid temporary increase in emissions can have a very long-term effect. Essentially, the only way in which appreciable amounts of CO_2 can disappear from the atmosphere in timescales comparable with a human lifetime is via uptake by plants followed by conversion into organic material, and via simple dissolution into the oceans and freshwater repositories. But that is not quick enough. Research based on modeling done in 2010 indicates that the global warming effect of CO_2 released now persists for more than 1,000 years.[24] No serious research has since contradicted that insight.

Why methane is a more potent GHG than CO_2 in the shorter term

First, CH_4 is IR active because it has bonds between more than two atoms of different identities (carbon and hydrogen). Hence for certain bending and stretching modes, there is a change in the overall molecular dipole (charge distribution of bonding electrons). CH_4 has four basic bond vibration modes (two bending, two stretching), two of which are IR active (similarly to CO_2). However, mass for mass, CH_4 is a more potent absorber of IR radiation because its two IR-active modes have more permutations (degeneracy) than CO_2's: counting degenerate modes, CH_4 has six IR-active modes, while CO_2 has three. Furthermore, a few of CH_4's absorption bands are at markedly lower wavelengths (i.e., higher radiation energy) than CO_2's, hence corresponding to absorption of higher energy and an unsuspected effect. Finally, CH_4 has more capacity for increases in radiative forcing for every ppm increase in concentration in the atmosphere compared with CO_2: CO_2's absorption at its two main peaks close to the ground is saturated (i.e., it can't absorb any further at these peaks by becoming more concentrated, only in the shoulders of the peaks); CH_4 is at a much lower concentration than CO_2 (approximately 1.7 ppm by volume) and hence has much further to increase before reaching absorptional saturation.[25]

Similarly to CO_2, even when CH_4 does reach absorptional saturation in the low atmosphere, rising CH_4 concentrations will continue to increase radiative forcing because the height of the atmospheric layer from which IR radiation *can*, finally, escape simply rises. Incidentally, Tyndall presented this general effect in 1862, but for some strange reason, his simple explanation was not appreciated for its fundamental importance by scientists of the time. Fortunately we no longer model the atmosphere as a single block of gas above the Earth, but instead as a system of layers that, together with energy transitions, explain the greenhouse effect very well.

CH_4 partly closes further "windows" in H_2O's IR spectrum. One of CH_4's crucial absorption peaks, at around 7.5 μm, had been known for many decades; two more had been overlooked until recently. CH_4 had been calculated to have between 28 and 84 times the global warming potential of CO_2, over, respectively, 100 and 20 years[26]: that is, 1 tonne of CH_4 released today would, over 20 years, cause 84 times as much global warming as 1 tonne of CO_2. The figure drops to 28 over 100 years because CH_4 (with a half-life of 7 years) is resident in the atmosphere for much less time than CO_2. Calculating net forcing is a complicated matter, superbly illustrated by the next point. CH_4's impact had hitherto been calculated by weighting its forcing effect (partly closing an H_2O window at 7.5 μm) against its *damping* effect (reducing overall forcing), but something had been overlooked.

Damping happens in the real atmosphere under incident angles of solar radiation and real-life water vapor concentrations (as opposed to laboratory experiments). It had already been observed for CO_2: the short-wave absorption of CO_2, by *overlapping absorption bands* (as opposed to closing windows) in the H_2O spectrum, actually *reduces* the total net radiative forcing at those points. The same was assumed for CH_4 because CH_4 has a relatively large overlap with the H_2O spectrum at around 3.3 μm, which reduces overall forcing. However, in 2016, a paper was published showing how even the Intergovernmental Panel on Climate Change (IPCC) had significantly underestimated CH_4's effect on the climate by missing something.[27] The research showed the opposite of net damping because it looked at even shorter wavelengths than conventionally studied: CH_4's short-wave absorption actually *increases* calculated radiative forcing because absorption bands around 1.6 and 2.3 μm close parts of the two windows in the H_2O spectrum at those extremes of IR. As a result, the calculated potency of CH_4 over 100 years rose to 32: that's 14% more than the previous estimate.

CH_4's cycle of production–destruction occurs almost exclusively between the surface of the Earth (land and oceans) and the troposphere (lowest atmospheric layer). CH_4 released at the Earth's surface (from microbial processes, digestive processes in animals, geological processes, etc.) diffuses into the troposphere, where almost all reacts with OH radicals (produced by the action of UV light on ozone in the presence of water). This produces formaldehyde and carbon monoxide (CO), which are rapidly oxidized to CO_2. That, and a similar reaction in the upper atmosphere, convert roughly 540 million tonnes of CH_4 annually. For

CH_4, the atmosphere is the sink, but human activity can prolong CH_4's existence there: for example, increasing concentrations of anthropogenic nitrogen oxides (NO_x), which reacts with OH, lead to lower concentrations of CH_4-destroying OH radicals. Though CH_4 is a much more potent GHG than CO_2, it is much more reactive, and hence its cycling time between source and sink is much shorter. CO_2 is an altogether different matter: it produces long-term deposits of carbon-containing compounds—a process with interesting consequences.

CO_2 cycles and sinks: How the inorganic and organic Earth work together

CO_2 is removed from the atmosphere by a variety of mechanisms.[28] All major mechanisms are well known, but they are part of a complex network of interconnections. Furthermore, the behavior of living organisms as a very important part of the global carbon cycle must also be understood (i.e., how much carbon is sequestered in them, where they are, *what* they are, how quickly they absorb/emit CO_2, and how they are reacting to global warming and increasing CO_2 levels). Here arguably lie the greatest scientific challenges in understanding climate change. Recent estimates suggest that the carbon of living biomass on Earth is distributed mainly among plants (450 Gt C, primarily terrestrial).[29] Animals possess a mere 2 Gt of carbon (mainly marine); bacteria 70 Gt; archaea (cousins of bacteria) 7 Gt; and fungi 12 Gt (Figure 3.6). About 6 Gt is sequestered in the total life in our oceans. It was only in this quite recent publication[29] that a fundamentally more accurate idea of the balance of terrestrial versus marine biomass was presented: organisms on land represent almost 100 times as much mass

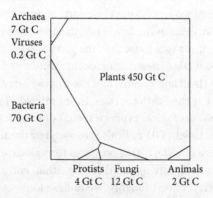

Figure 3.6 Distribution of biomass of carbon on Earth in gigatonnes (Gt). Redrawn from Bar et al.[29]

of carbon as organisms in the oceans. Even more striking: the mass of carbon in humans and livestock is approximately 23 times the mass of all wild mammals on the planet combined. Reflecting even briefly on how many of these humans live and the resources they use, we need not wonder why humanity is currently unsustainable.

Big blue: The largely mysterious workings of Earth's oceans

The complications of biomass calculations might conveniently be distracting us from the elephant in the room, the oceans, that greatly understudied part of the planet. The truly remarkable thing is that the Earth is, via its oceans, probably absorbing much more of our additional (anthropogenic) CO_2 than previously thought: 2020 estimates suggest up to 10.6 Gt net total per year[30]—a little less than the land, which absorbs around 30%, or 12.9 Gt, annually.[31] That leaves behind around 19.5 Gt. The worrying thing is that we have a very incomplete understanding of where most of the 23.5 Gt of absorbed CO_2 is going and what it's doing there. Basically, our planet is "trying" to equilibrate the carbon cycle—not surprising for a system embodying negative feedback (i.e., tends to react in the opposite direction to the force pushing it away from equilibrium). But the anthropogenic CO_2 is not being incorporated innocuously into long-term deposits because geological processes take millions of years. The oceans are admirably buffering CO_2 in terms of quantity, but pH values are falling consistently ("acidification").

Merely 6 Gt of organic carbon are sequestered in living organisms in the oceans,[29] but, over evolutionary and geological time, the tiniest marine organisms are responsible for creating today's most visible geological features: enormous expanses of sedimentary carbonate rock that were once sea beds and now fold and rupture out of the land in impressive formations such as the French Jura (a mountainous area after which the Jurassic period in Earth's history is named: roughly 201–145 million years ago [mya]). Almost all carbonate rock is either calcium carbonate ($Ca(CO_3)_2$) or calcium magnesium carbonate ($CaMg(CO_3)_2$), the latter being known as dolomite, after the Dolomite mountain range in Italy, where it was first identified.

The oceans harbor a little over one-hundredth of the biomass of the terrestrial biosphere, but they are the largest "sink" of CO_2. That is, they take more CO_2 out of the atmosphere and convert it to long-term deposits (carbonate rock) than any other environment. But they can't be hurried: increases in global CO_2 concentrations are not speeding up the incorporation of that CO_2 into marine organisms. If anything, the reverse is happening. Oceans will sequester around one-quarter of all "excess" CO_2 for millennia (i.e., into geological deposits). The "math" of this gigantic geophysicochemical equation was done back in 2009, by Indian mathematician Samar Khatiwala,[32] at that time working at the Lamont-Doherty Earth Observatory of Columbia University, New York. But Khatiwala and co-authors did more than just provide status quo figures. They plotted trends and revealed something horrifying: the capacity of the oceans for absorbing CO_2 is falling. The reason for that is one of the perilous positive-feedback mechanisms that threatens to set CO_2-driven global warming on an uncontrollable runaway trajectory. And, yes, this is also a complex system, so we are unlikely to be able to predict it beyond a certain time window.

Around 97% of carbon in the oceans exists as inorganic carbonate—basically dissolved CO_2. That is why oceans sink so much CO_2, despite containing much less mass of living organisms than the land. Seawater doesn't have nearly as high a concentration of CO_2 as a fizzy drink or sparkling mineral water (otherwise it would fizz), but the chemistry is the same. CO_2 dissolves in water quite well: at 25°C and normal atmospheric pressure (sea level), you can get around 1.45 g of CO_2 into a liter of water before oversaturating it (an unstable situation that doesn't last long); oxygen is almost 20 times less soluble than CO_2 in water under the same conditions.[33] In an interesting twist of physical chemistry, it's the same molecular dipole—imbalance of charge via "polar" C=O bonds in CO_2—that makes it relatively soluble in water and also absorb infrared radiation at wavelengths where many other molecules don't. This bond polarity leads to something interesting when CO_2 meets water: it "dissociates," as chemists say. At the turn of the 19th century, what becomes of CO_2 when it dissolves in seawater was largely unknown. Then, in 1904, the Danish physiologist August Krogh (1874–1949) published an equilibrium[34] indicating that most CO_2 just floats around as CO_2; but a small proportion grabs a hydrogen (and oxygen) from water and turns into hydrogencarbonate and H^+ ions, otherwise known as *carbonic acid* (Formula 3.1): this acidifies the ocean (it's the hydrogen that makes the acid).

Formula 1:
$CO_2 + H_2O \rightleftharpoons HCO_3^- + H^+$ (the larger backward-facing arrow indicates the balance of equilibrium)

This is exactly the same "carbonic acid" as in fizzy water. If you have a fizz-maker and can compare the acidity of normal tap water with water that you've carbonated (e.g., using litmus paper or a liquid pH indicator that you can get from a pharmacy) you'll see that the carbonated water is more acidic (has a lower pH) than the plain tap water. This is exactly what happens in the sea, except that the sea contains substances that make it slightly alkaline to start with. Carbonated drinking water has a pH somewhere between 4 and 6, depending on how "fizzy" it is. pH 7 is neutral, the sea is around 8, so whereas your mineral water can fairly be called "acidic," we can't really talk about acidification of the ocean: it's still slightly alkaline. However, life in the ocean is so sensitive to pH that even if it fell to neutrality, most organisms in our oceans would long since be gone: no CO_2-absorbing phytoplankton (algae) that make calcium carbonate skeletons (e.g., coccolithophores) could make these intricate wonders of nature anymore because calcium carbonate—at its low concentration in the organisms—simply remains in solution at pH 7. Disappearance of phytoplankton means reduced CO_2 uptake (via photosynthesis) and no food for animals higher in the food chain, right up to the magnificent whales, sharks, and other vertebrates.

Surface ocean water is, on average globally, somewhere around pH 8.1, but how, why, and where it is changing in response to anthropogenic CO_2 emissions and global warming is a matter of debate. Some studies report lower pH between the tropics compared with the Arctic[35,36]; others identify massive fluctuations within Arctic waters that don't fit with the larger model. This is telling, because the lowest pH values yet measured in seawater are, in fact, in areas within and around the Arctic (the Canadian Basin and the Bering sea): the research paper presenting these data was one of the last that Taro Takahashi and his group at Columbia University published, and it caused a major stir.[37] The Bering Sea henceforth became known as the most acidic sea on Earth, at pH 7.7 (even lower than the catastrophic scenario predicted by the IPCC in Figure 3.7). It fittingly demonstrates the synergistic effects of two facets of CO_2-induced global warming: (1) the Arctic ice melts and dilutes the seawater, reducing pH; (2) the increasing concentration of atmospheric CO_2 causes more CO_2 than previously to dissolve into this water, which—being colder than between the tropics—absorbs more CO_2 than warmer water. More dissolved CO_2 means lower pH.

The crucial point here is that while the regions studied represented extremes, far outside the average, they are connected with other ecosystems. Takahashi, an oceanographer, had been studying changes in oceanic constituents for more than 40 years and was particularly concerned by the ability of organisms to produce structures made of aragonite. Aragonite is a crystalline form of calcium carbonate, and it is rather special because at room temperature and pressure it is unstable: it readily turns into the more common calcite, which is chemically identical. However, at lower temperatures and higher pressures, and also in biological contexts, aragonite is stable enough to be used to build biological structures. A major class of corals (the so-called scleractinian corals) use exclusively aragonite for their exoskeletons. Sea snails and some other mollusks also make aragonite in their shells. The biological reason for this choice over calcite (which is used by other organisms) is still a mystery. However, the practical implications are obvious: aragonite has a significantly higher so-called *solubility product constant* in water than calcite. This means that aragonite dissolves more readily than calcite as the pH of the oceans falls (a consequence of its relative instability compared with calcite): larvae in particular struggle to make it into adulthood, where they are then less at risk from demineralization. Naturally, these organisms are food for others, and so a population collapse at this relatively low level of the food chain can have catastrophic consequences higher up.

Already in 1996, the IPCC had predicted a sudden and large fall in ocean pH as result of CO_2 taken up from the atmosphere (Figure 3.7). The plot showed the CO_2 concentrations in sea water increasing from 10 mmol m^{-3} in pre-industrial times to 30 mmol m^{-3} by 2100: a 200% increase. As a result, the pH of the oceans was predicted to fall from 8.2 to around 7.8. That might not seem like much, but all living organisms (aquatic or not) are very sensitive to pH changes, as we saw earlier: 7.4 is the normal pH of human blood; a decrease of only 0.05 pH points—to 7.35—is classified as "acidosis," a condition that can lead to loss of consciousness, coma, and even death if not treated. In fact, 7.35 is not even an *acidic* pH. If we take just the top 10 m of sea water, we have a relatively simple system because temperature can be measured easily, and the pressure varies from one atmosphere (1 bar) to two atmospheres (2 bar) (10 m of water produces a pressure of very close to one atmosphere, coincidentally; in the sea, it's a bit more than in plain water because the dissolved salt makes the water denser).

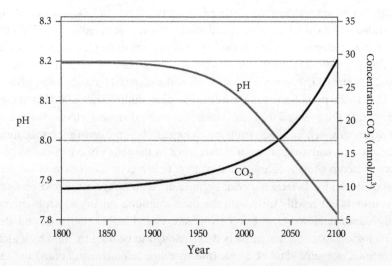

Figure 3.7 Projected change in atmospheric CO_2 concentrations and ocean pH, assuming that anthropogenic CO_2 emissions followed predictions made at the time (1996). Plot made by the IPCC in 1996, redrawn from Watson et al.[38] (Climate Change 1995, IPCC 1996).

The solubility of CO_2 in water at different temperatures and pressures—essentially depths—was calculated very accurately in 1940.[39] This is very relevant to our present-day challenge of understanding the ocean's buffering effect on CO_2 from the atmosphere because the deep ocean clearly has a much higher pressure than the shallow ocean. At 500 atmospheres (i.e., 5,000 m depth and at close to 0°C), the solubility of CO_2 is around 2.5 times as great as in the surface 25 m of ocean (at 10°C).[39] Hence the cold, high-pressure depths of our oceans draw CO_2 down into an invisible and vastly understudied realm. Deep ocean temperature is very constant compared with surface waters, where we see the effects of global warming. What appears at first sight to be a simple system of temperature, pressure, CO_2 concentration, and pH turns out to be very complicated. For example, as surface water warms, its capacity for absorbing CO_2 falls, hence limiting the fall in pH. However, counterintuitively, the Arctic contains areas where surface waters have among the highest measured pH values (8.25) *and* the lowest (7.9),[35] hence representing the ocean with the largest pH variability: understanding this phenomenon will aid our general understanding of the ocean's pH buffering capacity.

As the largest carbon sink on Earth, not only do oceans contain more readily exchangeable carbon than the land, but they also exchange more carbon with the atmosphere each year. Global warming is happening at the level of the behavior of our atmosphere; however, the conditions in the world's seas are arguably changing in more complicated ways. Furthermore, we understand oceans much less well than the atmosphere. Unfortunately, complex systems—and the more complex the more this holds—can move out of the dynamic area of variables (e.g., temperature, concentration of gases, etc.) where they exhibit negative feedback and enter unstable areas where positive feedback emerges (i.e., runaway behavior). This is the basic mathematical explanation of the often quoted "tipping point," after which global warming and its consequences will be out of reach of human intervention.

Where does carbon go in geological cycles, and for how long?

The net increase in carbon in the atmosphere in 2019 was around 5 Gt, mostly in the form of CO_2 (19 Gt CO_2). How much is that in comparison with the amount of carbon "on the move" each year? In Earth's short-term carbon cycle (at the scale of years or decades), around 260 Gt of carbon are exchanged between the atmosphere and the oceans/land per year. How much carbon goes into very long-term cycling parts of the total global carbon cycle on the scale of the Anthropocene (a few hundred years)? It's quite negligible, but clearly not zero, otherwise the Earth would not boast massive mountain ranges of limestone (calcium carbonate from minute organisms) or have accumulated such enormous reserves of fossil hydrocarbons and coal (i.e., the transformation of organic matter from dead organisms). Researchers currently estimate that between 10 and 100 million tonnes of carbon move through the slow parts of the carbon cycle (thousands to millions of years) annually,[40,41] representing between 1/500th and 1/50th of the annual increase of 5 Gt from human emissions.

Limestone is easy to explain: it's essentially unchanged in its chemical composition (mainly calcium carbonate) from when it was adorning a living organism or came out of solution as a precipitate. It can form in under a million years and is part of the long carbon cycle, which takes between 100 and 200 million years on average to return carbon to its source (e.g., CO_2 in the atmosphere). Some limestone formations are much older, witnessing fossils from more than 500 mya. How fossil oil and gas deposits were largely made is a more controversial question, but we have some fairly good ideas. The basic principle is one of organic material—containing, as we saw earlier, a great deal of carbon in its dry mass—turning into a different form of carbon and carbon compounds. These escaped most of the oxidative processes that happen on the surface of the Earth. Remember, it is the reduced nature (the opposite of oxidized) of hydrocarbons that gives them useful chemical energy (mostly stored in their C-H and C-C bonds, as opposed to C-O bonds). Basically, the anoxic environment (i.e., lacking oxygen) is a prerequisite for almost all types of fossilization process.

Creating coal: Of ancient plants and parallels to cosmic carbon

Most coal was formed from peat bogs, where the conditions are known to be acidic and low in oxygen—ideal environments for minimizing microorganismal decomposition of organic matter. The early plants monopolizing this soggy habitat were mosses (as they are today). They were in an almost endless cycle of growing and dying, the dead parts descending into the damp underground, decomposing to a small extent, and forming peat. Early shrubs and trees (types of fern and horsetail) joined them to make swampy forests. By studying today's peat bogs (some of which can be 12 m deep or more), scientists estimate that at least 1 m of peat is deposited every 1,000 years. The Carboniferous period lasted around 60 million years (359–299 mya), and so, in some places where bogs remained essentially unchanged for most of that period, a very thick layer of peat would have resulted. Of course, the theoretical 60 km could not possibly accumulate in one spot: any natural basin in the land capable of accommodating a peat bog could only be a small fraction of that height. However, the deepest peat was already undergoing greater compression and turning into coal (Figure 3.8).

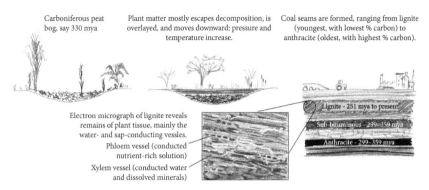

Figure 3.8 Schematic of coal formation. Sub-bituminous strata also contain bituminous coal (not shown); sub-bituminous coals, bituminous coals, and anthracite were all formed in roughly the same geological time window, between 299 and 359 million years ago (mya). Lower panel (electron micrograph of lignite) drawn from SEM micrograph in Wang et al.[48]

Clearly, a prodigious quantity of peat was laid down in some places if present-day coal seams are anything to go by: these tend to be between around 5 and 10 m thick, on average; the thickest known seam in the world, in Wyoming, measures a little over 24 m at its maximum.

Estimates of the compaction factor from peat to coal vary: the highest is between 10:1 and 11:1. Much lower estimates are claimed by researchers who have found quite well-defined dinosaur footprints in the roofs of coal seams. These indicate that the peat on which the dinos trod was already compact enough to preserve detail and clearly didn't become very much more compact upon transformation to coal. Dry peat has a density of 400 kg/m^3; coal on average around 870 kg/m^3; hence, by mere compression, a ratio of roughly 2.2:1 results. If the average compaction ratio lays somewhere between these estimates, say 6.5:1, then the peat bog that made the Wyoming coal seam produced 160 m worth of peat! During the Carboniferous, plants took so much CO_2 out of the atmosphere and exchanged it for oxygen via photosynthesis that the O_2 concentration reached 35% compared with the present-day 21%. Arthropod animals (e.g., insects, millipedes, centipedes, spiders, crabs etc.), which rely more on diffusion of gas to breathe* than do animals with lungs (e.g., vertebrates), reached enormous sizes: for example the 2-meter-long giant millipede (*Arthropleura*) or the giant dragonfly (*Meganeura*) with a 65 cm wingspan—the largest flying insect ever to have lived.

---

* Research published in 2003[42] showed that, contrary to previous assumptions, the tiny air passages in insects do actually move slightly to pump air in and out of tissues.

Soils in which ancient trees on Earth grew are known to have been both acidic and low in oxygen, ideal environments for preventing microorganismal decomposition. That is not to say that *no* microorganisms grow in low-oxygen or even completely anoxic environments, but such microbes (so-called *anaerobes*) are very "slow": they rely on fermentative metabolism to degrade their food source. So, to cut a long—very long—story short, this is what happened to an ancient tree fortunate enough to be preserved after death: having survived most of the decomposition by anaerobic microbes in its boggy tomb, it became buried by successive layers of other dead plants, then by inorganic sediment (e.g., brought by a river delta). Millennia passed as the "wood," peat—and anything that had escaped microbial destruction—descended into the Earth's uppermost geological layers (the lithosphere) and remarkable chemical changes took place. No joke, it's called "coalification"![43]

Brown coal, or "lignite": We can still see the wood

During their descent (basically caused by deposition above and gravity), the pressure and temperature rose. Heat in the lithosphere mainly results from radioactive decay in the rock, and it increases gradually with depth. The first geologically relevant product of peat compression that we would recognize as "coal-like" is *lignite*, or "brown coal." It has a dry-mass percentage carbon content of between 25% and 35%. Lignite derives its name from *lignin*, a constituent of wood (from Latin *lignum* = wood) and reflects the fact that a component of the plant matter in the ancient wetlands was woody (either woody shrubs or early trees). With an electron microscope one can even see the remnants of wood in lignite. Having undergone some degree of compression, but at relatively low temperatures (up to 100°C), lignite has less than half the percentage carbon of true coal and is, therefore, a rather low-energy fuel. It's also dirty because of its relatively high percentage of other elements (notably sulfur, from sulfur-containing amino acids, and also from seawater ingress[44]). Lignite is close enough to the surface to be mined by *open cast* techniques: removing the earth (around 10 m) covering the lignite seam and then simply scooping it out. The largest single deposit of lignite is in and around the Latrobe Valley in Victoria, Australia, where seams range from around 100 m to an astounding 330 m thick—the result of a deep Tertiary basin basically having filled up with peat. These seams together constitute around 25% of world's lignite deposits. Lignite is mainly used to fire older coal-fired power stations. Despite massive public protests, many supposedly ecologically aware countries still mine lignite: in Germany, the figure for 2021 was 126 million tonnes,[45] almost half the tonnage of the world's leading lignite producer, China, which produces 260 million tonnes per year.[46]

Sub-bituminous and bituminous: On the way to high-grade coal

Next, we identify sub-bituminous coal, which is also a lower-grade coal, containing around 35–45% carbon. Both lignite and sub-bituminous coal still contain quite a lot of water (around 30% by mass), hence making them rather inefficient fuels. Older still, geologically speaking, the next deposit, bituminous coal, varies in hardness quite a lot. The softer varieties contain relatively high proportions of tar-like (but distinct from tar) bitumen or asphalt, which we use on roads. Bituminous coals were formed in geological epochs spanning the Carboniferous to the Cretaceous (359 to 66 mya). They belong to the larger group of "hard coals" and have a chemical composition as follows, by ash-free dry mass: carbon, 84.4%; oxygen, 6.7%; hydrogen, 5.4%; nitrogen, 1.7%; sulfur, 1.8%.

At this point, the last remnants of the original plant cell structures are lost, and we're into the business of true coal, but not of the highest grade. Bituminous coals (of which there are several types) are burned in power stations, and they were the main fuel for steam trains and ships. They are also roasted to produce coke (during which the bitumen for road-building is collected). This process removes most of the ash and sulfur, making the coal suitable for the metal industry—primarily steel production in blast furnaces—where most of it is burned.

Anthracite: The hardest, densest, and highest-quality coal

Finally, at a depth often exceeding 1,000 m below the present-day surface, and "matured" at temperatures of 120–300°C and typical pressures of around 25 MPa (almost 250 atmospheres), we find anthracite, the hardest coal. The deepest actively mined anthracite coal face in the world is probably in the Jindřich II Mine in the Rosice-Oslavany coal basin in the Czech Republic, at almost 1,430 m below surface. Anthracite is typically around 90% carbon by mass, but it can be higher. Often deep black in color, its luster can even make it appear metallic. It is, indeed, the hardest of all coals and has the highest carbon content and the least impurities, notably sulfur. The sulfur did not disappear somehow, but rather became concentrated into nodules, such a pyrites, that can easily be separated from the coal. Anthracite is not pure carbon, but it's the closest burnable substance to it. Anthracite is used primarily for firing domestic heating. It is relatively expensive, being the least "dirty" coal to burn and having the highest energetic value per unit mass.

In aggregate across the different coal types, sulfur impurities create major pollution problems, principally by producing sulfur dioxide (SO_2) when burned (an irritant when breathed in, and the major cause of acid rain). The sulfur in coal is a remnant of the sulfur in two amino acids (methionine and cysteine) and reducing compounds (glutathione derivatives, which protect the living cell against oxidative damage) in the ancient plants.[47] The average elemental composition of plant cells varies somewhat between plants, and parts within plants. However, in terms of carbon content, it is around 49% of dry mass (similar to that of animals), most of the rest consisting of oxygen (39%), hydrogen (6%) and nitrogen (2%). I have omitted the sulfur for the sake of simplicity.

Still some mysteries surrounding the chemistry of coal formation
Though one might expect everything about coal to be established knowledge by now, the geochemical process of coalification (the proper scientific term) is far from a settled matter. We know the chemical nature of the starting materials and of coal, but what happened in between is still a topic of research and debate. The typical carbon-based molecules found in anthracite give us some clues (Figure 3.9).

Some of these look uncannily like the carbon compounds floating around in space, as reviewed in Chapter 1. Wherever it exists, carbon has a natural tendency to form cyclic structures, given half a chance. In fact, under particular conditions of temperature and pressure, these structures are the lowest-energy (most stable) form of carbon—as, for example, diamond is the lowest-energy form of carbon under the extremes of pressure found at the boundary of the lithosphere and the molten mantle of the Earth. They are, therefore, expected to become the predominant form, given time (and coal had plenty of that). What biology created above the Earth's surface, via the far-from-equilibrium energetic systems of life, inanimate geology then took back and returned to a form dictated largely by physical equilibrium.

How did the variety of biological molecules become simplified into the polyaromatic structures of coal? In metamorphosing into coal, it is clear that plant matter becomes greatly concentrated in carbon content and loses a lot of oxygen. One would expect that, too, because coal is a less oxidized substance than wood and hence gives off more energy per unit mass when burned with oxygen. Crudely put, a combination of heat and pressure chemically changes the organic molecules; less crudely put, oxygen-containing plant-derived polymers are split and recondense to create C-C bonds where previously C-O-C bonds were present, and chemical side-groups, mainly -OH, are removed. To cut a long story short, a lot of organic chemistry happens as cellulose breaks down, then lignin (from wood) gets involved, forming a kind of very stiff but reactive organic porridge—one that gets stiffer and stiffer as compression proceeds and water is eliminated. The reactions that cause cyclization and growth of smaller anthracenes into larger ones mostly remove oxygen and hydrogen from the molecules, raising the percentage carbon toward the levels of anthracite.

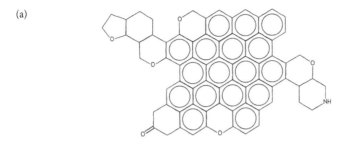

Figure 3.9 The intriguing chemistry of coal. (**a**) An example of an extended pattern of hexagonal aromatic groups (the essential chemical motif of which is benzene without its hydrogens) containing the odd ring-inserted or double-bonded oxygen and NH (amine) group (all of which break aromaticity locally); (**b**) an example of an extended network of different chemical groups containing aromatic and non-aromatic rings, and various side-groups that are ubiquitous in organic molecules (in living and non-living contexts). Note in particular the presence of sulfur (S), originally derived from sulfur-containing amino acids (methionine and cysteine) and cellular antioxidants (e.g., glutathione). Aromatic rings are depicted with a dissociated, circular bond structure: this is equivalent to the three double bonds (=) shown in other contexts. Image (**a**) drawn with inspiration from Chemistry Explained[49]; (**b**) re-drawn from Wikipedia.[50]

The origins of oil: Of cell membranes and the largest wax deposits ever identified

At this point you might be wondering what happened to organic matter that built up on the sea bed and became part of the geochemistry there. Did it also become coal? Not quite. The bulk of the biomass sinking into the marine sediment consists of single-celled organisms and tiny multicelled creatures: phytoplankton and zooplankton, respectively. The deep oceans certainly sequester a great deal of carbon from once-living organisms (and they support large communities of organisms that feed on large vertebrate carcasses that end up there); however, it is not in these depths that oil is thought to have formed. The story of coalification might be complicated and full of open questions, but that of oil is wide open at a much more fundamental level: whether oil is mainly produced by organic, biogenic mechanisms, or whether most of it is abiogenic in origin. First the biogenic theory.

Theory 1: The biogenic origin of oil from ancient cell membranes

Dead microorganisms also settle on the ocean floor in relatively shallow areas near sources of sediment (mainly river estuaries) and in so-called sedimentary basins where inorganic sediment is washed in (e.g., lakes with rivers flowing into, or through, them): the organic matter is covered by sediment faster than it can rot. In the sediment, anaerobic decomposition by the very slow-growing microbes is outstripped by geological processes, and so most of the organic material is preserved. It undergoes various chemical processes (known as *diagenesis* and *catagenesis*: basically breaking down of larger molecules to smaller ones), approaching pyrolysis (cleavage at high temperature) as it moves down with the sediment. At depths of around 1 km below the sea bed or lake bed it forms compressed deposits. For an overview, see *Petroleum*, National Geographic Society, Education.[51] It's known from sedimentary core sampling that a substantial proportion of the cell debris is from photosynthesizing algae. The clue that they ultimately turned into oil was the discovery by Alfred Treibs, a German organic chemist (1899–1983), of an organic compound with four nitrogen-containing rings in petroleum samples: a so-called *porphyrin*. It bore amazing structural similarity to chlorophyll, except that chlorophyll's central magnesium ion is replaced by vanadium (Figure 3.10).

From what we know about the depletion of oxygen from organic material that turns into coal, it is easy to believe that geochemical processes turned algae-derived chlorophyll into this vanadium porphyrin and other parts of the algal cells into oil itself. The big difference between coalification and petroleum formation is that whereas the raw materials for the first are largely ring-containing cellulose and lignin, the substrate for the second is mainly the cell membranes

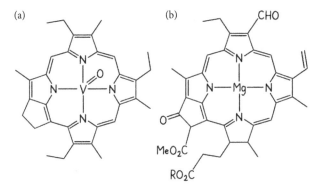

Figure 3.10 The compound that likely betrays oil's organic origin. (**a**) Vanadium-containing porphyrin from petroleum; (**b**) chlorophyll from green plants. The central metal ion in chlorophyll is magnesium. V = vanadium; Mg = magnesium. MeO_2C is a methyl group (CH_3) esterified to the next carbon; R indicates a variable organic group, which is also connected via an ester bond (O–C=O).

of unicellular creatures (phytoplankton and zooplankton). These membranes are overwhelmingly made up of straight-chain lipids rather than cyclic organics (Figure 3.11).

Lipids are basically biological fats and oils, and the geochemical transformations that make them into petroleum start with them being transformed into a substance known as *kerogen*—a process that happens as the organic matter descends to 1 km below the surface, through temperatures of up to 100°C. Kerogen (named after *keros*, ancient Greek for "wax" → modern Italian "cera," pronounced *chera*) is a hard, waxy mixture of a large variety of organic compounds and has no unique chemical formula. One type contains, for example, carbon, hydrogen, oxygen, nitrogen, and sulfur atoms in the ratios 215:330:12:5:1. There is clearly more hydrogen in kerogen than in coal because the carbons in lipid chains have more hydrogens attached to them than the polycyclics in coal. These lipids are not cyclized with themselves, and they have hardly any double bonds (C=C), thus leaving more positions for hydrogen bonds. Some cyclization does occur during kerogen formation, but not as much as in coal formation. A simple calculation based on the atomic ratios and published data suggests that the Earth has to date accumulated around 1.24×10^{16} tonnes (12.4 quadrillion [thousand trillion] tonnes) of kerogen.

Kerogen is partly defined by its insolubility in water and most organic solvents; a component that can be extracted with certain organic solvents is bitumen. Kerogen is the most abundant organic substance in the Earth's crust, or on Earth as a whole, for that matter. And much of it is on its way to becoming oil and gas, depending on where, exactly, it is in the Earth's crust: in the geological temperature

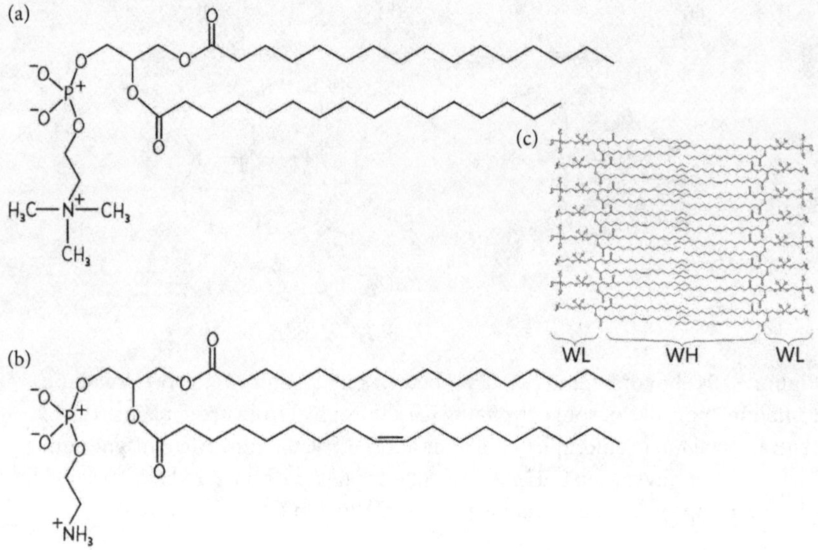

Figure 3.11 Lipids that are abundant in cell membranes: two examples. (a) Dipalmitoylphosphatidylcholine (DPPC), which contains two molecules of palmitic acid; (b) phosphatidylethanolamine, which contains one molecule of palmitic acid, and one molecule of a mono-unsaturated fatty acid (FA); note the double-bond in the carbon chain. Such membrane lipids (which are phospholipids because of their PO_3^- group) can also contain two FAs that are partly unsaturated. Note the three-member OC–CO–CO motive at the top of each molecule: this is derived from the glycerol (a triol) that we met in Chapter 2 as the carrier of three FAs in fats and oils. In membrane lipids, the remaining binding position on glycerol is occupied by a phosphate-based group, often with nitrogen, and it is variously positively or negatively charged. This makes the "head" of phospholipids polar and therefore water-loving (hydrophilic); (c) the water-loving heads associate with each other, as do the water-hating (non-polar) tails, hence making a double-layer (bilayer) with FA tails in the middle and head-groups on either side—the basis of 99.9% of biological membranes.

window of 50–150 °C, petroleum is the decomposition product; between 150 and 200 °C kerogen produces natural gas. The chemical transformations from material of lipid origin to gas and oil is quite easy to envisage because gas and oil are essentially a mixture of straight-chain hydrocarbons (Figure 3.12), very reminiscent of the chains present in biological lipids.

There are, as you might imagine, all manner of other constituents of petroleum and natural gas, but that is not surprising: chemistry rarely produces a

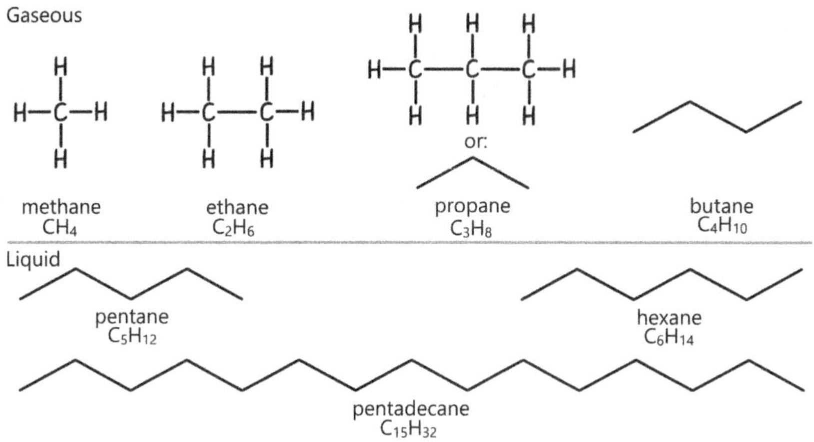

Figure 3.12 Alkanes that are gaseous or liquid at ambient temperature and pressure. Natural gas consists of short-chain alkanes, predominantly methane. Pentadecane is a constituent of kerosene/aviation fuel.

single class of compounds when faced with a large organic substrate, and certainly not when faced with the bag of organic material that was once a living cell.

Theory 2: The abiogenic origin of oil from geochemical processes and meteorites

This theory proposes that much, or even most, of the petroleum reserves on Earth were produced exclusively by geochemical processes (i.e., no once-living organisms were involved). Samples from a variety of kerogen deposits verify (via differential proportions of the carbon isotope ^{13}C) that a small proportion of it was, probably, formed only by geochemistry. "Probably" is a necessary qualification because another source may be the accretion disc surrounding our Sun— the material that made all planets, asteroids, and comets in the solar system. Here broils a hot debate, made even hotter by the possible discovery of kerogen on Mars by the analytical tools of NASA's Curiosity Rover (2018).[52] Those who believe that kerogen can only be formed in appreciable quantities from dead organisms cheer that evidence of past life has been found on the red planet. However, much more research needs to be done in order to exclude an *abiogenic* origin of that kerogen as well.

The presence of organic carbon compounds in meteorites has been known since at least the 1960s[53,54]: carbonaceous chondrites (rock-based meteorites containing mineral granules) have since been classified into at least eight groups, depending on their likely origin in the solar system. The C-chondrites contain higher than trace amounts of carbon and are therefore very dark in color. In 1969, a truly landmark meteorite impact occurred and was well observed. At 10:58 AM local time near Murchison, Victoria, Australia, a bright spot was seen searing through the sky, leaving a smoke trail behind it, and splitting into three fragments. The final impacts spewed debris over an area of 13 km^2. They were also heard and felt. The Murchison meteorite continues to provide researchers with a treasure trove of cosmic secrets. It was a CM-condrite, the "M" denoting "of the Mighei type"—after the famous and chemically fascinating *Mighei* meteorite that impacted in the Ukraine in 1889. In 2020, scientists reported finding silicon carbide crystals in the Murchinson meteorite that pre-date the formation of the Earth by around 2.5 billion years.[55] The meteorite also contains small quantities of amino acids, constituents of nucleic acids (DNA and RNA bases), and a variety of hydrocarbons. Some of these—as is generally the case with CM group chondrites—are chemically and physically very similar to kerogen; indeed, they may as well *be* kerogen.

Though opinions differ on this point, and proponents of life elsewhere in the solar system would love it to be otherwise, these hydrocarbons (and organic compounds in general) are probably not directly related to life and living organisms in the solar system—either in terms of life's origin (seeding life) or the traces of past life. Hydrogen and carbon monoxide, which, as we saw in Chapter 2, are abundant in interstellar dust clouds, are also abundant in planetary systems in the making. The variety and complexity of carbon-based molecules that overlap with biogenic organics can, and do, arise spontaneously and go about their own, private chemical business. They do this in the absence of water, which is essential for life as we know it. Meteorites containing water as well as a variety of organics (these are increasingly being identified) might be a different case, but only *might*, as, too, might the sediments on Mars.

Larger and smaller rocky and dusty bodies accreted to form rocky planets in the nascent solar system. These processes produced conditions of great temperature and pressure, creating a small amount of organic compounds similar or identical to those found in living organisms or produced by their geological decomposition: we are no longer surprised to find such abiogenic organics on Earth and meteorites that land on Earth. The conclusion of this part of the story of oil is that the geology on Earth makes of organic carbon in massive quantities what asteroid and planet formation in space made of inorganic carbon to a small extent. The parsimonious explanation (that requiring the least assumptions) for the majority of the enormous quantities of kerogen in the Earth's crust is that of geochemical decomposition of dead unicellular organisms. That process continues to this day, but it's millions of times too slow to replenish oil at the rate at which we're taking it out.

Is peak oil now really in sight?
There have been many false alarms so the public seems to have become desensitized. New oil keeps being found, but possibly not as frequently as a decade ago. In its latest Statistical Review of World Energy,[56] the oil conglomerate BP presents the global R/P ratio (identified oil reserves divided by current production per year) as 53.5 years. That is how many years' worth of oil they believe we currently have. Interestingly, 10 years earlier, in their 2011 report, they quoted an R/P of 46.2.[57] So despite clearly growing global consumption of oil per year, the rate of discovery of new "proven reserves" exceeded consumption over those 10 years. That appears to be changing now because, since 2014, the figure seems to be plateauing out to somewhere in the mid 50s: the rate of new oil discovery is starting to equal the rate of consumption, and, in coming years, we may well see consumption outstripping discovery. We will probably only be able to pin peak oil in retrospect, but it will surely happen in the foreseeable future.

The genesis of graphite: Carbonaceous sediments that got hotter and denser

It's hard to estimate the total mass of graphite on Earth: the tonnage of accessible deposits regularly undergoes large corrections. For example, estimates of China's deposits (probably the largest on Earth) stood at 537 million tonnes in 2018, having grown 135% since 2010, a result of improved prospecting methods.[58] The principal means of natural graphite formation is metamorphosis of carbonaceous deposits (mainly ancient plants) under increasing temperature and pressure as they subside into the Earth's upper crust. Some graphite results from hydrothermal activity on carbonaceous deposits. A further proportion is formed directly from coal seams as they descend and also from crystallization from carbon inclusions in magma (molten rock).

Graphite comes in various qualities, and only superior material is worth using for making good batteries, for example. Even the best must be artificially processed to greater than 99% purity. This has put anode-grade graphite into an important economic position: it's the single largest component by mass of Li-ion and other rechargeable batteries. Many countries rely on China as a source of this anode-grade graphite for electric vehicle batteries, and trade disputes have held up battery production in a number of facilities as a result: more local solutions are being investigated, both for the primary graphite and its processing to anode-grade. Mines are likely to be dug in countries that hitherto have no history of graphite mining. Despite graphite's abundance in Earth's crust, its mining and processing have the attendant environmental impacts associated with many other mineral industries. *Synthetic* anode-grade graphite is more expensive than natural graphite and environmentally more damaging, being made from petroleum coke. Graphite that escapes human industry may be lifted up by tectonics to be weathered away, or subducted into the magma, from where it may issue again in the form of CO_2 from a volcano, or, if it's lucky, become diamond.

Conclusion: Thermally catalytic CO_2 is overwhelming Earth's recycling and buffering capacity, but much more than climate is at stake

The mechanisms through which increases in atmospheric CO_2 and CH_4 cause increased radiative forcing leading to global warming are complicated and non-linear. It is important that we understand them fully in order to build accurate predictive models. Radiation, absorption, and re-radiation have been emphasized in scientific communications and media reporting as the mechanisms of CO_2's radiative forcing, but ultimately something importantly different is the cause of temperature increase: the conversion by CO_2 of radiation-derived bond-vibrational energy, in a way very reminiscent of catalysis, into thermal energy by collision with nearby molecules—energy that the "windows" in the H_2O spectrum would otherwise allow to escape. This system embodies great multiplicative power and endurance for a small mass of CO_2 (catalysts are required in tiny quantities, massively facilitate reactions, and are unchanged by them, surviving to catalyze billions more). The picture of CH_4's role continues to develop and frame it—albeit over shorter timescales—as an increasingly serious contributor to global warming, one that is ultimately converted to CO_2, thus adding to the 36 Gt released annually from fossil fuel burning.

Back-of-an-envelope calculations can estimate how much faster humans are taking accessible fossil fuels out of the ground than they are being replaced by natural processes: whether it's 30 million or 1 million times makes no practical difference, and the very assumption that oil is still being formed roughly as fast as in the geological past is almost certainly wrong. We've seen how coal, oil, and gas reserves were formed across hundreds of millions of years, time periods over which carbon fluxes were essentially at equilibrium. We're now very far from equilibrium. A minuscule proportion of the CO_2 that the Earth is currently absorbing is being turned into long-term geological deposits; a very large proportion is simply accumulating, mainly in the oceans, as "buffered" CO_2. However, that CO_2 is having detrimental effects of its own, exceeding, and reducing, natural ocean buffering capacities. Finally, what hits the headlines is what's left, increasing at 2.5 ppm per year. Less in the public eye, but just as urgent, is the need to reduce CH_4 emissions from agriculture.

Massive amounts of carbon go through biological parts of the carbon cycle in overall neutrality: it's not a question of quantity, but rather recycling. I believe that humans can mimic this and thereby also neutrally cycle carbon in large amounts. The role of living organisms in the carbon cycle is at once complicated and incompletely understood, but inherently sustainable. Ecosystems impacted by other stresses and direct human activities, besides climate change, are even less able to absorb CO_2. The overwhelming concentration (often fired by the media) on the climate crisis to the exclusion of the larger picture tends to focus our attention on physical phenomena such as temperature, drought, acidification: things on which we can put hard figures. The oceans are largely acting as a massive *physical* buffering system, but the biosphere as a whole is buffering via terrestrial ecosystems containing much more biological carbon: protecting the world's forests and wetlands is imperative. Reactions to the physical cause and consequences of global warming must not produce different environmental degradation of their own, otherwise we will also destroy Earth's *biological* buffering (which goes far beyond climate effects): ecosystems and whole environments are at stake, as we are increasingly seeing.

As a complex connected system, the physical Earth has already started a variety of reactions to our actions. Ecosystems debilitated by climate change are even less resistant to other environmental insults, so in these times we need massively enhanced environmental consciousness in all areas. Though it is clear that we have used naturally occurring carbon-based energy carriers and materials up to now in highly unsustainable ways, it is not clear that we are behaving more sustainably with alternative technologies. I explore this and ideas for major improvements while still taking advantage of carbon as an energy carrier, in the following chapters.

4
Fuels, efficiency, and emissions

Understanding carbon-based energy carriers in the larger picture of sustainability

Let's now look at fossil fuels and their alternatives more closely. Fuels (mainly gasoline, diesel, gas, kerosene, and a range of minor products) represent 85% of petroleum products according to latest figures from the US Energy Information Administration.[1] And the really long-chain residues flowing from the large stainless steel fractional distillation columns of petroleum refineries—a thick black goo—are mainly used for making plastics. To understand how crucial the energy choice is for a variety of applications, now and in future, I need to describe a few things in detail: the combustion characteristics of hydrocarbon fuels, the thermal efficiency of their applications, and the emissions. In this chapter, I summarize the developments and prospects of carbon-based fuels for various applications, focusing on certain examples across the bewildering and ever-widening spectrum of fuel use and research. They are symbolic of important principles and present interesting socioeconomic "stories."

We must fully understand energy density and its consequences

Crudely speaking, the final energy density of a fuel (usually given in megajoules [MJ] or kilowatt hours [kWh] per unit mass) is strongly related to the energy that goes into making one unit of that fuel energy (either via natural or synthetic processes). You can either have a low-density form of portable energy, of which you'll need a relatively large weight to produce a given quantity of useful work (e.g., powering a ship, car, or plane for a given distance), or you can have a highly energy-dense type of fuel, which will, for the same weight, move your ship, car, or plane further. Energy concentration (i.e., making a high-energy-density fuel) has high costs, both energetic and financial. Science and technology can sometimes make quantum leaps, but they are unpredictable. For the foreseeable future we should acknowledge a variety of disadvantages and advantages of current energy carriers, from batteries to liquid fuels: instead of placing all our bets on a specific technological development with merely the *hope* that it might solve all of our energy challenges, we should use the precautionary principle, be modest with our use, and hedge our bets between several technologies.

For the carbon dioxide (CO_2)-neutral fuels of the future—and energy sources in general—we must still be acutely aware of potential environmental impacts: it's not OK, for example, to create a technology that promises environmental benefits during its use if, during its manufacture, impacts on climate and environment are neglected. But environmentally sustainable production is only half of the equation: products must also be recycled in sustainable ways. Even if net CO_2 emissions are no longer a concern in future, other environmental impacts (land/water/air pollution, ecosystem destruction, etc.) definitely will be. So, too, will issues of how available energy is divided between different "needs." If increases in consumption can't be halted and instead continue, there will be inevitable impacts of even the "cleanest" technologies. If we continue to need hydrocarbon fuels, we will have to *make* them rather than *mine* them. Those that are made with CO_2, hydrogen, and electricity are termed *e-fuels* (electro-fuels).

High energy density is costly, but it is variously desirable or crucial for certain applications. Furthermore, additional factors are important in defining equivalence of e-fuels with their fossil counterparts. We can't lump all e-fuels together as "equivalents" to fossil fuels, even within one chemical category. At least four qualities must be considered:

- *Chemicophysical nature* (which has a bearing on how easy it is to make the fuel, transport it, and harness its energy in a work-producing machine)
- *Combustion characteristics* (which tell us for which applications we can use it)
- *Combustion products* (which equate to the immediately undesirable products such as soot and other products of incomplete combustion)
- *Energy density* (which is essentially proportional to the amount of useful work that we can get from 1 kg of the fuel)

All of these influence the fuel type and hence its energy balance of manufacture: the ratio of energy expended to energy present in the product per unit mass.

Portable fuels: Applications depend on energy density, energy balance, and politics

Methanol, the first alcohol, and a promising auto fuel

Let's start with an organic compound possessing only one carbon atom. Methanol deserves a special mention because it is currently the most energy-efficient carbon-based fuel to make: energy balances of between 50% and 55% can be achieved (i.e., 50–55% of the input energy is manifest in the product).[2,3] Production capacity is already large: globally, 148 million tonnes (187 billion L) are produced annually,[4] which could keep 163 million average cars driving, European-style, for a year—not that that would, per se, be a noble aim. One large production facility can make 2.3 billion L of methanol per year.

Methanol is very easy to substitute into existing applications, notably internal combustion engines (ICEs), where—to give mixtures of higher energy density— it is mixed with ethanol and/or gasoline/diesel. Here we see a disadvantage of methanol: its comparatively moderate energy density. Methanol has 21 MJ/kg, compared with 44 for diesel, 45 for gasoline, and 28 for pure ethanol (average values between the higher heating value [HHV] and lower heating value [LLV]; see footnote*).

However, methanol is relatively easily synthesized and doesn't compete with feed and fodder crops, as bioethanol does, because it's not typically made via carbohydrate fermentation. Methanol is toxic in much smaller quantities than ethanol, but exposure at toxic levels is extremely unlikely in everyday, sensible use. Fruits naturally contain small quantities of methanol, and methanol is a side product of fermentation by gut microbes, hence humans have evolved a small tolerance to methanol and enzymatic mechanisms to break it down.

Methanol has production and engineering advantages compared with fossil hydrocarbon fuels or their e-fuel equivalents that outweigh its moderate energy density.[7] Furthermore, it is still roughly 20 times more energy-dense than a good Li-ion battery. Notably, a methanol-burning ICE using spark-ignition (SI) produces less nitrogen oxides (NO_x) than a diesel engine. It also has a higher compression ratio than a petrol or even diesel engine, hence giving greater peak efficiency (in practice, up to 42%; *theoretically* nearly 50%).

A mere peculiarity of motor sport?

For the aforementioned reasons, one might expect methanol already to be used widely in cars. Wrong: it has only been used in significant quantities in racing engines, involving a relatively easy adjustment to increase compression ratio (off-setting methanol's moderate energy density) and modified fuel tank and fuel

---

* Higher heating and lower heating values differ typically by around 10% (see Figure 4.3, Appropedia[5], and US Energy Information Administration[6]), which corresponds to the energy of water evaporation during combustion. If this water can be recondensed, much of the vapor energy of the water can be recovered, hence leading to the higher heating value; if not, the lower heating value should be assumed as relevant. My justification for using average heating values for fuels is two-fold: (1) it is often unclear whether the vapor energy of the water is, or can be, recovered, and (2) many studies do not explicitly state which value is used; hence, for comparing my results with others, the potential for error is reduced to a maximum of 5% by using the average between HHV and LHV.

lines. Regular car engines need similar minor modifications to run on methanol or methanol mixtures. So, why such restricted use? The main explanations seem to be political and economic. The largest controlled trial of methanol (typically as an 85% methanol/15% gasoline mixture) in ordinary cars was done in California from 1980 to 1990. The motivation for the trial was initially not environmental but rather fuel security in the wake of the 1970s fuel crises. However, air quality factors soon became very evident: the methanol mixture burned much more cleanly than fossil gasoline. The results were a technical success, but—also as gasoline started to become cheap again—commercialization failed: not enough fuel stations offered methanol mixtures, and drivers fell out of love with the single-carbon alcohol.

At the peak of the "methanol rush," 15,000 cars ran on methanol mixtures in California—enough to get a statistical impression of its benefits/problems (toxicity was not a problem). Unfortunately, renewability was not among the benefits at the time because the methanol was produced mainly from natural gas.[8] However, as an automotive fuel, methanol had made an important point. Prospects for methanol in California deteriorated even further, however, as, in 2005, Governor Arnold Schwarzenegger stopped its use in preference to ethanol from corn fermentation. One might wonder whether lobbying from farmers was a major factor in that decision because the environmental justifications for preferring ethanol from crop plants over methanol are questionable.

The Open Fuel Standard Act and why it failed to stimulate technology diversity

Flexibility to do experiments with fuel types is important at a time when unpredictability can lead us into technological and economic cul-de-sacs. Congressman Eliot Engel saw that when trying to get the Open Fuel Standard Act passed in 2008. This would have required car makers to make 80% (95% in version 1 of the Act) of their cars manufactured or sold in the United States capable of burning mixtures containing 85% methanol, ethanol, or biodiesel. In fact, the diesel was not restricted to "bio"(agricultural): synthetic (Fischer-Tropsch) e-diesel would also have qualified because the Act permitted any other *sustainable* technologies. Admittedly, the major intention of the act was to reduce the strategic importance of oil to the United States, but it arguably had the potential for environmental benefit. What happened to the Act? Unfortunately it suffered a major setback immediately. Car manufacturers raised their hackles at the percentage targets for mixed-fuel vehicle production: far too high.

The Act has since been modified many times, introducing phased percentage increases. It now also permits hydrogen, flex-fuel (cars that can run on a variety of fuels), natural gas, and even hybrid and plug-in hybrid technology under "new technologies." As environmentalists have noted, the Act still does not explicitly stipulate reductions in CO_2 emissions, rather considering them almost as incidental outcomes. That is, indeed, its major weakness: it's all very well to experiment with alternative energy carriers, but the aim of the experiment must be clear, justifiable, measurable, and environmentally sound. It is extraordinarily hard to pass bills based on principle alone; further, the Open Fuel Standard Act is a riot of interests. Among the top three justifications mentioned in the latest draft are money-saving at the gas pump and economic benefits in the form of job creation. True, health benefits (less harmful emissions) are in second place, but only two of the 10 reasons are related to environmental impact. 2013 seems to be the last mention of the Act.[9] There are no further developments to report at the time of writing. Legislation that would have promoted environmentally less-damaging fuels in everyday cars was stymied by financial considerations: the car industry won. As we will see in Chapter 6, the European Union tackled the goal of reduced emissions differently, legislatively leaving the car industry little choice than to canalize its production into electric vehicles. This is no better because it excludes technological diversity. Both the US and European cases show that scientific input and analysis can be overshadowed by legislative procedures that reach very extreme outcomes (all or nothing).

China's hesitancy to switch from diesel to methanol trucks: A matter of cost?

Looking further afield, public health was very much in the minds of Chinese policymakers who considered methanol as a replacement fuel for diesel trucks. Though trucks make up only 10% of China's 300 million vehicles, they produce almost all of the vehicle-related particulate emissions.[10] As with cars running on methanol mixtures, the adaptations to a truck engine and fuel system are modest and would cost around 3,500 USD per vehicle on average—costs that are recoverable in under a year. Methanol is relatively cheap, and China already has a large production capacity—not least as a feedstock for organic synthesis of plastics and other organic compounds. However, legislators have refused to change vehicle laws, hence making it impossible to register a vehicle that has been modified.[11]

A geological reason, compounded with economic competition in global markets, probably plays a large role in the politics of methanol: China's enormous coal reserves are the primary feedstock for methanol production (via synthesis gas and the Fischer-Tropsch process), and it is said that China's environmental concerns about using those reserves are behind the resistance to methanol as an alternative fuel. However, China's emissions from coal burning are currently growing by at least 1% per year: production of electric vehicle (EV) traction batteries and whole electric cars for export and internal sale account for much of that 1% increase. Until recently, it was cheaper for China to make methanol from imported gas or import methanol *produced* from gas. However, as the

Russian war against the Ukraine continues to push natural gas prices through the ceiling, China may start to rethink this strategy. Organic waste as a source of synthesis gas ($CO + H_2$: syngas) is also becoming increasingly economically competitive.[12] Might it catch up with fossil gas? Could it constitute a significant percentage of demand? Unlikely—so back to coal.

Meanwhile, media reporting of the Chinese car industry, particularly in Europe, focuses almost exclusively on EVs—a topic regularly trotted out by some journalists who sensationally disgrace Europe for being so backward in the mobility revolution. These reports rarely note the environmental impact of the EV economy. And they never mention that the world's largest controlled study of regenerative methanol in ICE vehicles (ICEVs) was done by a Chinese company (Geely) in Iceland (where geothermal energy with low environmental impact is plentiful): highly promising results were released in 2017. In 2022, Beijing reported accelerated deployment of methanol-powered ICEVs as part of its carbon-neutral drive.[13]

Global demand for methanol as an energy carrier is growing fastest, but renewable methanol is still scant

The 2021 report from the Methanol Institute notes "energy-related" uses for methanol as the fastest growing sector, constituting 40%.[14] But where does this leave methanol as a potential automotive fuel? The answer seems to be in a tangle of undecided legislation, vehicle manufacturer insecurity, competing economic interests, and global industrial realities. It is clearly possible to make methanol from non-fossil feedstocks such as biomass, municipal waste, biogas, and waste CO_2 (plus hydrogen). If we can make it from atmospheric or waste CO_2, its downstream applications would be CO_2-neutral or even CO_2-consuming. However, governments would need to intervene to promote methanol production in significant quantities from truly regenerative sources: currently less than 1% of global methanol is from renewables.[15] So there is a final, rather bizarre, twist in this tale: for the foreseeable future at least, methanol—as an ICE fuel—will remain the preserve of racing cars at one end of the spectrum, and massive freight ships at the other. Yes, that's correct: in 2021, Maersk ordered eight enormous freighters with ICEs running on methanol synthesized with renewable electricity.[16]

The story of methanol highlights a number of important considerations for any portable fuel. Next, I cover some other potentially important portable fuels, largely from the perspective of energy density, the energy per kg needed to make them, and their application characteristics. In developing environmentally conscious portable energy sources, we must be as sparing as possible with energy but also conscious of other side effects and the suitability of a fuel for its intended purpose.

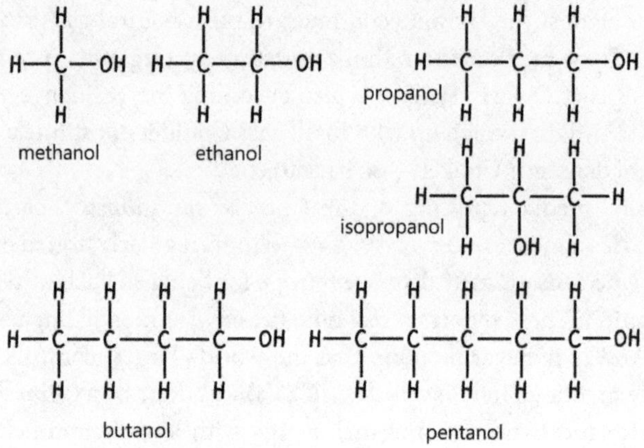

Figure 4.1 Alcohols, from C_1 to C_5.

More alcohols to consider: A large assortment, with large potential pitfalls

Artificially making a C-C bond (essentially in a large catalytic pressure cooker) to create ethanol from single-carbon substrates (e.g., methane) is energetically costly. Hence researchers have always looked to biology—specifically microorganisms—for enzymatic solutions to the challenge of making ethanol and longer-chain alcohols (e.g., propanol, isopropanol, butanol, pentanol). Higher-chain-number alcohols (Figure 4.1) have desirable properties for applications as fuel, such as lower evaporation, higher mass density, higher flash point (temperature at which spontaneous combustion occurs), and lower chemical reactivity: the last feature is important for use in existing technology because the corrosive aggressiveness of high-percentage methanol—and to a lesser extent ethanol—makes modifications to existing technology necessary.

Making longer-chain alcohols economically is quite a challenge, and so ethanol (two-carbon) from fermented sugars has, for decades, been the only widespread component of gasoline replacement, particularly in South America. In Brazil, for example, a minimum of 27% ethanol by volume is currently mandated,[17] and the average blend at gas pumps is 48%. Many European countries offer a choice of low-percentage blends (typically E10 and E5: respectively, 10% and 5% ethanol), but some have started offering E85 (85% ethanol/15% gasoline).

Replacing gasoline with bioethanol: How much could we achieve?

A calculation based on published data (see Chapter 4, Calculation 1, online at https://doi.org/10.1093/oso/9780197664834.001.0001) suggests 25% replacement, if we were to use a maximum advisable percentage of 25% of all waste crop material from the entire Earth's crop-harvested land (leaving most of the rest on the land, for good reason). Clearly, primary crops (seeds, grains, tubers/roots) cannot be considered as the main feedstock because the areas of land required are too large, and we would basically destroy all food/feed agriculture. This leads to two inevitable assumptions.

1. Most of the bioethanol for replacing fossil fuels must come from crop waste (1a) or biomass from non-crop plants, purpose-grown on land that is unsuitable for agriculture (1b).
2. However, the extent of 1b would be relatively small because (1) cultivation involves fertilizer provision, (2) monocultures of "energy plants" jeopardize biodiversity and (3) land use change can have unpredictable effects, releasing large amounts of greenhouse gases (GHGs) (e.g., draining wetlands for new cultivation areas).

Hence, it is safe to say that the mainstay of bioethanol production will be feedstocks from plant waste (cellulosic, hemicellulosic, and lignocellulosic). Europe and the United States have a long way to go, for example, because they produce, respectively, 90% and 94% of the bioethanol mixed with gasoline from food/feed crops, mainly sugar beet, corn, triticale, wheat, rye, and barley,[18,19] and largely corn (maize) in the United States.[20] The rest is from non-agricultural waste. Cellulosic biomass (e.g., crop harvest waste) has never crept above 1.1% of the total in Europe[19] and 0.2% in the United States.[20]

But (to avoid soil erosion and nutritional degradation via depletion of nitrogen, phosphate, and minerals), the realistic percentage of post-harvest biomass that can be removed is likely only 20–25%[21,22]. Around two-thirds should remain on the land, and 10% is needed for animal bedding and horticulture. This is not material that would just rot and be wasted.

And what does the standard (crop material fermentation) energy balance for bioethanol look like?

Enzymes achieve feats of kinetics (overcoming energetic barriers to reaction) that elude most industrial chemists. But microorganisms that contain the enzymes don't really like the ethanol that they produce: true, they are thousands of times more tolerant to it than humans, but, at around 13% ethanol concentration, most fermentation yeast varieties go dormant. If forced, they can produce more, and special strains can produce up to 19%. However, at that point, they're very sick, and, long before that, they become very slow. That's no good for a continuous-flow bioreactor, where one wants the microbes to keep producing alcohol at a fast rate. Hence, the typical concentrations of alcohol produced in such yeast bioreactors are only 2–3% by mass; selected or engineered microbes (among them, particular bacteria) may produce as much as 5% by mass of ethanol.

Ethanol for combustion in ICEs is either *hydrous ethanol* (94% ethanol and 6% water) or *anhydrous ethanol* (as close to 100% ethanol as one can get). Distillation technology can only reach 94% (hydrous) ethanol; to get anhydrous ethanol, dehydrating substances are used. The quantity of energy needed to evaporate water from an alcoholic fermentation mixture to produce 94% alcohol is quite large: reported values of around 30 MJ/kg are common with simple distillation technology that essentially boils the broth, very similarly to distillation of vodka and other spirits.[23,24] The energy density of 94% ethanol is around 26 MJ/kg, hence 15% more energy is required to produce the fuel than is contained within it. The Sun's energy that we get for free in the crop is outweighed by the energy that we need to extract useful fuel.

The main method of ethanol production for gasoline mixtures (termed "Ex") firmly remains microbial fermentation in aqueous solution, followed by distillation. But, depending on the exact methodology, degree of heat recuperation, and the means of calculating net energy expenditure, a vast range of energy balances can be reported. Some claim more net energy output than input; others claim the reverse. Are the fermentation residues used as animal fodder, and counted in kilocalories (kcal) of nutrition? Is the heat from these residues used for cogeneration of electricity that is calculated as an energetic positive in the larger calculation? Is the energy requirement for production of the crop included? If the fermentate is using waste biomass, how is the energetic cost of that material calculated? Are compensations made for additional fertilizer needed in the wake of biomass removal from a field? The list goes on and on.

Membrane separation (pervaporation) techniques that force the fermentation mixture through a membrane that only lets ethanol molecules through are less energetically costly. This is claimed to be the solution to the high energy demands of distilling ethanol. Improved energy recovery in such systems is sure to materialize (e.g., by using counter-current flow systems to transfer heat of products to incoming fermentate).[25,26] Feasible reduction to 23–24 MJ/kg final ethanol is claimed.[27] That is around 90% of the final chemical energy present in the ethanol. But if we now add the 4.5 MJ/kg for growing the crop (see Chapter 4, Calculation 3, online at https://doi.org/10.1093/oso/9780197664834.001.0001), we have 108% energy input for the energy in the ethanol output, thus giving a net energy ratio of less than 1 (more energy input than energetic value of the fuel).

An alcoholic muddle: If you you're not confused, you're underinformed!
There is an enormous number of facilities producing ethanol today via an enormous number of feedstocks and methods[28]: anything from crop plants, through bakery waste and sawdust (one study is entitled "Ethanol production from waste

FUELS, EFFICIENCY, AND EMISSIONS   137

pizza"). One of the largest problems of environmentally conscious liquid fuel developments is that efforts and money are split between too many initiatives and lack common benchmarks for comparison in terms of energy balance and GHG emission balance. A global commission to develop standardization and minimize waste of research money is needed.

Researchers have also been working on mixed fermentates, for example producing isopropanol/butanol/ethanol or acetone/butanol/ethanol (ABE). Trials of fuel mixtures containing various percentages of ABE (up to 100%) are ongoing.[29] ABE fermentates can, it is claimed, achieve distillation energies as low as 15% (4–5 MJ/kg) of final product.[30] The feedstocks for these fermentates are sugars and starches, which compete with food/feed agriculture; but also lignocellulose from woody plants (after enzymatic/chemical treatment), cellulose from crop waste, and even algal biomass. If these are grown in ecologically sustainable ways (e.g., algae can be grown in large commercial facilities) or if taking "waste" products away from farming soil is not problematic, then so much the better. Ethanol produced by fermenting algal biomass produced in photosynthetic bioreactors (and distilled via optimized conventional methods) is claimed to be able to reach net energy ratios of around 0.45 (energy in divided by energy out): the energy input represents 45% of the energy of the final product, so we get a net gain of 120% (100/45 = 2.2 = 220%, minus the 100% input energy = 120% gain).[31] However, as with all bioethanol concepts, it is highly questionable whether we could produce the volumes required to substitute large fractions of current gasoline use. It's also unlikely to be cheap if done with due respect for the environment. Moreover, the newer technologies are still some way from scale-up, and a variety of negative (as well as positive) things can happen during scale-up.

So, what is the biological feedstock for bioethanol in gasoline/ethanol mixtures at the pumps (e.g., E10), and how is the high-percentage ethanol made? You will struggle to find this out, and you might stumble across many policy documents where you would expect to find it. Even the European Renewable Ethanol Report 2020[32] mentions only "crop-based" biofuels and no concentration methodology: more information seems to be in annexes that are not available with the main report. You would be vaguely informed if you visited the ePURE website of the European Renewable Ethanol Initiative, where it states "Renewable ethanol is manufactured in a biorefinery by fermenting sugars into alcohol. In the EU, these sugars typically come from a variety of agricultural sources such as wheat, corn, barley, rye, triticale, and sugar beet. While the feedstock used typically varies depending on market conditions, the majority of renewable ethanol biorefineries are built to specifically process either grains or sugar beets."[33] So, primary crops. But do we at least have good energy balances? To find out more on the energetics, I contacted two German companies that make ethanol as a fuel additive. One of the companies replied that it was not prepared to send me any information on the topic; the other company did not even reply. Bioethanol production from primary crop, distilled in old-fashioned ways, cannot be considered a sustainable component of our vehicle fuel economy. "Biodiesel," by the way, is no better, originating, as it does in Europe, mainly from oilseed rape.

A struggle to compare: Policymakers can't judge different technologies properly, and industry can take advantage

Few policymakers (are able to) inspect the agricultural and environmental intricacies of distilled bioethanol for gasoline. The agricultural argument is easy (so no forgiveness there), but learning about energy balances of fuels is not; neither is finding out which energies have been offset against the production energy (as mentioned earlier) and whether they are justifiable. Some studies quote values per kilogram of product; others values per kilogram of substrate (e.g., straw, biowaste, etc. etc.).[34] Without reliable tools and great experience in the workings of scientific research, it is extremely hard to assess the relative merits of research projects competing for funding. In the call to develop more environmentally conscious fuel solutions, all sorts of research are being funded. I see little in the way of standard measures for designing projects, conducting the research, collecting results, and interpreting them in the context of potentially feasible and truly beneficial applications. Research scientists also have biases and hopes and need to make enough money to live. In this context, it is no less than astounding that the European Commission (a large funder of fuel technology development) only published its Energy Balance Guide for standardizing research in early 2019.[35]

To what do energy balances reported in studies actually refer? Do they always take account of the whole process from plant growth through to distilling 94% or 99.9% ethanol? First, we can easily calculate the theoretical energy input in the process of getting 94% ethanol from an imaginary water/ethanol solution (see Chapter 4, Calculation 2, online at https://doi.org/10.1093/oso/9780197664834.001.0001). My result emphasizes the challenge: with no energy recuperation, the requirement is 190% (i.e., 90% more energy input than present in the final distillate). Recent literature suggests achievable reductions to 92% of product energy (23.9 compared with 26 MJ/kg for the product),[27] a gain of energy on balance. However, adding the energy costs of the crop production (an extra 4.5 MJ/kg cost; see Chapter 4, Calculation 3, online at https://doi.org/10.1093/oso/9780197664834.001.0001) would increase this to 109% (28.4 MJ/kg, i.e., a slightly negative energetic balance). Recovering 70% of the distillation heat energy (a big challenge, but perhaps possible) would produce a net energy ratio of 75%: that is, an energy balance of 133% (100/75), that's 33% net energy gain. But what is model simulation, and what is reality? Even review articles rarely tackle the fiddly business of comparing full production chains and distinguishing between idealistic models and real facilities. One review concludes that basic energy indicators may vary up to 400% depending on considerations.[36]

Currently, fossil energy use in the production chain is a major concern; for the future it is land-use: in the United States, where 40% of field corn is used for bioethanol production,[37] a 2022 study concluded that profound advances in technology and policy are required to make biofuels environmentally acceptable and that "the carbon intensity of corn ethanol produced under the Renewable Fuel Standard is no less than gasoline and likely at least 24% higher."[38] Often, environmental goals mutate into plain financial incentives, even preserving outdated first-generation biofuels (e.g., see Cadillo-Benalcazar et al.[39]). This can easily happen because understanding the true, full-chain energy balance analysis of alternative fuels is hard. A few examples of the factors were given earlier. Alcohol synthesis from bio-syngas—for example, made from high-temperature treatment of plant material (cellulosic lignocellulosic by-products, e.g., crop plant and wood waste) with steam—is yet another possibility. These chemical syntheses are energetically more costly than fermentation, but they avoid much of the energy of distillation. Even hybrid routes with gasification of biomass followed by microbial fermentation of syngas are being investigated—bacteria of the "acetogen" type can use carbon monoxide (CO) and hydrogen (H_2) as "food"! Designated plantations of, for example, the fast-growing switchgrass have long been considered potential sources of biomass; however, biomass removal necessitates fertilizer addition, which also has energetic and environmental costs. It is unclear whether the sources of any of these materials is long-term sustainable.

In summary, primary crop produce can only responsibly be considered for minuscule proportions of bioethanol, if at all; most bioethanol should be produced from crop waste; and the most constraining factor is the area of land needed to produce it, hence setting an upper limit of 25% replacement of current global gasoline consumption. Bioethanol from crop waste and other waste should be part of our portable fuel solutions in my opinion.

Dramatically accelerating (photo)synthesis of carbon-based energy carriers: E-fuels

E-methanol from photovoltaic energy, H_2 generation and CO_2 capture

You read correctly "methanol." For this thought experiment, we need to switch to methanol because (1) methanol is energetically more efficient to produce via chemical synthesis (it's the first alcohol produced via Fisher-Tropsch synthesis), and (2) it can be integrated into an energy production and chemical synthesis facility that minimizes energetic and material losses (as described in more detail in Chapter 7).

First, a boundary-setting thought experiment: If we wanted to replace all currently used gasoline with synthetic methanol (e-methanol) instead of bio*ethanol*, how much land would we need to generate the necessary energy via photovoltaic (PV) panels? (i.e., artificial, silicon-based, photosynthesis). The answer: around 0.19% of the area currently used for crop-harvested land on Earth, or 3,230,000 ha: somewhere between the sizes of Moldova and Belgium (see Chapter 4, Calculation 4, online at https://doi.org/10.1093/oso/9780197664834.001.0001). This gives us some idea of the efficiency of photosynthesis compared with PV (a concept that, at a lower level of analysis, amounts to comparing apples with pears): efficiency is somewhere in the region of 260 times greater per hectare for PV than for photosynthesis (Calculation 4). Clearly, we would not place these installations on agricultural land. Furthermore, if we were to concentrate solar energy collection in arid deserts where sunshine hours per year are even higher and no crops can be grown, we would only need a tiny percentage of their total area to equal total current energy demands: around 2% of the Sahara desert would do it if we have very efficient PV panels that track the sun and adjust angle accurately (see Chapter 4, Calculation 5, online at https://doi.org/10.1093/oso/9780197664834.001.0001). If we then add a range of synthetic fuels with an average energy balance of 50% to replace the ones that we are currently getting almost for free, we'd probably need 3–4% of the Sahara (Calculation 5). If economics dictate less high-tech PV collector installations, we might need to double that. Still, divided between all Earth's arid deserts, this seems possible.

However, manufacturing all those silicon units with the present energy mix (10% renewable, worldwide) would release a sizable additional CO_2 bubble. This is a catch-22 situation, but we could envisage a solution based on quickly reducing car-driving by 50%. We could then use just half of that 50% CO_2 saving for fossil-energy–based solar cell production (i.e., "splitting the difference" between ourselves and the climate). How much PV collection area could we manufacture? And how much e-methanol could its resultant electricity produce? According to my calculation (see Chapter 4, Calculation 5b, online at https://doi.org/10.1093/oso/9780197664834.001.0001), in one year (using one year's worth of the 25% of current gasoline consumption in the form of industrial energy, assuming we could scale up production that fast), we could produce 6.6% of the necessary PV and chemical synthesis capacity for replacing gasoline with e-methanol (Calculation 5b): we would, after all, only need to replace half the existing consumed quantity of fuel. If we continued construction at this rate and kept our gasoline saving steady, in the next year, we'd be producing 13.2% of the necessary gasoline energy as e-methanol. In 15 years, we'd reach 100%. A great advantage, and incentive, for doing this is that we would be saving CO_2 emissions all the time, from the very beginning, and those savings would be growing further throughout the project.

You may ask "But wouldn't it be better to use PV electricity to charge electric cars?" Energetically speaking, maybe, though only moderately on the background of current gasoline consumption; however, if we cut gasoline consumption in half (which would, in any case, be good for the environment) EVs driven the same distance would on average significantly underperform ICEVs on total energy balance of manufacturing and driving (as explained in Chapter 6). Sustainable electrification of road transport involves many large challenges that have yet to be solved. In the context of a whole economy of manufacture and driving, considering all environmental impacts, it would not be better than e-methanol: the reasons are, briefly, that (1) for the foreseeable future, more than two-thirds of the energy used in car manufacture is fossil energy, and (2) the current EV battery economy produces very large environmental and human health impacts. I explain these matters in more detail in Chapter 6.

In the e-methanol scenario, at a stroke, we would be contributing to climate improvement and *building* longer-term improvement, whereas, when we buy an EV, the most immediate thing that has happened is that we have contributed to greater CO_2 emissions and environmental degradation. As we'll see in Chapter 6, the overwhelming majority of EVs will not break even with ICEVs in environmental/human health impact over their expected lifetimes. Establishing the PV/e-methanol solution (the renewable energy would also come from wind turbines and hydroelectric) would definitely take many years. However, the replacement of all cars on the road with EVs is also going to take a long time. And far from representing a genuine hope for environmental improvement, EVs will likely transpire to have caused more environmental degradation. An energy carrier such as methanol, which can be stored for long periods in easily and cheaply scalable holding facilities, is also very attractive for helping solve the problem of storing the wasted peak generation capacity that we currently have: make e-methanol while the sun shines and use it for energy generation in the winter.

Power-to-X fuels in general and a much cleaner diesel

Electricity can be used for making any type of e-fuel, and, from here on, I consider mainly non-alcoholic fuels, the group that contains most of the e-fuels (also referred to as power-to-X, and including Fischer-Tropsch fuels [Box 4.1] and power-to-liquid [PtL]). These are synthetic substances with higher energy densities than alcohols. As with e-methanol, and in contrast to *bio*fuels, the full-chain energy and material usage of completely synthetic fuels is easier to estimate: we have closed, highly controlled systems and can do small- to large-scale pilots with relatively low errors. The sustainable routes to e-fuel production largely involve extracting CO_2 from air or flue gases, generating hydrogen via electrolysis of water, and using these as the feedstocks for chemical synthesis.

Let's start with dimethylether (DME), a heavily researched potential replacement for diesel. Though DME has very similar combustion properties to fossil diesel, it is chemically rather different (Figure 4.2) and has around 30 MJ/kg

Box 4.1 The Fischer-Tropsch process

This process is named after the two German chemists who developed it, Franz Fischer (1877–1947) and Hans Tropsch (1889–1935). It combines hydrogen and carbon monoxide (often with small amounts of carbon dioxide), so-called *synthesis gas*, catalytically to form a variety of organic products from alcohols, short-chain and medium-chain alkanes (hydrocarbons) to long-chain alkanes: from gasoline-like compounds, through kerosene and diesel oils, to greases and waxes. Side reactions produce an even greater variety of organics besides alcohols, including acetone (H_3CCOCH_3) and olefins (cyclic and straight-chain molecules containing a C=C double bond). The F-T reaction typically takes place at temperatures between 150°C and 300°C, pressures of up to 25 bar (25 atmospheres), and in the presence of cobalt and/or iron as catalyst. Longer-chain and more diverse components are produced by longer reaction times, and the balance can be changed by tuning temperature and pressure. Individual components are separated by fractional distillation in columns. Patented by Fischer and Tropsch in 1925, the process became indispensable for Germany during World War II because the country could make petrochemicals and other organics from its own coal reserves and circumvent oil embargoes: roasting coal with too little oxygen for complete combustion produces CO; the necessary hydrogen was made by electrolysis of water. A few other countries have since built F-T plants to make their own fuels and lubricating oils from native coal and/or gas in the absence of oil reserves. Many other, organic, sources of synthesis gas have been identified to date, including the pyrolysis (high-temperature/high-pressure steam treatment) of waste, including wood and plastics, for example. F-T continues to be an invaluable process for the chemical industry, producing the ingredients of a massive array of consumables from lubricating oils through to waxes and oils for cosmetics, and components of adhesives. Most recently it has become the focus of attention for the production of CO_2-neutral fuels by taking its CO from CO_2 extracted from ambient air or flue gases.

energy density (fossil diesel is around 45). DME is the simplest ether ((CH_2)$_n$ groups joined via a C-O-C bond) and under normal conditions it is a gas; however, it can be compressed to a liquid under relatively low pressures (5 bar = 5 atmospheres, or twice the pressure in an average car tire), similarly to propane.

DME (Figure 4.2) is one of the most versatile synthetic organics. Currently, according to the DME Association, global production is a mere 9 million tonnes per year[40] (compared with 157 Mt for methanol). The fastest growing and largest

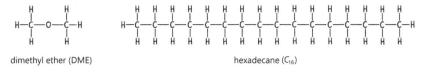

Figure 4.2 Dimethylether (DME) compared with hexadecane to emphasize the different chemistry, despite similar combustion properties. Hexadecane has a carbon chain length in the mid-range of diesel fuel alkanes (between 12 and 20).

market for DME is as a cleaner domestic fuel; it has also long been used in numerous chemical syntheses and as a refrigerant gas and propellant. Its production method—though well-established—is still improving in efficiency.[41]

DME can be synthesized in two practical and efficient ways, both of which work at 260–300°C, maximum 10 bar pressure (four times a typical car tire pressure)—relatively moderate conditions.

Method 1: First, CO is synthesized by reduction of CO_2, traditionally by reacting it with carbon (from coal) at high temperature and pressure; other methods use hydrogen as the reductant. More recently, nano-particulate aluminum surfaces and ultraviolet (UV) light-excited plasmon resonance at room temperature are being developed,[42] which, if scaled up, could cut the energetic and financial costs of making CO by an enormous margin. CO is then mixed with an appropriate quantity of H_2 to give syngas, which is then catalytically reacted (first step of Fischer-Tropsch) to make methanol (highly simplified in Formula 1).

$$\text{Formula 1: } CO + 2H_2 \rightarrow CH_3OH$$

Finally, a molecule of water is extracted from the methanol to give DME (Formula 2).

$$\text{Formula 2: } 2CH_3OH \rightarrow CH_3OCH_3 + H_2O$$

Method 2: The aforementioned steps of methanol synthesis and methanol dehydration are combined: "direct DME synthesis." This involves a single redox reactor and a different catalyst (usually containing oxides of copper, zinc, and aluminum, but improved catalysts include zirconium oxide). All catalysts are used in small quantities, are relatively easy to recover and recycle, and are relatively abundant on Earth.

DME has almost ideal combustion characteristics, burning evenly with extremely little incomplete combustion (notably particulates, NO_x, and CO). For automotive technology, DME is primarily a diesel engine fuel (spontaneous ignition); mixed with gasoline it can be burned in a gasoline ICE. Pilot plant models suggest full-process energy balances of 44–62% (energy out/energy in), depending on methodology and feedstock.[43–45] These figures seem reliable according to my calculations, and, if maximum energy capture (minimizing heat waste) is factored in, values of at least 65% may well be achievable (see Chapter 4, Calculation 6, online at https://doi.org/10.1093/oso/9780197664834.001.0001 and references therein). That doesn't beat bioethanol, but it avoids competition with food/feed agriculture and also CO_{2eq} emissions from land-use changes caused by expanding agriculture to meet demand.

Portable energy versus stationary energy: No universal solution

The jet airplane is an extreme example. Commercial airliners are designed and built close to the limits of their propulsion technology to get airborne and carry a full load to their destination. They therefore need a fuel of very high energy density—much energy for little weight. Jet aviation fuels (grades of kerosene) range between 43 and 45 MJ/kg.[46,47] Biofuels containing distilled oils from Jatropha and oil palm kernels mixed with fossil aviation fuel can produce energy densities of up to 44.3 MJ/kg.[48] But the idea of powering aviation with plant fruits is fraught with serious problems of land competition with forest, food, and fodder. Fischer-Tropsch synthesis can produce synthetic aviation-grade fuel of around 43–44 MJ/kg energy density from CO_2 and H_2.[49,50] True, the process is very energy-consuming per MJ of energy in the product, but do we have a choice?

Weight, transient power, and heat dissipation: Why hydrocarbons are the aviation fuel of choice

Movement requiring high power involves great heat production, and the higher the power, the greater the heat generated (and, currently, mostly wasted). There is no example more striking than the modern jet engine: the hot exhaust, rapidly expanding combustion gas *is* the propulsion mechanism (simplistically put). Many of us would rather see airliners taking off with the whirring of motors, but the peak energy needed to get 150 tonnes of machine and contents airborne is the largest challenge yet for electric aviation. Even at cruising speeds, a Boeing 737-300 requires 7.2×10^6 watts (7.2 megawatts) of power[51] (which is less than the nominal energy in the fuel because its engines are only around 35% efficient at converting fuel into thrust). In Chapter 5, I'll discuss battery power in this context, but first to battery *heat*.

At takeoff, the engines must produce 90% of maximum thrust, typically around four times the cruising power. Hence, at takeoff, an average 737's engines (together) will be producing around 30×10^6 watts of useful power—30 megawatts, or almost 500 acres-worth of silicon solar panels on a sunny day. With the thrust, a lot of heat is being given off (roughly 60% of the fuel's energy), which is conveniently "left behind" in a stream of gas that cools, producing con-trails. If the exhaust heat did not leave the engine, it would rapidly fail. It's an enormous challenge to recover any of this heat. It's not impossible, but far more difficult than thermal recovery technology for ICEs. Dissipating heat, though not wonderful, is sometimes necessary, and a battery-powered plane would have to do the same. How much, at 90% maximum thrust? A fairly easy calculation (see Chapter 4, Calculation 7, online at https://doi.org/10.1093/oso/9780197664834.001.0001) gives 6.75 megawatts, the equivalent of more than 5,600 typical kitchen kettles boiling within the fuselage (i.e., excluding heat from the motors), one-third of which would come directly from the battery and two-thirds from the power electronics. Hence, for the battery alone, 1,875 kettles-worth during takeoff and 470 at cruising speed (plus double that from the power electronics). For a constant 3D form (constant ratio of lengths of sides), the surface area of an object decreases in proportion to its volume as it increases in volume. Hence, in large batteries, with proportionally less surface area than similarly shaped small ones, rather special battery architecture and cooling would be needed, likely active (i.e., consuming further energy).

At a maximum usable energy density of 1.2 MJ/kg (current production-scale technology) for Li-ion batteries, we're looking at a plane that would, in fact, be too heavy to take off if it were intended to fly any appreciable distance (more details in Chapter 5). For intercontinental travel and trade aviation, energy density is a very strong potential deal-breaker: only fuels of 40 MJ/kg or higher are currently practicable. The degree to which synthetic fuels of this sort can be produced without major environmental impact will very likely rest on the extent to which we continue to fly in future: though we will definitely pay more for those fuels, simply paying more does not guarantee long-term sustainability. If we regard air transport as crucial for particular applications, we must assess energy consumption as a whole, within which flexibility exists, and consider tradeoffs with other sectors.

It's important to continue research on a range of aviation energy solutions, including electricity, and apply them in situations where they truly have full life cycle benefits in sustainability compared with existing technologies. However, we are in another catch-22 situation because the useful life-time of airplanes is relatively long (around 20 years)—and so it should be because of the large material and energetic costs in making them. Chopping and changing every few years on a whim of novel technology—that can't even be long-term proven to be sustainable—is likely to be very destructive for the environment. A massive techno-economic tension exists here because technology-based industries might well profit greatly from such strategies. It would arguably be better to make existing planes last longer, with less environmentally damaging fuels, and use them less, hence reducing total energy consumption (manufacture and fuel).

Why methanol or ethanol could also not replace aviation fuel

Could airliners use a fuel that is energetically less costly to make than synthetic kerosene? Methanol again? With an energy density of less than half that of aviation fuel (21 compared with 45 MJ/kg), we're already off to a bad start with methanol: in fact, few analyses seriously consider it as a potential aviation fuel. Let's go up one carbon atom to ethanol, at 28 MJ/kg (anhydrous). Here, at least one detailed study shows the consequences for aviation technology.[52] Planes are very sensitive to changes in shape, weight, and surface area—particularly in terms of drag. Hence, the 2006 paper concludes that not only would an ethanol-powered Boeing-737-equivalent need 50% larger engines and 25% larger wings, but that, on a 3,000 nautical miles journey (5,400 km), it would consume 26% more energy. The take-off weight and in-flight weight would always be greater than the conventional plane's, and the larger engines and larger wings would create more drag. Here is a thought-provoking tradeoff because the implications of using a lower-grade fuel are greater environmental impact in plane manufacture. We also have greater fuel use, the environmental consequences of which are likely bad: much depends upon the source of the ethanol, and we know that ethanol from fermentation would never be plentiful enough.

What are the high-energy-density options for aviation fuel?

Many alternatives, but which are environmentally sustainable?
Similarly to methanol and ethanol, DME is not energy-dense enough for widespread aviation use, and it has other characteristics that make it unsuitable; instead, we need something in the region of 42–45 MJ/kg that is liquid at operational temperatures of aircraft—something that is much more like the straight-chain hydrocarbons in crude oil. Aviation kerosene is between diesel and gasoline in average carbon chain length (10–16 carbons), and chemically similar fuels can be made in numerous ways (Box 4.2).

As with all fossil-fuel–equivalent replacements, the environmental impact of non-fossil aviation fuel depends strongly on both the ultimate source of the chemically reduced carbon, the energy balance of making the final fuel,

Box 4.2 Non-alcoholic fuels from a variety of sources

Biofuel can be made from lipids (oils and fats) contained in Jatropha seeds (a type of spurge), algal bioreactor cultures, animal fats (as by-products of the meat industry), waste oils (e.g., cooking oil or certain lubricants), palm oil (similar to that used in some cosmetics and food products), Babassu (a South American palm species), rape seeds, and Camelina seeds (a distant relative of *Brassica olracea*, or Brussel sprouts). Waste biomass (either solid or liquid) can also be used, via steam reforming (pyrolysis), to produce syngas (mainly CO and H_2), followed by Fischer-Tropsch processes. Syngas sources range from natural gas (and other deposits), through waste plastic, to biomass. Hydrogen on its own can be produced via electrolysis of water (into H_2 and O_2)—a technology that also has many forms ranging from electricity production via standard solar (photovoltaic) panels followed by anode–cathode electrolysis in a water cell, to direct production of hydrogen in photocatalytic solar panels fed with water. CO can also be made in a variety of ways, including from coal, but the environmentally most acceptable is by reducing CO_2 collected from atmospheric air or flue gases.

and the source of that energy. Oil from plants grown on arable land, or land "stolen" from tropical forests, is very unlikely to be environmentally sustainable; feedstocks from waste seems to be better in that regard. However, we must ensure that the waste truly *is* waste, that it isn't ecologically or agriculturally important, and that it would otherwise simply decompose, producing methane and CO_2 of its own.

As with gasoline or diesel replacements, researchers are not aiming to reproduce exactly the chemical composition of kerosene: that would be detrimental because, more often than not, a *homogeneous* synthetic fuel has more predictable characteristics (combustion, storage, transportability) and burns more cleanly because it also lacks impurities found in fossil-derived fuels. A straight-chain hydrocarbon with 12 carbon atoms and 26 hydrogens ($C_{12}H_{26}$; i.e., dodecane) is a good starting point (it's within the range of kerosene chain lengths), with an energy density of 46.2 MJ/kg. Blends of fossil aviation fuel with Fischer-Tropsch "paraffinic" kerosene are already starting to be used—albeit at very low percentages. Energy density per kilogram declines with every carbon in the chain (Figure 4.3); however, other characteristics improve, for example, lower volatility, higher flame point, particular combustion properties, absence of need to compress to liquid (above four carbons: i.e., butane), and therefore simplicity of distribution and combustion technology.

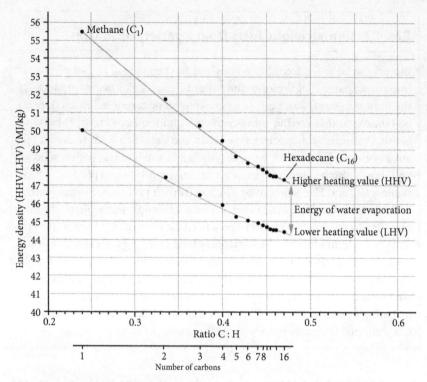

Figure 4.3 Inverse relationship between the ratio of carbon atoms to hydrogen atoms and energy density (MJ/kg) in fully saturated hydrocarbons (alkanes). Methane has the highest energy density; hexadecane the lowest on this plot. The difference between higher heating value (HHV) and lower heating value (LHV) is the heat that leaves the combustion process in the form of vaporized water. When water recondenses, it gives out this energy again. Processes that cannot recapture the latent heat of vaporization for reintegration need to calculate energy requirements on the basis of LHV.

Data from standard published sources.

And which are, in addition, economically sustainable?

Cost is a major factor because synthetics are still significantly more expensive than fossil fuels. A 5% compulsory blending quota for synthetic fuel applied to all flights departing from Germany, for example, would increase annual costs by €1.1 billion.[53] At around 850,000 departures in 2019,[54] and an average of 100 passengers per flight, that's an average of €13 more per flight. In this example, synthetic fuel cost six times as much as fossil at 2019 prices (€3,000 per tonne, compared with €500), but the cost of an average flight wouldn't rise by that much. However, by extrapolation, if 100% of aviation fuel were synthetic, at current prices, the passengers in the previous scenario would end up paying €260 more per flight on average. Could that be the price of sustainable air travel? The virtually nonexistent taxation of aviation fuel in most parts of the world certainly doesn't help in the switch to synthetic fuel.

The economic challenge with automotive fuels is smaller because, at least in Europe, their pump prices typically consist of 50% percent tax and duties[55] (in the United States, tax constitutes merely 9%). In 2021, standard 95+ octane gasoline in Germany had already increased to €1.40 per liter on average (US gasoline costs around €0.52 per liter), partly because of an additional CO_2 tax and partly because of inflation. Russia's war in the Ukraine (starting 2022) caused the price to rise to around €2—around the level at which synthetic fuel would (tax- and duty-free) compete on equal terms. How aviation has managed to escape most fuel taxation is worthy of a book of its own, but I believe that the days of cheap kerosene are numbered. Then we will realize the truly sustainable extent of air travel—ironically, not by grasping its environmental impact, but rather its impact on our bank account. Cost will surely fall as demand increases, capacities are expanded, and new technologies are developed, but true environmental sustainability will not be cheap.

The latest candidate for an alternative aviation fuel is triisobutane—basically three isobutane molecules stuck to each other and having the same chemical composition as dodecane ($C_{12}H_{26}$); it has a very similar energy density to aviation fuel, and, if current literature is to be believed, it could be synthesized at a cost-at-pump of €850 per tonne—only 1.7 times the price of conventional aviation fuel at 2021 prices.[56] What is the source of this wonder fuel? The type priced at €850 per tonne is produced via catalysis of butanol obtained from fermentation of "cellulosic feedstocks": that's short-hand for plant waste, hence competition with bioethanol for cars is clear, and the extent of replacement is likely to be relatively small because of agricultural constraints.

Energy carriers for stationary applications: Somewhat different considerations, but no less important

The question of energy density is less critical in many stationary energy applications, but supply infrastructure is still important. Factors such as practicality, affordability, and emissions may outweigh energy density for certain applications: here we might be able to transition from fossil fuels to completely different energy solutions without considering synthetic fuels. In some critical areas of industry, however, it is energy density that makes it so hard to replace hydrocarbons at present or in the foreseeable future. Hydrocarbon fuels turn out to be very hard to beat for high-temperature processes that must run at high throughput for long periods. Moreover, large depots of them can easily and safely be accumulated to smooth out supply irregularities and prevent industry stand-stills.

Industrial and domestic requirements: Big challenges, but immediate opportunities

Methane currently provides part of the necessary heat in many steel foundries, but it can also provide the source of hydrogen as reductant for the iron oxide in the ore: essentially methane is pyrolyzed using the molten metal heat of the furnace itself, making—if done efficiently—H_2 and C. Hydrogen can, ultimately, be the heat-energy provider *and* reductant in one that could replace coal and reduce CO_2 emissions from iron and steel manufacture. The challenges of H_2-driven steel production are currently being tackled.[57] Note, too, that methane is a convenient carrier of hydrogen, circumventing problems of compression, supply, and storage that beset pure hydrogen. Cement-making relies largely on coal and natural gas (mainly methane), but here, too, progress is being made to replace these with CO_2-neutral hydrogen.[58] That said, these two industry sectors are unequivocally seen as "hard to abate" in terms of removing their fossil fuel dependence. I deal with them (also in terms of their % $CO_{2eq.}$ contribution) in more detail in Chapter 5. A further challenge is making these industries more thermally efficient: almost 40% of input energy is lost in typical modern rotary cement-making kilns.[59]

Domestic energy presents lower-hanging fruit, technologically, but large costs that governments must help ordinary people to pay. Travel-related CO_2 emissions declined during most of the coronavirus crisis, but home heating-related ones rose somewhat: people went out less. The main heating technologies that can immediately help us reduce net CO_2 emissions in this area are:

- Solar heating (water-circulation)
- Air-source heat pumps
- Ground-source heat pumps
- Biomass heating system (from single houses to small communities)—but only if the biomass is from sustainable sources

Emerging technologies include domestic conversion of solar energy to hydrogen and storage on-site, and solar electricity production followed by storage in Li-ion batteries. Further discussion of these is outside the scope of this book. A carbon-fueled example is the biomass boiler. This might seem to be a sustainable (carbon-neutral) option, but its environmental balance depends on the source of biomass. If it's wood, we must deal with a serious misconception: yes, trees are growing anew all the time, but to generate the same heat per mass as from hydrocarbon fuels, we need to burn much more, and we produce much more CO_2 per tonne; moreover, if we don't specifically grow wood for the purpose, the principle is similar to burning fossil fuels but over a much shorter cycle: depletion. Biomass boilers that burn true waste that would otherwise—and, importantly, over short periods of time—naturally decompose to CO_2 and methane are worth considering among a range of biomass heating systems.

Why burning wood is not an ecologically sound large-scale heating solution

Let's disregard for a moment the problem of particulate matter (soot), which, per mass of fuel burned, is much greater for stove-burned wood than that associated with modern ICEs. With that caveat, there is a common perception that burning wood for heating your house is CO_2-neutral and therefore environmentally sound. After all, if that wood then grows again in the forest, what could be more obviously net emission free? Unfortunately this is a misconception. Furthermore, and equally seriously, the principle of wood-burning (in whichever form) does not compare well in terms of energy density and CO_2 balance with that of more concentrated fuel sources such as pure hydrocarbons. This will surely disappoint the many political parties and pro-environment movements worldwide that proclaim that nature's fuel—increasingly in the form of wood pellets or wood chippings—is the answer to environmentally sustainable heating. There is, indeed, nothing more "natural" than throwing a log onto a merrily blazing fire; however, some simple forestry science, chemistry, and thermodynamics show where it goes very wrong.

First to the question of CO_2 neutrality: one would need controlled plantations of trees specifically—and only—for fuel production in order to ensure that not more wood is burned than grows again. However, it's even more challenging than that because, mass-for-mass, burning wood has much inferior energy and CO_2 balances compared with, say, methane. Firewood contains around 20% by mass of water (which must be evaporated, hence using combustion energy); moreover, wood contains many compounds in which the carbon is partly oxidized already, or chemically bound with other elements that effectively increase its oxidation state (e.g., in cellulose, lignin, protein, and DNA). Hence the amount of oxidative energy per kg released upon combustion is substantially smaller than when burning a pure hydrocarbon. This is because in hydrocarbons there is very much more oxidation potential: all the carbon atoms are bound to hydrogen (in methane) or are bound to each other and to hydrogen (in longer-chain hydrocarbons). Hence, when we burn wood, we fully oxidize a lot of partly oxidized material, completing the oxidation process and releasing CO_2, but for a smaller release of heat energy than by burning the same mass of hydrocarbon

Table 4.1 Average (between HHV and LHV) energy densities of wood, methane, liquid natural gas (LNG), and heating oil

Fuel	Energy density in MJ/kg
Wood	13—15 for seasoned wood, i.e., containing 20% w/w water
Methane (at 1.013 bar pressure and 15°C)	53
LNG (compressed to 250 bar)	52
Heating oil	44

(e.g., methane). This is a rather awkward truth for those who think that fully organic (perhaps "natural") solutions to our energy demands are better than chemical ones.

The point here is that the energy that a plant soaks up from the sun is not simply used to make high-energy carbon compounds, but rather a whole mixture of things that make that tree in the forest. The tree cares naught for energy balances and oxidation states: it just wants to be a tree and reproduce itself. To put this all into numerical contrast, I give the figures for wood versus hydrocarbon fuels in Table 4.1. The value for wood includes wood pellets, which have a comparable density to many kinds of wood used for burning. Hence wood has less than half the energy density of typical hydrocarbon fuels.

If most households burned wood for heating, our forests would be gone in a few years

An average family with oil central heating in the United Kingdom burns around 2,600 L—2,200 kg—of oil per year,[60] releasing 94,600 MJ of heat. To produce the same heat from wood, one would need almost 7 tonnes of seasoned timber, equating to almost 12 tonnes of freshly felled wood (see Chapter 4, Calculation 8, online at https://doi.org/10.1093/oso/9780197664834.001.0001). If all families in the United Kingdom were to use wood for heating, local CO_2 neutrality would only be achievable if the same amount of wood were to grow each year the United Kingdom. How much is that? Given 28 million households in the UK,[61] we're talking about 325 million tonnes of freshly felled wood each year. Clearly the wood needs a couple of years to be seasoned, but that merely introduces a small time lag, which doesn't change the final balance.

Next question: Is enough wood for heating growing anew each year in the United Kingdom? No: 90% of the forests would be gone in 2 years (Calculation

8). There's no possible way for 325 Mt of wood to be regenerated for the next year's harvest: one doesn't need to be a forestry scientist to intuit that. At the level of individual trees, we're talking about somewhat more than a 200-year-old oak, or at least five mature pine trees, each year per household (Calculation 8). The problem is similar to that of CO_2 emissions from fossil fuel burning: trees just don't grow fast enough in our natural landscapes to use them as a general fuel supply. Other countries have higher ratios of forest to human population (Germany has almost three times the UK value), but, even there, we're still talking about the decimation of forests within a single generation. OK, so now that the United Kingdom's forests are gone, how did the CO_2 balance of this enormous incinerative deforestation look? In comparison with the most efficient natural gas central heating (i.e., also accounting for the thermal efficiencies of the boilers), heating with wood produces, on average, more than twice the mass of emissions of CO_2 for the same in-practice heating value (Calculation 8) (compared with oil, wood emits 1.7 times as much CO_2).

CO_2 molecules from combustion of wood or pure hydrocarbon fuels are indistinguishable to the natural cycle: we need not correct for difference in source. Wood releases substantially more CO_2 because it's a less highly reduced substance than pure hydrocarbons and hence produces less heat per mass upon oxidation during burning. Logistic costs and effects on soil also need to be considered because, for 1 tonne of natural gas, we would need to burn 4.4 tonnes of seasoned wood, which is 7.5 tonnes of freshly felled tree (Calculation 8). Clearly, one must be careful in touting the "advantage" of fossil fuel as a "superior" heating fuel: after all, it currently comes from geological reserves, and its CO_2 emissions certainly outstrip nature's absorption capacity. However, if we heat with gas, at least we leave large forest trees standing and capable of absorbing some of that CO_2; burning wood instead outstrips the forest's CO_2 absorption capacity even more. We must, instead, invent energy/material economies that can recycle CO_2 as quickly as it is produced.

Efficiency must be viewed from several perspectives, from fuel production to use

Why *expending* energy to *store* energy is necessary: A parallel with biology

The example of wood shows that there can be large differences between the energy used to make a substance and the energy that we finally extract from it via an energy conversion technology. The biological creation of organic material is full of side reactions that produce "things other than fuel" and

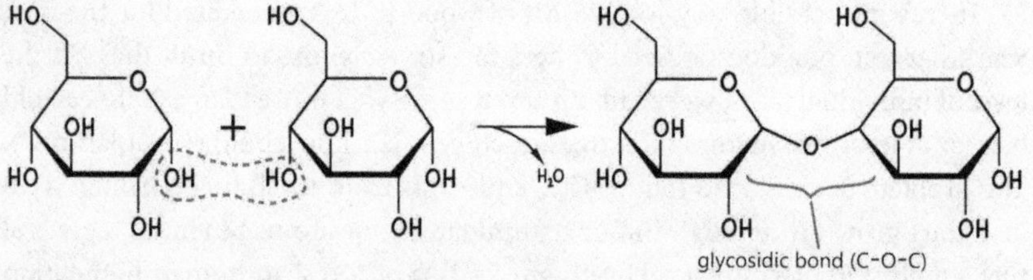

Figure 4.4 Polymerization of sugars into complex carbohydrates. Simple sugars (here, glucose) are polymerized into disaccharides and then longer-chain starches or glycogen in a two-step reaction involving a phosphorylase and an isoamylase. The glycosidic bond is formed by the overall loss of water upon joining of the two sugars.

lose energy to the surroundings; artificial synthesis of fuels also encounters these principles. It takes energy, some of which is infeasible to recover, in order to make an energy store. But this is a practical tradeoff that biology does all the time. Let's revisit the polysaccharides from Chapter 2 in a different context. Simple sugars are polymerized via a "glycosidic bond" (Figure 4.4), but this process is not simple: first glucose (or any simple sugar) must be "energized" by adding a phosphate group from the universal energy-carrier adenosine triphosphate (ATP) (via the enzyme *phosphorylase*); next, the phospho-sugar is joined to another sugar molecule (via the enzyme *isoamylase* [*amylosynthase*]), and so on.

Here we see it as the dimerization of glucose to make maltose (malt sugar). Wherever there is a chemically suitable OH group (the details are beyond scope), this "condensation reaction" can occur (the other product is H_2O). In living organisms, the activation energy of the reaction (needed to get it "over the hump," so to speak) is reduced by an enzyme (isoamylase) that perfectly positions the two substrate molecules so that they can react more easily. The actual reaction energy (this reaction is unfavorable thermodynamically) is provided by a high cellular concentration difference between ATP (high concentration, high energy) and adenosine diphosphate (ADP) (low concentration, low energy). ATP turns into ADP during many energy-requiring reactions in cells.

In Chapter 2, we saw the reasons why organisms from bacteria to humans go to the energetic costs of building long chains of sugars, ultimately to split them up again and burn them in central metabolism as single-sugar units. Let's now consider how each methylene (CH_2) unit is added to the growing chain of a fatty acid for energy storage in fat/oil. This is not a simple addition reaction, but rather takes place—similarly to starch—via activated intermediates of central energy metabolism. Furthermore, not one, but *two* carbon units are added at each step (thus making most fatty acids even-numbered in chain length). Still, the overall effect is one of polymerization, and the net reaction consumes energy. Most of that energy is not lost but rather stored in the potential of the molecules to react with oxygen during biological burning (cellular respiration). Some *is* lost, as heat. Furthermore, an organism must invest energy in making and replacing enzymes to do these material conversions, and *most* of *that* energy is, indeed, lost.

A reversible power station and chemical plant in one: Central metabolism unites material with energetic fluxes

Central energy metabolism is the hub where all basic ingredients for biological polymers (from DNA and proteins, through carbohydrates and fats) are made. By shifting the balances, and in the presence of the right enzymes, forward and backward interconversions often happen: for example, carbohydrates can be converted into fatty acids and vice-versa in many organisms—an exception is animals, which cannot make carbohydrates from fatty acids. That's one reason why losing weight is hard: we must run our carbohydrate stores down to zero, and then switch to fat burning, which uses a different metabolic pathway. However, when we start hydrocarbon combustion, we turn it into useful energy every bit as efficiently as when we burn carbohydrates.

The most metabolically versatile creatures transpire to be the simplest: bacteria and archaea—unicellular organisms from the dawn of cellular life on Earth. Via relatively simple genetic techniques they can be "reprogrammed" by humans to synthesize all manner of carbon-based compounds, including fuels with similar properties (often improved) to fossil-derived hydrocarbons. The principle that I hope to convey is that biological organisms expend energy, not all of which they can recover, in order to make energy-dense compounds that can be stored and mobilized easily without undesirable side effects: particularly in the case of oils and fats, they are highly stable but flexible depots. These are the virtues of hydrocarbons, not only in biology, but also in the human economy. The tradeoff for these useful characteristics is that one must invest relatively large amounts of energy to create the desired substances—albeit energy, a proportion of which is recoverable during synthesis, and most upon combustion.

How thermally efficient are living organisms?

Given that living organisms have a carbon-based energy economy, we might ask how efficient *they* are. Endotherms are animals that generate internal heat[62] and, when active (as opposed to hibernating), operate at a constant temperature of between 36°C and 38°C, depending on species, regardless of the external temperature. At rest (basal metabolism), and in cold external conditions, the energy needed to maintain such a temperature difference compared with the surroundings can be up to 100 times the energy consumed by an ectotherm (e.g., "cold-blooded" lizard, snake, crocodilian) of comparable size.[63] Endothermy, despite its seemingly horrendous wastefulness, brings distinct evolutionary advantages such as higher stamina and greater (and more quickly reached) peak physical performance on account of a more highly evolved apparatus of oxygen and fuel supply to skeletal muscles; ability to grow faster because of faster enzymatic and metabolic processing of food (enzymes and anabolic—body-building—metabolism are faster at higher temperatures); and the potential for intensive care, protection, and "education" of offspring, allowing for the occupation of more challenging ecological niches. An advantage in reproductive *output* (i.e., number of offspring per time unit) per se, is, however, far less clear.

Basal metabolic rate (BMR) tells us how much energy an animal uses at "tickover" (i.e., not doing anything). Human males have a BMR of around 1,600 to 1,800 kcal/day, and females between 1,300 and 1,500 kcal/day.[64] An (ectothermic) alligator manages with 60 kcal/day. Taking an average human value of 1,500 kcal/day, that's 6.3 MJ (1 kcal = 0.0042 MJ), or 0.15 kg diesel; or 1/6th of a liter: not bad![63] An alligator gets by on a mere 0.0068 L diesel (i.e., an amazing 6.8 mL)! A lizard or alligator (ectothermic) *can* reach high temperatures if it lays in the sun long enough; but its metabolism is not optimized for constantly high temperatures, and it is not able to regulate its temperature except by moving out of the heat. A lizard is intrinsically bound to the temperature fluctuations in its environment, and, if it's cold outside, the lizard is also cold and hence lethargic. In contrast, the largest evolutionary advantage of endotherms seems to have been their ability simply to remain active for longer each day, eat more, and mitigate environmental impacts and fluctuations better. For example, when the hypothesized asteroid impact that wiped out the dinosaurs (mostly ectotherms) happened, our proto-mammalian forebears, plus the ancestors of modern birds, were able to continue food foraging. They succeeded in mitigating the most extreme environmental impacts simply by being relatively mobile because they were warm.

Thought experiments reveal quite modest energy efficiency in humans

So, how much energy do we need to walk up a flight of steps, a task where we can attribute almost all of the *mechanical* energy requirement to raising the body's weight (our joints have as little friction as a skating boot on ice). A published experiment[65] provides a value that we can use in a calculation (see Chapter 4, Calculation 9, online at https://doi.org/10.1093/oso/9780197664834.001.0001), giving 0.011 kcal per kg weight per meter rise. The theoretical, physical minimum is 0.0012 kcal (Calculation 9)—roughly 10 times as little as the human body expends. A tailor-made machine for weight-lifting is doubtless much more efficient than the human, but it also has energy losses.

Where did most (90%) of the human energy go? Some into the machinery for getting oxygen into our lungs, then into our muscle cells (the heart pumps blood at an energy cost), processes that are themselves not 100% efficient. Much was dissipated in heat of chemical reactions in our muscles, including carbohydrate combustion, which makes us warm up and start to sweat. Most of this chemical energy is consumed by our legs and heart, but our torso and arm muscles are also working to keep us straight and balanced—energy that also doesn't directly go to lifting our weight. So, is that efficient? Perhaps not in comparison with a machine specifically designed for weight-lifting, though that's not the purpose of a human being. However, looking at the basic machinery of movement, we discover that skeletal muscle itself (our motors) is only 30–50% efficient in converting chemical energy into work[66]: not stunning compared with the 90-plus % of an electric motor. However, it is very sustainable as part of a cyclical system encompassing the whole biosphere, where energy and material are easily converted and leave no residue as pollution.

What about a 10-km bike ride compared with a car drive? Let's assume a speedy amateur cyclist, say 30 km/h on average, along a cycle path parallel to a country road on the flat. A simple calculation (see Chapter 4, Calculation 10, online at https://doi.org/10.1093/oso/9780197664834.001.0001) shows that the human–bike combo would be 17.5 times as efficient at transporting the human 10 km, but the car would do it much more quickly, safely, conveniently (indeed, why do people drive cars?). If we look at weight/energy comparisons, the human–bike combo (weighing, say, 82 kg) would use 0.013 MJ per kg, and the human–car combo (weighing, say, 970 kg) would use 0.018 MJ per kg. So, there is not a massive difference in energy consumption per unit weight. In terms of mechanical efficiency, the human–bike combo is around 38% more efficient than the human–car. Much lighter cars would contribute greatly to reducing energy consumption and hence environmental damage, not least because their manufacturing impact would also be much smaller.

The point of the comparisons is this: endotherms such as humans have high resting energy consumption and hence heat waste, but that goes hand in hand with great flexibility of operating conditions, evolution, and food supply. And, because they achieve this with carbon-based compounds that are easily recycled, it's ecologically sustainable. Furthermore, all animals, whether warm- or cold-blooded are not particularly thermally efficient: they burn carbon compounds and experience the typical energy losses: bonito tuna fish muscle can be up to 14°C warmer than the surrounding water.[67] Relatively large amounts of heat loss are a tradeoff against the "ease" with which animals can acquire and convert carbon-based fuels, store them (related to high energy density), mobilize them quickly (also related to high energy density), and burn them when needed. They can do this either gradually or suddenly when peak exertion is required: a very versatile system, with strong parallels to human-made heat engines in terms of flexibility of fuel and recyclability of that fuel.

Minimizing avoidable losses but making sensible compromises: How much heat can we reasonably recover?

Though many have insulation, organisms generally don't or can't recover emitted heat. However, much of human technology produces markedly greater quantities, and so we must strive to recover/use most of that energy. A ubiquitous example is the ICE: for a diesel engine at peak efficiency the useful mechanical work that it produces via the expansion of the combusting air–fuel mixture cannot exceed 45% of the energy from the fuel it burns.[68] The remaining energy is dissipated as heat. Other examples are thermic power stations, with typical thermal efficiencies of 37–38% for coal-/oil-fired and 56–60% for combined-cycle gas-powered plants[69] (without inclusion of waste-heat uses).

When the expanding steam, after passing through two turbines at most, can do little useful work it leaves the power station—traditionally in large cooling towers that condense the water to be reused for steam generation. Increasingly, this steam travels in insulated underground pipes to residential areas or businesses for heating and warm water. This can be calculated in favor of the power plant's efficiency. We must recover as much energy as feasible from any energy transformation that we perform but not fool ourselves with recovery strategies that themselves have high ecological impacts. The heat dissipated by ICEs has long been a thorn in the side of automobile technologists: around 33% of the combustion energy simply exits via the exhaust manifold as hot gas—energy that can partly be recovered in two ways, through direct harnessing of thermal energy into mechanical energy, or through thermoelectric mechanisms.

Direct harnessing of thermal energy into mechanical energy

This is usually done via the *Rankine cycle*—a type of steam generator fitted to the exhaust and driving either a piston engine or a turbine. The Rankine engine can help drive the crankshaft of the ICE or power an alternator to generate electricity. Research suggests values between 2% and 10% fuel savings with Rankine technology,[70–74] depending on the size, fuel source, and compression of the engine. Perhaps the practical "truth" at present is somewhere in between, say 6% fuels savings. Several companies have started R&D projects in this area.

Rankine cycle systems for ICEVs are thought to be capable of recovering more energy than regenerative braking in EVs, where the following consideration arises: under normal operating conditions, a battery's maximum charge rate is lower than its maximum discharge rate. A 2019 study using a pure-electric dump truck concludes that energy consumption of the vehicle can be reduced by between 1.06% and 1.56% via regenerative braking.[75] Calculations based on theory and ideal conditions and values indicate braking-related energy recuperation of between 2% (motorway driving) and 14% (urban driving).[76,77]

Thermoelectric mechanisms of energy recovery
The second means of energy recovery in ICEs is generation of electricity from exhaust gas heat, which could be attractive for hybrid vehicles, but probably only of goods-vehicle size (because of the relatively low efficiency of the process). The German-based company Mahle, for example, is experimenting with such a system in conjunction with the 48-volt electrics of heavy goods vehicles, where, they claim, it can reduce fuel consumption by up to 5%[78]; simulation projects suggest between 2.5% and 3.7% fuels savings[79]: small, but not insignificant, given the enormous amount of freight on our roads. This technology is certainly feasible for large vehicles, but not necessarily for everyday cars.

Some machines must dissipate heat in order not to break or degrade unacceptably fast. An ICE will simply seize up if not cooled; an EV battery will degrade very quickly or even explode if not cooled when large currents are drawn from it; the circuitry and even the motor itself require cooling to prevent damage. EV technologists strive to reduce the internal resistance of the battery, the resistance of the power electronics and motor so that they produces less heat. In ICEVs, the compression ratio, density of fuel in the cylinder, and subsequent expansion of the combustion gas determine thermal efficiency. Reducing heat generation in the first place is an important goal in any technology, but no work-producing system can avoid it.

Material is also "lost" in open processes. If we want to make carbon-based CO_2-neutral fuels, we should try to capture as much CO_2 as possible at high concentration, rather than letting it mix with ambient air. It's not impossible to capture CO_2 from car tail pipes, but the larger the vehicle, the more it makes sense. CO_2 can also be collected pre-combustion (see Chapter 5): an experimental ship that extracts the hydrogen from fuel to feed into the engine, and captures the CO_2 before combustion, has already been designed[80]; similar systems are proposed for diesel engines in trucks,[81] filtering exhaust CO_2 out via a thermo-swing filter and using the heat from the exhaust gas to run a compressor to put the CO_2 into a tank for off-loading. Pre-combustion could also be applied to industrial processes such as cement and steel manufacture.

In summary, car technology is rapidly addressing major energy losses, but many stationary processes have great room for improvement. The energetic waste from heating, power generation, and industrial heat-intensive processes—basically into the atmosphere—is more significant in terms of total energy waste than that from transport. Here enormous savings could be made, but, as we will see in Chapter 7, this requires full integration of energy and material economies. Attempts at recovering significant amounts of energy from "harsh" environments have so far been largely unsuccessful, but we must persevere.[82]

Combustion engines: Not a technology at the end of the road

Much achieved, but still great scope for improving the ICE

Breaking news at the time of writing: an Austrian company announces a zeolite-based technology that is ready for roll-out to all kinds of combustion engines (and combustion processes in general, including domestic heating and industry)[83]: it is capable of reducing hydrocarbons, carbon monoxide, nitrogen oxides (NO_x), sulfur dioxide, and fine particulates by between 90% and 100%. The cartridges can also act as collectors for valuable substances that can then be "desorbed" and used as industrial feedstocks—an important concept in the framework of cyclical economies with minimal waste. For ICEs, the big advantage is that they work very efficiently from cold.

Engine optimization could still reduce fuel consumption by 30%; hybrid drives and light-weighting may achieve 50% reduction, according to a recent paper.[84] There are still many developments in ICE technology that can improve efficiency of combustion and reduce emissions, and it seems likely—contrary to what many technology spectators say—that we are still some way off the end of the developmental road: it seems premature to prepare the funeral of the ICE wholesale, given the myriad applications in which it is found and the comparably low base from which alternative technologies are starting.[85] Moreover, the environmental and social dangers of rolling out replacement technologies too fast without due diligence on their full-cycle consequences are already evident. Even older ICEs can be modified relatively easily, but customers seem to prefer to buy shiny new cars instead of spending a fraction of the price on a refit. The car industry certainly supports that preference!

The oil crises and stricter emissions standards of the 1970s finally forced car and engine manufacturers to concentrate more on economy. From the 1980s onward, a series of developments greatly reduced fuel consumption.

FUELS, EFFICIENCY, AND EMISSIONS   161

- *Lambda sensor* (Lambdasonde, from Robert Bosch GmbH, a German engineering company), first used in Volvos and Saabs, to optimize fuel–air mixture
- *Engine management systems* (EMS) that detect when the driver is slowing down using engine-braking, and cut engine fuel supply
- *EMS developments* to control fuel–air mixtures, timing, and profile of fuel delivery (in small packets) and ignition
- *Direct fuel injection* into the combustion chamber
- *Turbochargers* that increase the useful work from each power stroke
- *Smaller engines* producing the same power as larger ones, hence saving materials (ongoing)
- *Large improvements to car diesel engines*, capitalizing further on the diesel combustion cycle's greater thermal efficiency than gasoline (unfortunately largely offset by manufacturer cheating with emission values)
- *Start-stop automatic engine control* when idling
- *High-compression gasoline engines*, approaching thermal efficiency of diesel (Mazda Skyactive) (ongoing)
- *Reduction of friction* in all parts of the drive-train, from engine to wheels (ongoing)
- *Increasing use of plastic parts* in motors, hence reducing weight and energy of recycling (ongoing)

Of course, car manufacturers are still building large engines for certain customers, but, on average, engines are getting smaller for the size of car that they power—significantly. Some are even losing a cylinder and managing very well on only three: this saves materials and makes the engine easier and less energy-intensive to build. Larger car models increasingly contain much smaller engines than before because of improvements in engine design and management and in turbocharger technology. A 1.2-L gasoline engine that produced 55 brake-horsepower (bhp) in the 1970s can now pump out 113 bhp: more than double. This is accompanied by other trends,[86] such as reduction in fuel consumption per ccm. Compared with 1970, power-to-displacement has increased by up to 150%; fuel consumption per ccm has decreased by almost 60%. Even greater saving could be made by car models becoming lighter and smaller, hence requiring even smaller engines, but many customers want the opposite.

The graphical plots of efficiency-increasing trends in cars have not suddenly "plateaued" out, demonstrating stagnation: improvements are still being made. Nothing can improve ad infinitum, but there are so many aspects of wheeled locomotion that are amenable to technical evolution—and revolution—that we would be foolish to rule out further significant ICE improvements. Cold-start efficiency and energy recovery are big areas of improvement yet to see their heyday, for example. Hybrids that combine the virtues of ICEs with those of EVs may also be very pragmatic solutions. During development we need to assess the full-cycle wisdom of the product—from the mining of minerals, through making of components, all the way up to the use, death, and recycling of the object. Different technologies and combinations of technology will continue to have different optimal spheres of use, and one-size-fits-all propositions are unlikely to be sustainable solutions. A development that has, unfortunately, not yet occurred is an increase in average load transported in a private car: usually it is only one person, rarely two or more. Hence, the greatest potential for improving car efficiency is not technological at all: it's sociological.

Jet engines: Heading for 20% reduction in consumption by 2040

Jet engines and other turbines are an extreme case. These heat engines are one step removed from a rocket that simply blasts out a burning mixture of fuel and oxygen to produce thrust. That said, the jet engine is anything but crude, and, because the combustion of the fuel–air mixture happens inside the very mechanism that drives the air in and compresses it, the jet engine is also a form of ICE. It works rather like the turbocharger of a car: the expanding combustion gas powers a series of turbines mounted on a shaft. In front of the fuel nozzles—at the cool end of the shaft—there is one or more fans bound to the shaft (hence the name "turbofan" for some engines). Thus the engine sucks in, and compresses, the very air that is needed for the combustion process. Jet engines also produce a thermodynamic graphical plot of heat versus pressure: the *Brayton thermodynamic cycle*. For large aircraft and long distances jet engines are more efficient than piston-driven aero engines. The pressures on flight technology caused by World War I concentrated the minds of two men in particular on the task of making aircraft that could fly much faster than propeller planes of the day. Frank Whittle (a British RAF cadet) was granted a patent for his design in 1932; the German Hans von Ohain—initially unaware of Whittle's work, probably because the RAF also had no interest in it at first—designed his prototype in 1935. It was first envisaged for use with hydrogen as fuel, but gasoline was finally used for practical trials, and the first jet-powered aircraft lifted off on August 27, 1939, with Erich Warsitz at the joystick.

Both turbojets and turbofan engines power planes, but let's concentrate on the turbofan because it is the more commonly applied technology (most of the principles apply to other jet engines). The hotter the combustion process, the more efficient it is; however, the turbines of a jet engine can easily reach temperatures upward of 1,700°C: engines need robust cooling and special oils. Even higher temperatures, for improved efficiency, require very special blade coatings, as recently developed by researchers at the German Fraunhofer Institut[87] and used in the Airbus 350-1000, lowering fuel consumption by up to 10%.

FUELS, EFFICIENCY, AND EMISSIONS   163

Here is a summary of the efficiency-promoting developments in jet engines:

- Turbine blade coatings, enabling hotter combustion
- Combustion component heat-resistant coatings to produce better combustion control
- Improved turbine fluid dynamics (airflow)—speeds can reach an amazing 9,000 km per hour in such turbines!
- Improved fuel–air mixing

These, and other modifications, notably in material weight and aerodynamics, have led to a significant decline in aviation fuel consumption and pollution since the 1960s, when jet engines started to be used routinely for passenger and freight airliners. Taking 1960 as 100%, International Council on Clean Transportation figures indicate a reduction to around 50% fuel consumption in 2021, and further reductions are projected on a curve that crosses the 40% mark in 2040 (i.e., 20% reduction from present levels).[88] Clearly, this curve will flatten at some point, and, as it does, we will have to assess the environmental costs of reaching further efficiency in comparison with the environmental benefits, as always, comparing with alternative technologies. Part of the solution for reducing environmental impact to an acceptable minimum may involve reduced mobility: technology alone is unlikely to solve the challenge.

From emissions to healthcare: Mismatches between public perception and the realities of challenges and progress

One of the arguments for abandoning carbon-based energy carriers is pollution. Next I summarize some of the most important challenges and the progress toward addressing them. I do not consider reduction/capture of CO_2 emissions here because that is dealt with extensively elsewhere, mainly in Chapter 5.

The transport sector shows greater success than most others

The iconic catalytic converter: Now achieving 95%-plus reduction of by-product gases

The standard exhaust gas treatment device is the *catalytic converter* ("cat"). The first were fitted to cars earlier than you might think: 1975, in compliance with the US Clean Air Act (1970). Europe lagged far behind, making cats obligatory on new cars as of 1993, with the introduction of the EURO 1 norm (at the time of writing, the EURO 7 norm is due for release). Via catalysis on honeycomb-like surfaces coated with rare metals, cats accelerate chemical reactions in the exhaust gas[89] that

1. Promote oxidation of CO to CO_2 (with residual oxygen, and oxygen produced in Point 3)
2. Promote oxidation of unburned hydrocarbons to CO_2 and H_2O (with residual oxygen, and oxygen produced in Point 3.)
3. Reduce NO_x to N_2 and O_2

Three-way cats only work with EMSs with a lambda sensor (i.e., ones that compare O_2 concentration in exhaust gas with concentration in the air and adjust the fuel–air mixture). At designed operating temperatures (400–600°C during normal operation), a three-way cat will be 99% efficient when new, but typically needs replacing at between 150,000 and 200,000 km. Cats have been moving ever closer to the exhaust manifold of engines, allowing them to reach optimal operating temperature faster. A further exhaust gas technology relies on feeding part of the exhaust gas back into the combustion cycle to reduce concentrations of unburned hydrocarbons and NO_x. This also helps the engine reach optimal operating temperature—and hence maximum combustion efficiency—faster.

Cats are now fitted to all ICEVs and increasingly to ships' engines. The platinum, palladium, and rhodium in most cats is recycled to a large degree (around 60%), and more than 90% should be chemically feasible.[90] Reconditioning of cats can be done to some extent, for example with the kinds of detergents that one uses to de-soot a wood-burning stove and 100°C steam.

Particulate filters: Removing solids formed during combustion

Improvements in combustion technology reduce emissions of particulates greatly but cannot eliminated them completely. Moreover, the size distribution changes to ever smaller ranges with improvements in combustion technology. Concern over fine dusts (particulate matter [PM] ranging from 10 down to 2.5 µm in diameter) in general has grown massively over the past 30 years or so, and rightly so. PM25s (with a diameter of 25 microns—40th of a millimeter—or less) were the original focus of attention: more recently, PM2.5s (10 times smaller) have aroused concern. PM10–PM2.5 in the air that most of us breathe outside come mainly from combustion processes, the largest producers globally being domestic heating (particularly fireplaces), coal- or natural gas-fired power stations, industry, and transportation. Depending on where one lives, the balance of sources differs. For a recent review of the health risks from PM10–PM2.5 from short-term and long-term exposure, see Orellano et al.[91] and Chen.[92] Regardless of source, the basic particles are very similar and are termed "black carbon" (i.e., they are spheroids of almost pure carbon).

Particulate exhaust filters in diesel vehicles (known as diesel particulate filters [DPFs]) became mandatory in Europe with the EURO 5 norm in 2009. For many older cars, retrofittable filters were developed. DPFs act as holding containers for PMs until they are burned off in the exhaust gas. They most commonly consist of a honeycomb matrix made of cordierite or silicon carbide—the same material used for catalytic converters. A massive area for filtration is achieved (hence keeping down the pressure against which the engine has to work in expelling the exhaust gas) by the long and fine structure. The walls are "doped" with a small amount of platinum, which acts as a catalyst for accelerating the next step: at intervals, the collected soot is burned off by the exhaust management system injecting a small quantity of fuel into the filter. Older and retrofitted filters contain vortice-generating metal baffles and rely on just getting hot by a long drive at high speed; some have metal filtration meshes, which are heated by electricity to burn off soot.

Ceramic filters can remove more than 95% (by mass) and 99% (by number) of exhaust particles, down to sizes even below 100 nanometers (that's 0.1 μm; i.e., 10 times smaller than PM1s). The filters covered so far typically remove 85–99% of particulate matter from exhaust gas, but there are cheaper solutions (for example, for retrofitting vehicles in developing countries or in situations where normal filters are infeasible, e.g., coal mines). Special types of paper are one solution: these are simply replaced when clogged. For older vehicles, there are various options to reduce particulates—though admittedly not to the 85–99% mark of the factory-fitted devices.

For ship engines, *selective catalytic reduction* (SCR) systems for NO_x reduction, together with DPFs, can also be used. Here legislation has lagged behind road-traffic regulations. However, it is every bit as urgent because large ships typically burn the lowest-grade heavy fuel oil, which contains a relatively high percentage of sulfur (sulfur content is positively correlated with PM emissions). Synthetic fuels with improved combustion characteristics are in development, but interim measures to curb maritime emissions are a must. It was not until 2016 that the International Maritime Organization (IMO) for seagoing vessels set Tier III NO_x emission limits in its Emission Control Areas (ECAs).[93] Emission of particulates is not yet covered by any legislation in international waters, and it would be surprising if research on PMs in marine environments revealed no ill effects. In coastal waters and inland waterways, national and international legislation exists to regulate PM emissions, and DPFs are being developed and gradually introduced into ships, particularly in catalytically coated forms combining filtration with NO_x reduction.

What is prescribed for diesel engines also works for gasoline engines, where particulate filters are also becoming routine factory-fitted exhaust components. Where economic conditions preserve an older population of cars on the roads, local ingenuity can also help: in 2019, M. Ravishankar of Tamil Nadu in India was celebrated on the Better India website for his invention of an ingeniously simple exhaust component that, without any particular optimization, could already reduce emissions of all kinds from a gasoline engine by 40%.[94] The device is a steel cylinder containing activated charcoal and coconut fibers, kept in position at either end by steel wool—and it is cheap. The patent is pending. 100% reductions would be better, but, as the saying goes, perfection is often the enemy of the good. Doing something good in a situation where economic, social, and geographical constraints are strong is something to be celebrated.

Uncritical proponents of EVs might consider EVs the perfect solution to air pollution, but their superior environmental sustainability overall compared with ICEVs has not been demonstrated, and, given that we are discussing PM in this section, the fact that an EV has no PM coming out of its nonexistent tailpipe does not mean that it produces no PM during operation, as we will see next: all cars produce tire dust, road dust, and brake dust—particularly in towns, where braking and accelerating are frequent. ICE cars built to conform to EURO 6 norms have particle filters that reduce PM to 0.0045 g/km or less.[95] A catalytic converter and particle filter can be fitted to almost all cars, new or old, though producing differing percentage efficiencies (a pre-engine management car cannot benefit from a three-way cat, but it certainly can from an "unregulated" cat: it's just a modestly priced bolt-on). Brake pad composites are constantly responding to research findings (e.g., replacing copper with graphitic carbon nitride).[96]

Non–ICE-related particulates are now taking center stage

To contextualize ICE-related particulates, I've decided to mention some that are not products of combustion. Brake pads also contain carbon, and wear produces not only carbon PM, but metallic PM. Values vary somewhat (but certainly not by an order of magnitude): a 2020 study, drawing on other published data, notes that for an average car driving in a town in average conditions, PM from brake wear is 4.7 mg/km for PM2.5 and 11.7 mg/km for PM10. So, a total of 16.4 mg/km for PM10 and lower.[97] That's three and a half times as much PM as that car would, on average, emit from its exhaust if it were a modern diesel. The study concludes that switching all EURO 6 cars to EVs would have no benefit because (the study's calculations show) their emissions would not fall below those of EURO 6 (let alone EURO 7), even with regenerative braking. Essentially, we should not encourage cars of any type to be driven within towns.

Brake wear is a more constant contributor to PMs[98] but tire wear can—depending on road surface and road wetness—be considerable, too,[99] and rubber particulate emissions can exceed those from brakes. According to an EU study, an average car tire (roughly 7 kg) loses 10% of its mass in around 55,000 km.[100] For four tires, that's 2.8 kg. In Chapter 6, we'll go into this microplastic pollution in more detail. Research into better rubber compounds is important for all road traffic, but particularly for EVs, because the greater weight per unit area of tire-road contact, the greater torque upon acceleration and braking, and higher forces during cornering cause significantly greater wear.

Air sampling on the road can present a surprising real-world picture. In one urban area of Germany, a recent study concluded that "only 22% of [automobile] particulate matter emissions are exhaust emissions, while 78% are produced due to the abrasion of tires, brakes and the road [surface]."[101] Warnings about non-combustion traffic-related dust are already making headlines.[102] Indeed, one study on these underresearched sources of PMs concludes that the metallic particles present in brake dust (brake pads are typically a compressed composite of powders of iron, steel, copper, and graphite) might be just as cumulatively detrimental to human health as PM from ICEs.[103] No human being can (yet) live forever, so we must also consider ambient hazards relative to each other: yes, humans have, since their invention of fire, been breathing in the products of anthropogenic incomplete combustion, but they have not (as far as I know) been breathing in microparticulate and nanoparticulate metal, rubber, road surface, and resuspended road dust until the advent of the motor car. Again, a noble aim would be to reduce the traffic of *all* types of cars in town and city centers. Ironically, recent studies have shown that in European cities, the concentrations of fine and ultrafine particulates in subway air (on platforms and in many of the trains themselves) are higher than at road level[104]: metallic particulates (from train braking and track friction) are a major component.

Political developments versus the truth of the present road-transport situation

Whether relativizing sources of particulates will impress the European Commission is debatable: at the time of writing, the European Commission's EURO 7 emissions norm is due. This will be so stringent with emissions that it will—according to ICE experts and representatives of the car industry—amount to a de facto ban on the production of ICEVs. Then we will be almost entirely reliant on battery technology for most civilian road transport because it is very unlikely that hydrogen will make major inroads quickly enough. Given the many unsolved problems of which we know—let alone the ones that may yet emerge—surrounding battery production and recycling, this appears to be a very short-sighted and risky course of action. And one might fairly ask whether it's necessary, given that 1. access restrictions and large improvements in combustion and exhaust technology have already reduced air pollution well below acceptable limits in many cities; and 2. the energy balance comparisons between modern ICEVs and EVs that have, conveniently, failed to reach the public on the whole (see Chapter 6). Or does the Commission want towns and cities to be repopulated with as many cars as before, as long as they're electric ones? Where did "towns are for people" get lost in all of this?

Might we be tipping the baby out with the bathwater? Is the persistent criticism of the ICE—often being referred to as a "filthy old machine" (in German *alte Dreckschleuder*—old filth-spreader)—really doing good? I certainly haven't invested time and effort in these pages to support the car industry, and I certainly don't consider it one of the more desirable components of modern industry. However, plain truths about technology and technological achievements must be disseminated and understood if we are to reach sustainable forms of industrial production and consumer behavior. The truth about ICEs today is very, very different from that in the 1960s. In towns and cities of developed countries where emissions regulations are strongly enforced, PM emissions from ICEs are now usually markedly below those from traffic-related non-combustion sources. If the ultimate aim is to reduce PMs that people inhale to a minimum in towns, the answer is clear: restrict cars of *all* kinds to the town periphery and provide efficient, clean, non-automotive public transport within the city.

Emissions from jet engines: A very hard nut to crack

Now for one of the largest technical challenges in hydrocarbon-fueled transport: exhaust emissions from jet engines. It's not surprising that recent research has shown that PM from jet engines can be damaging to human health,[105] generating similar oxidative stress in cells lining our airways as does PM from other sources or from smoking. Present progress is centered on the combustion technology and the loading of the engine. One study shows that a typical turbofan engine produces least PM when operating between 40% and 60% of maximum thrust.[106] This could, for example, be achieved by having four engines, two large and two small; at cruising speed, where a jet typically needs only a quarter of the thrust of take-off, one would shut down the larger of the four engines, hence making the other two work more. Whether this model works in other respects and what its tradeoffs are is anyone's guess; PMs released at cruising altitudes of jet planes have not captured the attention of regulators as much as ground-based emissions.

Research conducted between the mid-1990s and 2014 indicates increased incidence of respiratory and cardiovascular diseases in populations residing close to airports: peak particulate emission occurs during take-off and climb.[107-110] As well as the extremely small particle size (all below 1.5 μm) in the studied air samples, the suspected most serious contributor is heavy metal particulates. Jet engines also produce appreciable quantities of NO_x. The necessary high speed of exit of exhaust gas from the turbine rules out a particulate filter or catalytic converter on grounds of immense resistance to the thrust of the engine; however, that has not stopped human minds inventing other possibilities. Basically, one idea is to inject urea and ammonia (a chemical relative of urea) into the exhaust stream to achieve non-selective reduction of NO_x—similarly to the solution for diesel vehicles. That will not deal with heavy metals, however.

Healthcare-related aspects of carbon emissions: We might be surprised by the realities

The first surprise is that despite the widely reported and increasingly substantiated health risks from fine PM, the exact mechanisms of their immediate and long-term harmfulness to us are far from understood. Most mysterious are the ultra-fine particulates (UPFs, ≤0.1 μm), for which we may only speculate a toxic role for lack of data linking them directly to morbidity/mortality.[111] The prevailing paradigm is that, in our tissues, PM produces reactive oxygen species (ROS) (many of which are "radicals," against which antioxidants are supposed partly to protect us) and hence oxidative stress and damage to biological molecules and structures. However, the picture is transpiring to be much more complicated than a simple chemical damage model. ROS doubtless plays a key role, but, in addition to direct damage, complex internal communication networks related to redox (balance of reduction–oxidation, i.e., "redox signaling pathways") are involved.[112] Adhered contaminants—largely metallic—may be significantly more damaging than the core particles themselves.[111]

Stationary sources: Your kitchen might be quite a significant one

Let us first consider immobile sources of emissions (i.e., furnaces, heating systems, cooking apparatus, and natural sources). In homes where gas is used for cooking, use of extractor fans is a must. The PMs produced by incomplete combustion of gas can reach concentrations every bit as high, or even higher, than those on a busy road—though only temporarily, during cooking, as a 2001 study showed.[113] However, of equal concern is the size range of particles produced, most of which are in the nanometer range (15–40 nanometers), about 50 to 250 times smaller than the PM2.5 that we were discussing earlier in connection with ICE emissions. This is not something that we should be inhaling in appreciable quantities because these particles can reach the smallest of our airways, remaining there for very long periods, suspended in the mucus coating the epithelial cells. Our bodies have three, albeit slow, ways of expelling such particles: (1) the flow

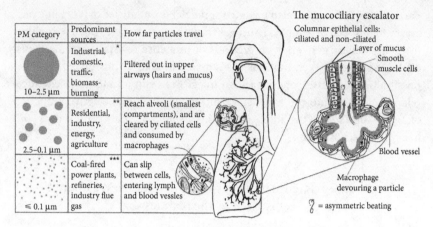

Figure 4.5 How far do particles go when we breathe them in? Particles smaller than 2.5 μm diameter reliably reach the finest endings of the microscopic airways in our lungs, the "alveoli." The finest particulates, those less than or equal to 0.1 μm (100 nanometers), are able to pass between cells lining the alveoli and enter the lymph and blood vessels. The smallest airways are patrolled by macrophages, specialized cells of the immune system, which devour foreign particles and pathogens. Particles are also wafted away from alveoli to larger vessels by ciliated cells (i.e., bearing tiny beating hairs) and ultimately to the bronchi and trachea to be coughed up or swallowed in sputum.

Data for the sizes of particles and their passage through the respiratory system taken from Kwon et al.[111]; predominant anthropogenic sources around residential centers, globally, from Junkerman et al.[115]; McDuffie et al.[116]; Kargulian et al.[117], and Supeni et al.[118] Note that PM2.5 from traffic is a major source in developing countries, though not in wealthy ones.

of mucus driven by tiny hairs (cilia) on epithelial cells, transporting the mucus up to our trachea and esophagus to be coughed out or swallowed; (2) phagocytes of the immune system, which ingest particles; and (3) passage between the cells into the lymph fluid (Figure 4.5). Worse still, the ciliary cells are slowly damaged by long-term exposure to PM above certain concentrations (as smokers know), losing ever more cilia and working less effectively. PM size and number appear to be more important than mere mass (weight) of particles breathed in: the smaller, and the greater the number, the more harmful. Those interested in more details can read a fascinating paper on the analysis of PMs from Chinese restaurants, representative of the diversity of cooking styles.[114]

In Europe, as noted in a 2020 publication,[119] the ROS-generating potential of PM is particularly high in "fine-mode secondary organic aerosols largely from residential biomass burning and coarse-mode metals from vehicular non-exhaust emissions." In the European Union, data analyzed from 1990 to 2017[120] show consistent decreases in all notable anthropogenic air pollutants (the largest being in sulfur oxides, on account of cleaner oil, gas and coal); particulate emissions show a similar decrease to NO_x, for example.

PM2.5s: The smallest routinely measured particulates

Today the sources of PM2.5 (mainly black carbon) to which average people are exposed in the European Union is broken down as follows[121]:

- Commercial, institutional, and household: 53% (mainly heating and cooking)
- Manufacturing and extractive industry: 17%
- Road transport: 10% (mixture of combustion products, brake dust, tire dust, road dust)
- Waste: 8%
- Agriculture: 6%
- Energy supply: 3%
- Non-road transport: 3%

Seen globally, although the potential for error is larger, the pattern—if not the exact percentages—is quite similar[122]:

- Other sources: 24% (including commercial and institutional)
- Residential energy: 21% (heating and cooking)
- Industry: 18%
- Power generation: 15%
- Agriculture: 9%
- Transport: 8%
- Open fires: 5%

Local regulations and practices can cause massive differences. However, it is likely that the health hazards from PMs are generally greater in cities, where there is less air circulation and a higher concentration of fossil fuel-burning heating and vehicle soot concentration, than in surrounding areas on average. Indeed, forms of carbon (elemental [EC]; organic [OC]) are the largest single contributor to PM2.5, both OC and EC originating primarily as products of incomplete combustion (i.e., anthropogenic sources). OC incorporates mainly cyclic structures, the variety of which we have already discussed; EC is essentially graphite.

Let's take a medium-sized European city that experiences large temperature fluctuations between summer and winter, such as Krakow, Poland.[123] In winter (when heating is turned up and cold temperatures can trap air close to the ground), total carbon mass present in PM was more than five times the summer value (4.7 µg/m^3 and 22 µg/m^3 respectively). But there is much more to this because appreciable quantities of elements other than carbon were found significantly increased in the city's winter air (see Samek et al.[123] supplementary material online): chlorine at between 850 and 1,600 ng/m^3; iron at between 370 and 1,370 ng/m^3; and lead at between 14 and 18 ng/m^3, to name a few. Though these are between 10 and 600 times less concentrated than the carbon particles, the fact that they are present as nanoparticles likely implies a much higher reactivity—and hence toxic potential—compared with the bulk elements. This is a complicated area of research to say the least.

Domestic heating: One of the largest current challenges

Domestic heating the world over produces a large proportion of PM10–PM2.5, and, in certain rural areas, people are absolutely dependent on coal-fired stoves for heating and cooking: Northern China is one example (most others are in Asia, South America, and Africa). While better energy sources are being developed and introduced, simple measures such as decreasing the size of the pieces of coal burned can greatly increase combustion efficiency, hence lowering PM and polycyclic aromatic hydrocarbons (PAH) emissions.[124] Though evolution with the occasional barbecue probably gave us certain tolerance from paleolithic times, it didn't fit us for standing next to an indoor smokey cooking hearth every day for prolonged periods. This is a massive, often ignored problem: it is estimated that more than 2.8 billion people worldwide cook with solid fuels.[125]

Technologies that have reduced particulate emissions by more than 90% in ICEs have not touched this sector and are unlikely to under current policies. If we truly care about the health of all fellow humans (as well as people in our First World cities), supporting technology developments of this nature is a responsibility. Replacing dirty domestic combustion technologies should have a much higher priority. Even in First World cities, heating dominates the emissions of PM2.5 in the colder months of the year and is a pressing target for "cleanup." It's worth considering that an average household with oil central heating can easily burn around 2,500 L of diesel heating oil per year and usually without flue gas filtration; by contrast, an average modern diesel car driven a European average 12,000 km per year would burn roughly 660 L and with very effective exhaust gas treatment.

There's more to emissions than meets the microscope: Complex particulates

Epidemiological evidence certainly supports the harmfulness of PM—particularly PM2.5 and smaller. In the case of metallic nanoparticles, the mechanisms are believed to be directly related to the production of ROS, which degrade cellular structures and can mutate DNA, leading—respectively—to cardiovascular disease, neurodegeneration, and cancer if present regularly and in sufficient concentration.[126] But how can something as seemingly harmless as a tiny piece of carbon do damage to the human body? It has long been known that PAHs—making up a significant proportion of the carbonaceous PM from combustion processes (of fossil fuels or of biomass), aluminum and steel production, and cement manufacture—are present in our air: there is no simple type, but rather a wild mixture of gaseous forms (small molecules) and solids adhered to other tiny particles or aggregated to make such particles. Graphite-based PM will gladly embellish its surface with PAH present in the combustion gas.

PAHs are absorbed by ingestion, inhalation, and skin contact.[127] Because they are lipophilic (fat-loving), they accumulate largely in adipose tissue and may therefore also be present in a mother's milk. The largest health concern related to PAHs is their carcinogenicity. Acute exposure at sufficiently high concentrations will produce symptoms of irritation; long-term exposure to high levels statistically increases one's risk of a variety of cancers (e.g., skin, lung, liver, bladder, and intestinal tumors) as witnessed, for example, by studies on firefighters, who are exposed to PAHs.[128] Note that these include very particular PAHs (e.g., from combustion of synthetic materials) and in unusually high concentrations: not risks to which the population at large is routinely exposed. Two principal mechanisms are believed to account for the carcinogenicity of PAHs: (1) an oxidation reaction that produces a radical cation—a positively charged version of the PAH that has an unpaired electron floating around and hence easily reacts with biological molecules (e.g., DNA); and (2) oxidation to produce a type of diol epoxide—a PAH with two OH (alcohol) groups sitting next to a triangular C-O-C construct. This diol epoxide is also pretty reactive. Both of these PAH derivatives, which form in the human body, can react with DNA bases, producing "DNA lesions." If a lesion can be repaired by the cell's DNA repair machinery (which evolved already in bacteria and archaea and is present in some form in all cells), then all is well; however, it's not always spotted, and the repair machinery isn't faultless: some lesions escape. The higher the "bombardment" from reactive PAH derivatives, the higher the chances that DNA lesions will accumulate.

When the cell comes to divide and replicates its DNA, the lesion basically leads to an incorrect base (letter) being produced in one of the DNA strands: the coding sequence is permanently changed in one of the daughter cells, and this change will be inherited by all subsequent cells in that lineage. If a change occurs in a gene that regulates the division of cells (cell cycle), and other conditions are propitious (foremost the "microenvironment" of the cell), further carcinogenic mutations may arise and be "fixed," bringing the cell ever closer to the cancer state. However, mutations are probably only half of the picture because cancerous behavior has also been observed in cells without obvious mutations, and it is known that constant irritants (countless chemical compounds fulfill this criterion) can induce cancer: notable in this group are aldehydes. PAHs can mutate DNA, and many are proven irritants, and yet, in Chapter 2, I mentioned that aldehydes may be emerging as more important carcinogens. We still lack critical biological insight about what is happening in human cells in real time.

PM also acts as a carrier for a wide variety of other things that can potentially do us harm—many of them not present in paleolithic times. Carbon-based PM has the ability to absorb other molecules (activated charcoal filters for air and water are a useful application of this property!). Carbon PM2.5s generally have a core of elemental carbon (essentially graphite) to which, apart from PAHs, secondary sulfates (derived from sulfur in fossil fuel), redox-reactive metals (e.g., iron, copper etc.), soluble hydrocarbons, and secondary nitrates can adhere.[129] Some of these may also synergize with PM in their oxidative damage, potentially contributing to the accelerated progression of cardiovascular disease and a variety of degenerative diseases (though the full mechanisms have not been completely elucidated). PM has also been suggested to be a significant carrier and concentrator of respiratory allergens[130,131] and even viruses.[132]

Reaching a realistic perception of risks and potential for improvement

I don't wish to downplay health risks from organic PM and other emissions, but (1) they are highly dependent on dose and (2) different PMs and their absorbed substances have very different properties, making generalizations senseless. Furthermore, many realities manage to escape deeper analysis, media coverage, public attention, criticism, and political measures. For example, the upper limit of NO_x concentrations to which a factory worker may be exposed in the European Union is around 20 times higher than the limit of NO_x on the street in a town. Permissible maximum concentrations of PMs in industry are also markedly higher than those tolerated on our streets. Research into long- and short-term (acute) effects of harmful substances is certainly not a finished art, so we cannot assume that all of these regulations make perfect sense (either in terms of being "good enough," or in terms of being "excessively restrictive"). The "natural" environment also presents us with numerous physical and chemical dangers. It is for this very reason that we and other air-breathing creatures have, for example, evolved the "mucociliary escalator," an epithelial mechanism that transports mucus and trapped particles out of the lungs and smaller airways.

When we hear or read figures of so many tens of thousands of people dying each year from fine dust, we should consider that in many countries these days, a lower percentage of the population than ever before in history dies directly from fine dust–induced diseases. In others, regrettably—particularly in and around the major cities—fine particulates from combustion processes of all kinds (including the burning of wood and coal in domestic stoves) are a major, and growing, health concern. Strict regulations, laws, and stupendous fines have forced the car industry to go the extra mile and explore more technology than they might otherwise have had to. Now we must focus similarly on other industries—the ones that continue to produce the larger fractions of PMs—particularly UFPs in the air we breathe: coal-fired power stations have long been known to produce UFPs containing metallic contaminants, which are likely more harmful than those from automotive fuel combustion,[133] and industrial stacks are, in general, major emitters.[111] We must also take responsibility ourselves and apply a certain measure of realism—both privately and in conjunction with purchasing decisions.

The truth of it is that even a candle will produce combustion products that are classified as harmful,[134] including relatively large concentrations of NO_x and PM in the 7–15 nanometer range. That is 200–300 times smaller than the routinely scrutinized PM2.5s discussed earlier. However, I know of no plans to introduce regulations forbidding the use of candles in homes or public places. A further truth is that the manufacture of an EV that produces no combustion particulates in a wealthy First World city gives rise to plenty of fine particulates and other noxious substances during battery production elsewhere in the world. That is a tradeoff that we should not accept: all efforts to ensure cleaner production of consumer goods outside the countries of purchase and use are vital components of better global sustainability.

Conclusion: Increasingly feasible and acceptable replacements for fossil fuels, more efficiently produced and burned

We have the capacity on Earth to generate some of our fuel feedstock from photosynthesis as starches, hemicelluloses, celluloses, and lignocelluloses, mainly from crop waste: "biofuels." All of our fuel needs could be satisfied via "artificial photosynthesis" (i.e., photovoltaic electricity production followed by chemical syntheses drawing on CO_2 and H_2 to make "e-fuels"). However, biofuels always run a risk of competing with food/feed agriculture, and primary crop produce cannot be envisaged as a significant feedstock. E-fuel production requires very small land areas in comparison, but e-fuels are energetically more costly per liter. E-fuels burn more cleanly than their fossil fuel counterparts, but all fuels—whether fossil or synthetic—produce combustion heat that can be hard to recover. These are tradeoffs that we must seriously consider living with because rapid and total replacement of carbon-based energy carriers has its own, severe climatic and environmental consequences.

Methanol and ethanol (mainly for use in ICEs) have convenient properties for application in existing combustion technologies, distribution, and storage infrastructures; DME could be a practical replacement for diesel fuel in trucks and ships, along with methanol; for long-haul jet planes, a very high-density fuel is mandatory, hence e-kerosene is the most feasible solution. All alternative fuel development requires much more standardization of experimentation, planning, assessment, and reporting, such that different methods and production facilities can be compared. Production capacity is growing, but how should we use it? Let's imagine a future in which the only source of electricity is non-fossil, and we have choices about how to use it. Carbon-based regenerative fuels and battery technology may both be used to power vehicles, industrial processes, and store energy: each technology is completely CO_2-neutral on the scale of the atmosphere and the carbon cycle. The only factors distinguishing the technologies in terms of economic suitability are their relative energy efficiencies in full-life-cycle analyses and the costs of the necessary infrastructure. Here, the high energy densities that make portable fuels so practical (an extreme example being the jet airplane), the existence of distribution networks and associated technology, residue-less recycling, and small mineral mining impact, must be taken into serious consideration. We might consider using e-fuels in mobility and trading off their modest energy balances against energy savings in the largest industry sectors.

Biology provides us with important insights into tradeoffs between relatively low-efficiency energy economies (from photosynthesis to muscle contraction) and enormous advantages of using carbon-based fuels for energy storage and mobilization. The parallels with human industry are obvious because we must distribute and accumulate energy stores in convenient forms, bridge energy fluctuations, and prepare for energy shortages. And because of the large quantities of energy involved, we must strive to recover it wherever possible, reduce energy losses, and capture and store as much of the renewable potential as possible. Energy storage is a major problem at present, leading to enormous wasted potential. The argument that e-fuels are energetically costly should be viewed in that context: we could already be using the wasted energy to develop manufacturing facilities and make e-fuels as stores for the regenerative energy that is currently slipping through our fingers. We must seriously consider carbon-based fuels for this purpose. Biology is literally built on systems of energy storage in carbon compounds, and we have learned very little from it.

And yet, the concept of potentially harmful material leaving combustion processes in exhaust gas has little in common with biology. We are starting to understand more about the mechanisms by which air-borne pollutants of all kinds harm us in the short- and long-terms, but we are still not sure exactly which ones pose the greatest danger and how; we have evolved over hundreds of thousands of years with exposure to some of them (notably from ubiquitous organic compounds) but not to less "biological" ones. It is wise to aim for reductions across the board. Here, enormous advances have been made, primarily in the automotive sector, hence the majority of particulates in the air in First World towns increasingly comes from other sources. Other harmful components of combustion technology have also declined greatly in recent decades, and it is now time to concentrate much more on industry-related and domestic emissions.

Ever more emitted CO_2 should be captured at source, where it is concentrated, hence reducing the costs of its recycling into e-fuel. The concept of complete recyclability of fuel/material-in-one is important: critics of combustion technology note the disadvantage of both energy (a lot) *and* material exiting in the exhaust gas. However, the material–energy composite can be just as much an advantage if we understand and use it properly because of its recyclability in one: we get material *and* energy back together, without troublesome residue. Furthermore, we have substances that are key feedstocks for the chemistry industry, hence enabling great flexibility between material use and energy use. Flexibility has an energy price, but it fits us for uncertainty and future challenges. A further tradeoff—one that could genuinely represent some hope in our darkest hour—is the following: markedly, and quickly, reducing consumption of fossil fuels, leaving one half of that saving as non-emitted CO_2, and using the other half of the saving to produce the energy needed to build the infrastructures of the regenerative fuels and energy storage of the future.

If renewable carbon-based fuels are to be part of our future energy, we should also prioritize taking CO_2 directly out of the air and converting it to fuel, instead of using green plants as an intermediate. After all, we are currently emitting greatly more CO_2 per year than all plant life—including agricultural plants—can fix into plant matter. Most of this biomass, as the example of wood shows, is a much inferior source of heat energy than hydrocarbons. As if that weren't bad enough, anthropogenic impacts are reducing natural biological CO_2-fixing capacities yet further. We should invest at least as much effort in halting destruction of existing forests as in planting new trees, and we should gear up our own CO_2 collection technologies. These are major focuses of the next chapter.

5
The call to "decarbonize"
Public perception, hard-to-abate carbon positives, and hard-to-achieve carbon negatives

In this chapter, I address the concept of "decarbonization" and how feasible and environmentally sensible it might be in a variety of sectors. In a sense, decarbonization is yet another in a series of developments that humans have been making increasingly quickly to distinguish themselves increasingly from the "natural" world. The separation between human economies of energy and material and natural cycles is not only increasing in terms of quantity, but also in terms of quality: the materials that we are using in increasing amounts. True decarbonizing signifies the need to build recycling economies that have increasingly less in common with biological and geological recycling. Can this work? Even if "decarbonization" refers only to *net* anthropogenic carbon dioxide (CO_2) emissions, we appear to be failing: the global growth in fossil energy production continues to outstrip the growth of truly CO_2-neutral energy.

Decarbonizing public perception: Of idealistic politics, bizarre consumer behavior, and harsh reality

As politicians the world over—particularly in Europe—proclaim the bold aim of "decarbonizing" economies, it pays to reflect on this and its messaging from two perspectives: (1) What exactly does "decarbonizing" mean? (2) Should we really decarbonize as much as possible? Furthermore, are we aware of the proportions of CO_2 that stem from the variety of everyday human activities and the activities of industry? Is there sufficient awareness in the general public of how ubiquitous carbon-dependent substances are in our lives? Do we realize that many of them are extremely difficult to "decarbonize" or that the confidence radiating from politicians when communicating these goals often belies the reality of not being able to solve one problem without creating another? As we strive (or *should be striving*) toward net zero CO_2 emissions, some strange developments in consumer behavior and thinking are emerging. Indeed, it sometimes seems that the most successful political and industrial efforts today are in decarbonizing public *perception*. At the same time, a tiny proportion of Earth's population produces CO_2 footprints that can only be regarded as obscene.

Killer tomatoes on the rampage: A public relations experience to learn from

In the year 2000, giant tomatoes with menacing toothy grimaces were to be seen parading the streets of some European towns. These were the days of the "Killer Tomatoes," and increasing numbers of the public were in the grips of a fury at what industry had started to do with crop plants for human food and animal feed. Nongovernmental organizations (NGOs), some of which presented themselves as consumer-protection agencies, emphasized the dangers of genetic modification (GM; the acronym also covered "genetic *manipulation*"; rather less flattering!). Industry was discredited, and perhaps that, rather than the mere technology, was the aim of these NGOs' activities; researchers working both in basic research and applied research in molecular biology—a large part of which deals with understanding and changing the genetic "code" of cells—were utterly unprepared for the backlash that these developments would have. It would impact their research, the regulations under which they did it, and their funding. The so-called Swiss Referendum of 1998 narrowly averted a disastrous outcome for basic and applied research in the life sciences alike, but it obviously had to come to such a massive debate to solve the matter.

Whether one agrees with genetic engineering technology or not, the referendum's outcome also helped preserve applied genetics research in Europe and GM technology in the form of companies based in, or with operations in, Europe. As some observers of these developments commented, that was a good thing from the perspective of keeping a regulatory eye on the business: "If you're not playing the game (anymore), you can't make the rules"—that is, if you abandon the development of a technology unilaterally, you might still need it, but you won't be involved in deciding the form that it takes. At present, in Europe, we see probably the hardest political stance on CO_2 emissions, particularly from the transport sector: the aim is, indeed, to decarbonize that sector, starting with cars, and seeing how far we can go. In the midst of this political fury, there is a tendency to demonize hydrocarbon fuels of any kind, regardless of whether they are from fossil reserves or not. Certain environmentalist organizations even claim that to preserve the internal combustion engine (ICE) and its fuel supply infrastructure would be dangerous because it would help prolong, rather than curtail, fossil fuel mining. But have they thought enough about the potential for *positive* activism *for* e-fuels and hydrogen? Europe may not embrace e-fuels in large measure, but others surely will, and there are already signs that e-fuels are contenders in the race for a CO_2-neutral portable energy source for many sectors of transport.

Many consumers back in the 1990s were gulled into thinking that one could make gene-free vegetables, and that those were better than the "gene vegetables," as the engineered variety had become known in certain quarters. In 2000, I had been living in Germany for one year, and I had observed the behavior of consumers and sellers as part of my job at the European Molecular Biology Organization. Strolling through the aisles of a local organic supermarket one day, I spotted a sign sticking out of a heap of healthy-looking tomatoes stating "Gene-free tomatoes" (German: *Genfreie Tomaten*). I smiled and wondered how long they would stay so healthy-looking compared with the genetically engineered Flavr Savr tomatoes. I then asked an assistant whether the store also sold "natural" tomatoes (i.e., ones *with* genes): he looked at me as if I was tired of life and said "what could be more natural than gene-free?" Here we see an interesting parallel in the perception of clean energy: decarbonized anything is seen as better than carbon-containing, even though carbon-based energy carriers are the most natural thing in the world. As with genetic engineering, we can abuse the existing science, or we can use it reasonably and with restraint. The last concept, restraint, is particularly important, for no energy source on Earth, regardless of its technology, will slake a thirst for ever increasing industrial production, travel, heating, etc.

So, genes were OK as long as they didn't find their way into our food, and carbon is OK, as long as we don't use it as an energy carrier. Incidentally, the "Gene-food? No thanks" sticker that proliferated in Germany had a predecessor in the form of an anti-nuclear one from the 1970s (Figure 5.1).

Figure 5.1 Anglicized version of the 1970s anti-nuclear-power sticker in Germany. The original read "ATOMKRAFT? NEIN DANKE."

However, the harsh reality of our present energy requirements, urgent need to reduce net CO_2 emissions, and slowness at developing regenerative alternative energy is now bringing nuclear energy back: perhaps we will need to rely temporarily on a certain percentage of nuclear energy simply because—despite its environmental problems—it has very low CO_2 emissions per kilowatt hour (kWh) of energy generated.

Psychological games with carbon: How we are persuaded and persuade ourselves

In a psychological sense, if we "decarbonize," we remove the object that we have "abused," similarly to imagining that we could have gene-free vegetables: if they have no genes, then we can't have abused anything within them. However, decarbonization goes even further in the public mind: carbon is seen as plain bad. Whereas the Swiss Referendum on genetic engineering was provoked by the Initiative to Protect Genes (*Genschutzinitiative* in German), few people wish to protect carbon, regardless of its origin, and that threatens to produce an even more radical outcome. But some people—similarly to those who find animal slaughter off-putting but will happily buy a plastic-wrapped steak at the supermarket—will gladly use carbon if it doesn't *seem* unsustainable (e.g., the example of wood burning explained in Chapter 4); alternatively, they don't *realize* that they are using carbon.

Wrong-thinking can have serious side effects, particularly when cemented by devious advertising, political, and industrial communication strategies. Probably the most absurd is the claim that battery electric vehicles (BEVs) are CO_2-emission-free, period. In almost all BEV advertising that contains this claim, it is made as an absolute statement rather than qualified in terms of "as measured by tail-pipe emissions," or "provided that you recharge with renewable electricity." If there *is* some qualification, it's linked with an asterisk (*) and some small print at the bottom of the advert. I recently read that a particular BEV of the SUV variety produced 114 g CO_2 per kilometer if charged with the average electricity mix in Germany;[1] interestingly, there are many small-size ICE vehicles (ICEVs) purchasable that produce less than 100 g CO_2 per kilometer, and several of them weigh in at around 0.4 times the weight of this SUV (less material required) and have a better range. Could consumer behavior here be partly a victim of a value judgment that is hard to justify? I won't elaborate on the matter of battery and drive-train manufacture here because it's laid out in Chapter 6. Instead I'll explore the carbon balance of large parts of the modern car that are unrelated to the drive-train. The reason why I focus on the car is that, second to a house, it is the largest object that most people buy in their lifetime. As the steel industry struggles to reduce its dependence on carbon, most other parts of a typical car *consist* of carbon compounds. Far from being able to survive without carbon, the car industry is in a clinch with the element: it goes all the way from the crude oil needed to make automobile plastics through to the emission of CO_2 when these components are burned at the end of a car's life.

The main plastics that we are talking about are:

- *Polypropylene*, mainly in bumpers, fuel reservoirs, and carpet fibers. It is a thermoplastic, cheap for its strength, easy to mold, and has high resistance to chemicals, ultraviolet (UV) light, and impact.
- *Polyvinyl chloride* (PVC), in dashboards, door linings, and other interior moldings, but also for certain exterior parts. It has good flame-retardant properties and can make rigid or flexible structures.
- *Polycarbonate*, mainly in bumpers and headlight lenses because of its impact strength and UV and chemical resistance. It is also light for its strength.
- *Acrylonitrile butadiene styrene* (ABS), mainly in steering wheel moldings and dashboards, but also in body parts because it is tough, and, in the event of an accident, it deforms and absorbs energy rather than suddenly snapping.

In 2020, the average car was estimated to contain 350 kg of plastics, having increased from 200 kg in 2014. Carbon fiber usage (with plastic resins) is also increasing in response to the need for lighter vehicle structures: the 2030 prediction is 9,800 tonnes per year, almost three times the 2013 figure.[2] Carbon-fiber–based composites are notoriously hard to recycle: most of this waste is burned or buried. After the packaging and the construction industries, car manufacture is the third largest consumer of polymers. This sector is certain to grow in demand per year, particularly as EVs become more popular and the pressure to reduce their total weight increases.[3] Even now, across the whole fleet, fuel savings of lighter vehicles are clear: a 10% reduction in weight equates to a decrease in fuel consumption of between 5% and 7%[4] (incidentally, EVs are harder to "lightweight" than ICEVs—not only because of the hefty battery, but also the metal frame needed to keep it safely in place in the event of an accident).

The state-of-the-art method for dealing with plastics from old cars is basically a combination of recycling and burning. Heavy plastics are easily separable from lighter synthetics and can largely be remelted into pellets for reuse in a variety of products, including new cars. Foams, rubber, and light plastics are converted to gas via pyrolysis and burned to produce energy in steam turbines. Metal recycling from cars is very efficient[5-7] and very desirable because these particular metals are more costly to mine and smelt than recycle; the same logic does not apply to most plastics, however. Depending on location, 75–90% by mass of a conventional car is recycled (aluminum and steel to 90%); most of the remainder is shredded and buried in land-fill sites, and almost all of that is plastic and other synthetics (automotive shredder residue accounts for up to 10% of hazardous waste in Europe[8]). A growing challenge is separating polymers from metals present in the increasing amount of electronics found in cars. Anyone who thinks that the car is on the way to sustainability via decarbonization of materials had better think again, because the car industry is facing massive sustainability challenges.

Waste incineration: A sizeable percentage of anthropogenic CO_2 emissions

The CO_2 footprint of plastics: An ever-increasing challenge

Despite what we see in certain defined industries, the overall picture of plastic recycling is not rosy. Estimates have it that, in 2019, the United States burned six times as much plastic as it recycled:[9] around 40 million tonnes of plastic waste are produced each year in the United States, and around 5.5 million of those (13.5%) are burned with recovery of energy (electricity and heat); most of the rest—70%—is land-filled.[10,11] The burned plastic produces 20.3 million tonnes of CO_2 (see Chapter 5, Calculation 1, online at https://doi.org/10.1093/oso/9780197664834.001.0001). Most of the polyethylene terephthalate (PET; see Figure 5.2) in bottles (ranging from drinks to washing liquid) is recyclable, but, despite reports of greater than 90% recycling, much, even in Europe, becomes "downcycled" to lower-grade products that are ultimately burned or buried.[12]

From a variety of sources, the current annual anthropogenic CO_2 emissions in the United States are around 5 Gt, around half of them resulting from electricity generation and industry combined. Hence the CO_2 from plastic burning is approximately 0.4% of the total. Not very much, you might say, but plastic is only part of the waste that most countries burn to generate electricity and heat. Globally, plastic waste production is predicted to develop from around

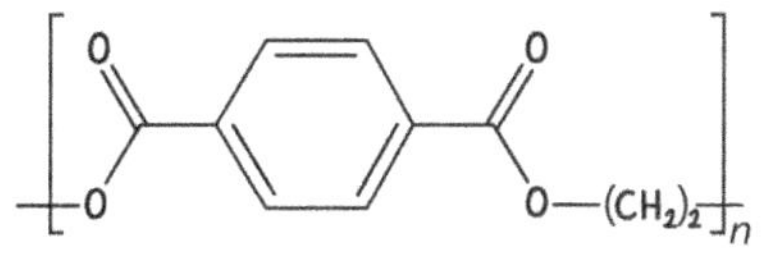

Figure 5.2 The repeating unit of polyethylene terephthalate (PET), where "n" denotes an undefined, variable number of units.

220 million tonnes in 2016 to 420 million tonnes in 2040, and the proportion burned as compared with recycled is not expected to change.[13] Humanity is so dependent on plastic products that experts predict 20% of fossil oil being used for their manufacture by 2050,[14] equating to 930 million tonnes. That's more than 60-fold greater than in the 1960s, representing an increase from 5 kg to 90 kg per person per year.

Domestic waste incineration produces at least 2% of CO_2 emissions

Not all waste produces as much CO_2 per tonne as plastic: it's estimated that combustion of general domestic waste makes net emissions of between 0.7 and 1.7 tonnes of CO_2 per tonne. Studies on combined heat and power stations running on domestic waste suggest an average value of around 1 tonne of CO_2 per tonne of incinerated refuse. The European Union is not entirely blind to the contribution of waste incineration to climate change,[15] noting that, in 2017, around 70 million tonnes of so-called municipal solid waste (MSW) were burned in the European Union—a rising trend. Even if it's still 70 million tonnes today, that translates to around 70 million tonnes of CO_2 per year. Of the 3.5 billion tonnes per year in total emitted by the EU countries, that's an amazing 2%—not much less than the CO_2 produced by burning kerosene in passenger jet engines in 2019. Incidentally, this figure only accounts for *municipal* waste: we haven't even considered industrial waste, not to mention waste from defined end-of-life-cycle processes such as scrapping of cars. We might be tempted to consider CO_2 emissions from burning waste as merely a part of the combined anthropogenic and natural carbon cycles—and possibly marginally better than land-filling it. Tipping waste into land-fill sites and waiting for slow decomposition brings many other problems. Tapping into the methane released during the process is a possibility, but any scenario involving deriving energy from waste must confront a reality: burning waste, or the methane from waste, amounts—in large part—to burning products made with fossil energy/material.

Are there environmental arguments for waste incineration?

The environmental argument for incineration—particularly the now-common high-temperature fluid bed combustion technology—is that if tipped into a land-fill site, not only would the waste create local pollution of earth and groundwater, but much of it would be decomposed microbially, releasing an amount of methane equivalent in greenhouse effect to 1.35 times the CO_2 emitted by combustion. One could find fault in this comparison because a given quantity of waste is combusted relatively quickly, essentially releasing the greenhouse gas (GHG) almost instantaneously into the atmosphere, where its half-life is greater than 100 years; the land-fill site, by contrast, releases methane slowly, and that methane has a half-life of about 9 years.[16] However, because methane's climate forcing potential over a 100-year period is 30 times that of CO_2, perhaps we don't want to take the risk—moreover, methane in the atmosphere is ultimately converted to CO_2! Releasing methane is always worse than releasing CO_2.

Students of atmospheric methane will be used to the controversy surrounding the sources. In 2020, methane emissions were reported as having been vastly underestimated.[17] According to this article in *Nature*, the geological emissions of methane (as judged by the percentage of methane in our atmosphere containing the ^{14}C isotope) may be 10 times more than previously acknowledged. The researchers believe that most of the increase in geological methane since the late 1700s results from "leaks" and other emissions associated with fossil fuel mining. Other researchers favor benthic volcanoes in our oceans and lakes as the main source. Regardless, we need to identify and reduce methane emissions urgently because they exacerbate an already desperate situation.

Tying us into a new, unsustainable energy system: Waste supports both legal and illegal economies

The thing about burning refuse is that it has long since graduated from a means of making it disappear to a means of generating much-needed electricity and heat: ergo, we need these incinerator power plants. But we also need to be doing something that we hitherto haven't: factoring the CO_2 that is emitted from these carbon sources into the other sources of CO_2 produced by burning fossil-derived products. The reason is clear: much of the incinerated rubbish has been produced either directly from fossil oil and/or gas, or it has been grown (fruit, vegetables, animal products) with the help of fossil fuel—the whole fossil-fuel–driven agricultural planting, fertilizing, growing, harvesting, and distribution system. Seen yet another way, every time that we waste something that could have been eaten, we contribute to unnecessary fossil-fuel–derived CO_2 emissions.

As suggested earlier, waste incineration for power or heat generation is largely fossil fuel burning with a small delay introduced. Yet more worrying: the energy needed to create much of the products that, in their own time, end up as waste, is mostly of fossil origin. The bottom line is that it's not OK to consider "energy recapture" by burning waste, or burning the methane from waste, a good practice in general: it acts as a kind of "excuse" for quantities of material that we should never have generated in the first place. As we will see later in this chapter, this has turned into an enormous blessing for the cement industry. There is no way to justify the "combustible waste" concept because it goes hand-in-hand with levels of production and consumption that are intrinsically unsustainable (in terms of how many Earths we currently use).

Recycling of materials can currently be done in innumerable ways, and we're very far from the end of our technical imagination: you might already have guessed (if you didn't already know) that technologies developed for turning methane into liquid fuels can also turn methane into plastics. If it's not fossil methane, the product will simply be more expensive. Indeed, even downcycling, such as making mixed unrecyclable plastic waste into railway sleepers (the beams that carry the tracks), is not profitable, and companies that do it survive largely on government subsidies. Most plastic waste (yes, most plastic, by mass, that we put into recycling bins) seems so worthless for producing new objects that a different industry has long since taken off—taking waste for money: as many countries officially close their doors to this "trade," others stay open. Furthermore, large networks of illegal and semi-legal plastic disposal have arisen to "help" get rid of massive quantities of mixed, un-recyclable plastic: you pay me, and I'll take care of your trash . . . often by burning it in the open or tipping it into a landfill.[18] Any number of customer-facing labels on plastic packaging blind us to the realities that most of this material is not recycled or even downcycled.

As in the global economy of electricity generation, competition for the cheapest method of cycling synthetic materials in some useful way—rather than tipping them into land-fills—is a race to the bottom: economic competition, or might we already call it "warfare," between the United States and China leads to some very sobering realities: to survive in a world where an economy can so easily fall into the hands of a politician who cares naught for the environment, other economies are naturally obliged to apply the same measures. Recycling *is*, in very many cases, more expensive than de novo production from newly mined or created raw materials, and almost everything that we consume *should* cost more if it is to be produced with sustainability in mind.

If burning synthetic waste (e.g., for cement production, heat generation) is considered the techno-economic solution for certain parts of industry, then we will *have* to recycle the CO_2, because (1) we can't let it contribute to increasing atmospheric parts per million (ppms), and (2) we must transition away from fossil sources of the raw ingredients: hence we will "need" high-concentration CO_2. Whether recycling is at the level of old-plastic-object → new-plastic-object or at the level of CO_2-from-burned-plastic-object → new-plastic-object-via-chemical-synthesis matters less than the concept that it is *full recycling with respect to net CO_2 emissions*. Hence in a CO_2 recycling economy, burning waste (and collecting the high-concentration CO_2) is not per se bad, but it must be judged in efficiency against *material recycling* and decisions that we make as to which synthetic substances we use and in which proportions (i.e., ones that we *know* will be recycled and ones that we *know* can only be burned because they're very hard/impossible to recycle satisfactorily). This must be a very conscious,

THE CALL TO "DECARBONIZE" 187

systematic, and technically thought-out economic development: nothing can be left to chance.

How risk-blindness and complacency develop from ubiquity

No precautionary principle here!
GM organisms appeared, at least to the general public, as a very sudden and unusual development. An important consequence of the great GM debate in Europe and elsewhere was the development of a practical implementation of the "precautionary principle" (PP). The PP draws on risk assessment to advise on behavior. The risk assessment part is basically a multiplication of the risk of an event by the severity of the event. If an event is very unlikely but has very severe effects when it *does* occur, the product of the calculation is a medium value; an event that is very likely, but has small consequences, would likewise produce a medium value. If an event is quite likely and has quite serious consequences, it produces a relatively high score. One perception at the time of the GM debate was that certain GM technologies stood a palpable chance of "escaping" from the confines of the plant, field, or test area and that the consequences, while not utterly disastrous, would be quite bad for the environment and/or human well-being (e.g., antibiotic resistance in pathogens). Such a scenario produced a relatively high risk rating, hence the PP was implemented in a relatively strong form.

The renewable energy and sustainable technology revolution, as I will call it, is a much larger leap than most genetic engineering, the roots of which go back thousands of years to the first human selection of favorable qualities in cultivars. And yet few people in rich countries perceive their lives changing much compared with 20 years ago. In the next 10 years, we will still be surrounded by massive energy-intensive objects such as houses, road vehicles, ships (even more of which will be cruise ships), airplanes etc. Anyone who has read even half of the literature describing the energy requirements of production (let alone use) of such objects cannot avoid feeling queasy: that in radically changing energy sources, we are embarking upon something with likely environmental impacts, some possibly even larger than conventional technologies. Material fluxes are a major factor, but we have been distracted from those by the matter of energy production and CO_2 emissions. However, energy production and material fluxes are intimately linked: the more we artificially "churn" material around between the Earth's surface layers and the biosphere, the more energy we need. The consequences of "getting it wrong" in terms of sustainability are horrendous: the greatest environmental disaster we have ever seen. So where is the PP in *this* technology scenario? Has it slipped out of the picture because we are so attracted to keeping our lifestyles pretty much as they are? And if we are really going to try to decarbonize as much as possible, are we aware of the scale of the challenge?

The ubiquity of synthetic carbon-based products in our lives

Basically, almost everything we touch contains synthetic carbon-based compounds. Let's start with clothing: though an increasing proportion of clothes contains, or is even entirely made from, recycled synthetic fibers, the overwhelming majority comes from chemically modified components of crude oil. Not wishing to belabor a point that might be obvious, when we say "synthetic," we basically mean derived from chemically modified components of crude oil. Of course, the recycled proportion of what is mostly referred to as "plastic" or "synthetic polymers" originally came from crude oil, too.

Now let's inspect the house or flat in which we live: to be fair, if it's mainly built from wood or concrete or bricks, then by mass it doesn't contain much fossil-derived hydrocarbon-based substances. However, for important parts of the interior and exterior, most houses have synthetic materials. Under your floors run pipes, increasingly made of polyethylene (PE) or polyvinyl chloride (PVC), and all of the wiring in your house has insulation made from PVC, PE, or polypropylene (PP) (Figure 5.3). Most floorings that are not stone, tile, or natural wood are made of some form of polymer: 97% of carpet by the meter is, on average, synthetic[19] (nylon, olefins, polyester, and acrylic, often in mixtures). So-called "click parquet" for flooring is made of several layers of wood bound with synthetic glue. If the customer wishes a very quiet floor, the parquet will be stuck to the underlying cement with synthetic glue as well.

Domestic glues are increasingly water-based, hence making them safer; however, the part that does the sticking is just as much a product of the oil industry as was the smelly solvent-based adhesive of the past. The components of adhesives, sealants, and paints that give them their crucial properties are almost exclusively polymers of carbon compounds from the petrochemical

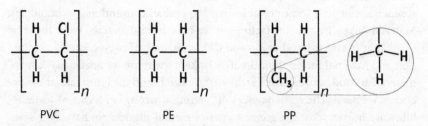

Figure 5.3 Polyethylene (PE), polyvinyl chloride (PVC), and polypropylene (PP). "n" stands for an undefined, variable number of repeating units.

industry. Interesting exceptions are silicone-based sealants and rubbers, but, as we saw in Chapter 1, the combinatorial chemistry of silicon is very limited compared with that of carbon, and so in the world of polymers, carbon is king—by a long way.

The paint on the walls of your house (inside and out) is most likely an emulsion of a synthetic oil-based component that, after evaporation of water during drying, is no longer water-soluble. The paint on wooden parts of your house's exterior has always been a product of the petrochemical industry, polyurethane (PU) in particular having been used for decades because of its excellent weathering properties. PU is increasingly being replaced by water-based paints—often alkyl resins—but, as with water-based glues, the crucial component is still a product of the petrochemical industry. Acrylic paints have expanded in applications enormously over the past 20 years, and many exterior and interior wood varnishes are now water-based acrylic: essentially an acrylic polymer emulsion that, when dry, is no longer water-soluble. Acrylates, so popular today in a variety of applications from paints and adhesives to transparent paneling, are one of the simplest synthetic polymers (they are made by reacting an acrylate ether with a methacrylic acid). Your house may have PVC cladding on certain parts, PVC-framed windows, and/or guttering made of PVC. Finally, you might even have a thick thermally insulating layer of expanded polystyrene or PU foam on the outside of the concrete or brick walls.

We take the fossil-oil origin of other objects in our lives for granted, and perhaps we don't even know that they came from oil: almost all plastics; detergents and soaps; antiseptics; medicines and drugs; creams; solvents; lubricants; candles; car tires; coolant fluids; almost all kinds of glues, paints, and sealants; dyes and other colorants—the list is enormous. We have become utterly dependent on products made from carbon that currently comes from fossil deposits. Can we get away from them? Very unlikely; instead, the oil and petrochemical industry will have to find non-fossil equivalents.

How can we reduce and replace the fossil origin of raw ingredients?
It's not only the global economy that's in a clinch with carbon. A large proportion of almost every consumable product that we buy contains carbon compounds that risk entering the atmosphere sooner or later as CO_2 or methane. Hence, each of us must try hard to reduce that eventuality to a minimum. There are only three ways to do that, and two of them largely overlap.

1. Reduce consumption of absolutely everything.
2. Maximize the life span of everything that we use, hence overlapping with 1.
3. Choose disposable substances (e.g., packaging of daily items) that have the smallest impact when they end up in refuse.

The reality of our relationship with carbon as a useful element is that it will be with us for as long as we are around. We must realize this as an immutable truth, otherwise we will not rise to the challenge of living sustainably but rather live with the ridiculous illusion that one day we will become a carbon-free civilization—through technological innovation. However, as consumers, we can, through our purchasing behavior, put pressure on industry to use what I call the "anthropogenic carbon cycle" (recycling through closed, net-zero-escape, pathways) rather than the geological carbon cycle as a source of carbon compounds for its products. As the global population rises, this becomes all the more important in the context of the increases in flux through the human side of the larger carbon cycle. It will be no cheap solution, and the rich countries of the world, which have—to a substantial extent—become so rich because of cheap oil, will find themselves not so wealthy.

In the United States, 80% by weight of a barrel of crude oil goes into making fuels; the rest goes to the chemical industry[20] (Table 5.1). Though the figures in Table 5.1—which, in total are 8.8% higher than in 2020—are for the United States, they are roughly representative of most industrialized countries. The largest combined sector is combustion fuels for transportation and gas for domestic and industrial heat-requiring processes (HGLs). Interestingly, although jet engine kerosene is hailed as an urgent candidate for replacement by synthetic liquid hydrocarbon fuel, it amounts to less than one-third as much as the diesel fuel and heating oil burned in industry and homes. Note that asphalt and heavy oil for roads consumes almost 2% of the crude oil, and petrochemical feedstocks (components used for making everything from plastics and rubbers to paints and pharmaceuticals) around 1.5%. The total percentage of the crude oil that is not "burned"—be that in transport, heating, or industrial production—is around 5%, or, seen another way, around 95% of mined oil is burned in one way or another. However, we know that quite a lot of the 5% that is not ostensibly burned *is* actually burned as waste. Hence, in comparison with demand for fossil fuel as energy sources, the material (feedstock) demand is relatively small. We should, therefore, be able to accommodate it in a strategy involving the following components:

1. Longer use spans for all products of industry
2. Massively more recycling of all human-made items
3. Feedstock production from syngas (via methanol, methane, and other hydrocarbons) from a variety of sources, including steam-reforming (pyrolysis) of waste plastic and other refuse
4. Substantially lower consumption of short-cycle products

Table 5.1 Petroleum products by type and volume consumed per day in the United States (2021)

Product	Annual consumption (million barrels per day)	Liters per day	Percentage of total
Finished motor gasoline[a]	8,795	1,398,405	44.58
Distillate fuel oil (diesel fuel and heating oil)[a]	3,943	626,937	19.99
Hydrocarbon gas liquids (HGL)	3,410	542,190	17.29
Kerosene-type jet fuel	1,371	217,989	6.95
Still gas	642	102,078	3.25
Asphalt and road oil	370	58,830	1.88
Residential fuel oil	313	49,767	1.59
Petrochemical feedstock	289	45,951	1.46
Petroleum coke	269	42,771	1.36
Lubricants	104	16,536	0.53
Other petroleum products	211	33,549	1.07
Special naphthas	42	6,678	0.21
Aviation gasoline	12	1,908	0.06
Waxes	6	954	0.03
Kerosene (domestic)	5	795	0.03
Total petroleum products	19,728	3,136,752	100

[a] Includes biofuels. Data source from U.S. Energy Information Administration.[21]

I cannot overemphasize Points 1 and 4 as the components on which we must work most, and most urgently.

Oil's geological cousin, coal, ends up as CO_2 even more completely than oil (>99%): the proportion that finds its way into longer-term products is minuscule. Almost 60% of global coal is burned for electricity generation, and here renewable electricity production must make great inroads. However, 16% of coal is used mainly to convert iron ore into iron and then steel, and 12% in the cement industry. Steel and cement are the two largest so-called *hard to abate areas* of net CO_2 emissions, as we will see in the sections that follow.

Hard-to-abate areas: Sectors where fossil fuels are very difficult to replace

Steel manufacture: The quest for an affordable energy source and chemical reductant in one

The problem here is two-fold: (1) steel manufacture from iron ore (primary production) requires, in a single source, a chemical reducing agent and a lot of energy, both of which are conveniently provided by coal (coke); (2) to replace such a quantity of energy and reducing power from renewable energy sources is an enormous challenge involving massive new infrastructures. I'm now going to highlight the German situation, because Germany is quite representative of a medium-sized wealthy country with one of the highest proportions on Earth of renewable energy in its total electricity mix (upward of 40%). It is also Europe's largest producer of raw steel, at a primary production (i.e., excluding recycling) of around 30 million tonnes per year (2020 figure).

In 2021, Germany produced an estimated 42% of its electricity demand via renewables.[22] However, that is just electricity; when seen as percentage of the *entire energy* demand, it shrinks to 6.8% (225 MWh[23]). Other renewable energy brings the total "renewables" up to around 16% of total primary energy demand. Germany's primary energy consumption in 2021 was 12,193 petajoules (12,193 billion megajoules)[24]; its 30 million tonnes of primary steel cost 720 × 10^9 MJ (720 billion megajoules) of energy, based solely on the quantity of coal consumed per tonne of steel (see Chapter 5, Calculation 2, online at https://doi.org/10.1093/oso/9780197664834.001.0001). That represents 6% of all energy (recycled steel requires only 20% of that energy, but global demand for steel is still outstripping several-fold the quantity that we can recycle).

Now, 6% doesn't sound like much, but we basically took that energy, in high-density form, almost for free, out of the ground as coal. Steel's 720 billion megajoules equates to almost one third of the renewable energy in Germany, so the question is, if we replace coal with something else, will that something else create a new, 720 billion megajoule demand on renewable electricity? Not quite, and to appreciate why, we must break the process of steel-making down into two chemically different steps.

1. Reduction of iron oxide at high temperature, typically in a blast furnace (Figure 5.4), which produces pig iron
2. Incorporation of tiny percentages of carbon into the pig iron, which produces steel of various grades, depending on the percentage; this must also be done at high temperature, hence requiring substantial amounts of energy

Step 1 can be done with any suitable high energy content reducing agent that can produce heat in an appropriate reaction (most conveniently, combustion with

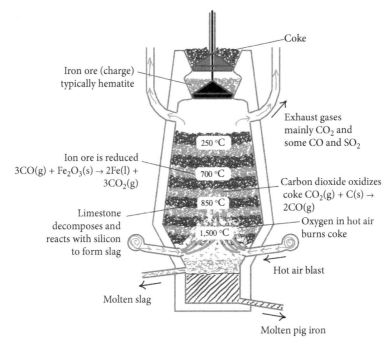

Figure 5.4 Essential principle of a blast furnace. Extremely hot air is blown from below through layer upon layer of a mixture of coke and iron ore, both of which are loaded periodically from the top of the furnace stack. Oxygen burns the coke to produce CO_2, which, at such high temperatures, reacts with the coke to form CO. CO, being a stronger reducing agent than Fe_2O_3 (iron ore) reduces the iron ore to iron (Fe) and is in turn oxidized to CO_2. (g) = gas; (s) = solid; (l) = liquid.

oxygen). Hydrogen is chemically and environmentally very suitable because it is a good reducing agent and it burns to produce enough heat for the reduction reaction, producing only water as a by-product.[25] However, present-day hydrogen generation technology makes H_2 anything but suitable for large-scale iron production compared with coal: apart from requiring large amounts of electricity to produce, it also needs a massive new storage and distribution infrastructure. Step 2 requires less carbon and less energy, and the solutions are somewhat easier: for example, the "biochar" process uses waste wood as a carbon source,[26] and there is even talk of using old tires, although their combustion would produce air pollution challenges of its own. Still, Step 1 is by far the larger challenge, making it very hard to transition iron and steel production to zero net CO_2. Calculating the hydrogen needed to create the energy to make 1 tonne of primary steel can't be done using the energy involved in the coal-fired process because the reactions involving hydrogen are different and require quite different types of furnaces.

Theoretical calculations indicate that hydrogen-powered steel production could be significantly more efficient than coal-fired because hydrogen releases more energy per reaction with oxygen and has a higher reducing power than carbon monoxide.

The—currently theoretical—lower energetic costs of hydrogen-fired steel production are around 2.8 MWh per tonne of steel (2.04 MWh hydrogen plus 0.75 MWh electricity).[27] However, we have to make that hydrogen—electrolysis of alkaline water being the most practical large-scale method at present, with a maximum energetic efficiency of 80%. Hence to make 2.04 MWh of hydrogen, we would need 2.55 MWh of electricity (or 3.6 if we decided to cool and compress the hydrogen for transport). The total energy of the hydrogen-driven process (i.e., including electricity delivered directly to the smelter) would be 3.3 MWh per tonne of steel with gaseous hydrogen (see Chapter 5, Calculation 3, online at https://doi.org/10.1093/oso/9780197664834.001.0001), and 4.35 if we use liquid hydrogen. A survey of the literature on pilot projects (i.e., not just theoretical) suggest a range of 3.6 to 4.7 MWh per tonne of steel. Taking an average value of 4.1 and multiplying by the 30 million tonnes of steel, we would need 123 TWh, which is a little more than 53% of present regenerative electricity in Germany.

The bottom line is that we can, purely in terms of process energy, make steel production more energy-efficient by using hydrogen instead of coal; however, making that hydrogen places an enormous new draw on existing electricity generation infrastructures. To get steel manufacture into proportion, the chemistry industry in Germany (one of the largest in the world as a proportion of total industry) consumes a greater mass of carbon-based compounds per year than does steel production; however, 22% of that consumption is not for energy, but rather the material of the products made. That is an enormous quantity of material that we need to synthesize from sources other than fossil. If we take the chemical industry's 22% materials out of the equation, it raises metal production and processing (mainly steel) to the largest user of primary energy in Germany.[28]

Cement manufacture: Transitioning toward waste as fuel?

Germany (home to the world's fourth largest cement company) produces about 35 million tonnes of cement each year and much more abroad. Public pressure to reduce the cement industry's environmental impact has led to the collection of unusual amounts of data on German cement production and an unusual degree of technological innovation to reduce energy requirements and CO_2 emissions. German cement production consumes (using low-energy technology) around 4,430 MJ (4.43 GJ) of energy (combustion heat and electricity) per tonne of clinker produced.[29] Hence Germany's annual cement energy budget is around

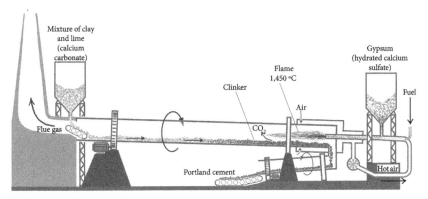

Figure 5.5 Principle of a rotary cement kiln. A mixture of clay and limestone (calcium carbonate: $CaCO_3$) is fed in at the top, dries while moving down the slightly inclined rotating pipe (the kiln), and upon reaching the flame at around 1,450 °C decomposes to calcium oxide (CaO) (the main constituent of "clinker") and carbon dioxide (CO_2). The fuel for the flame can be gas, oil, or pulverized solid fuel (e.g., coal and/or waste plastic). In Portland cement manufacturing, finely powdered gypsum ($Ca[SO_4] \cdot 2H_2O$) is injected with the fuel and combines with the clinker to form a cement that has excellent water-resisting properties.

43 TWh. Cement production (Figure 5.5), similarly to steel production, is extremely hard to electrify because of the high temperature needed to turn calcium carbonate into calcium oxide—typically above 1,400°C. Making electric heating elements to withstand very-long-term use under such conditions is an enormous challenge. Here the concentration of energy in hydrocarbon fuels comes into its own because they can deliver enormous amounts of concentrated heat via relatively simple and robust technology. This is, of course, the principle that makes steel and cement production "hard to abate." Natural gas is already used in some cement works, and (bio)synthetic methane would potentially be a replacement. How much electricity would we need to make the 43 TWh-worth of methane that German cement-making would need each year? Taking a realistic value of 40% energetic process efficiency for methane production from cement factory flue gas and hydrogen made from electrolysis of water, we would need 123 TWh ($100/40 \times 43 = 108$ TWh) electricity—around half the renewable electricity currently produced annually in Germany.

Pure hydrogen is being piloted in some cement works, but combining H with C to make methane instead is advantageous because it creates an energy carrier for H that is much more convenient to deal with (liquid or easily compressible gas). Whether we can economically break from hydrocarbon fuels for cement-making is far from clear so we must make as much use as possible of the heat produced, for example for other industrial processes and heating of buildings.

An alternative source of methane for cement-making is biomass decomposition in dedicated facilities. Here, Germany currently produces enough biogas to generate 7% of its electricity needs, or 0.47% (57.3 petajoules or 15.9 TWh) of its *entire* energy needs. And indeed, almost all of the German biogas, as in all other countries, goes into electricity generation. Cement-making is as much a thermal process as that of energy generation from burning gas. Hence we can estimate that the biogas in Germany would potentially power 37% of the present 43 TWh cement industry demand. The annual increases in biogas production in Germany—that were very steep between 2005 and 2011[30]—are now flattening out strongly. The end of the biogas boom seems to have come.

If it's going to be so tough to reduce CO_2 emissions from hard-to-abate sectors that burn natural gas (merely one sector of fossil fuel consumption in Germany), how will other countries with much lower proportions of regenerative electricity ever manage, let alone by 2050? Furthermore, a massive industry of thermal recycling of waste is supporting the cement industry world-over: un-recyclable (or "uneconomical-to-recycle") waste is used in increasingly large proportions to fire cement kilns. As a recent documentary revealed,[31] in Germany, for example, 70% of the cement industry's fuel is from waste of one sort or another—much of it plastic, for which the industry even receives payment from waste companies. How convenient: at once being paid to make heat for one's cement production, and, simultaneously, giving the impression of being less dependent on fossil fuel. Unfortunately, in terms of fossil-derived CO_2, there is no difference because 99%-plus of all of the plastic burned is produced from fossil oil.

Transport sectors: The larger the vehicle, the harder it is to abate

Jetting around the world: A disproportionate environmental impact

Estimates vary within the range of 2–4% of global CO_2 emissions stemming from the aviation industry (excluding military flights) in recent years up to 2019, 2.5% being a credible average.[32] That is a relatively small figure, but aviation is the fastest-growing transport sector, as is continually observed (e.g., in the report issued by 11,000 concerned scientists in November 2019[33]). CO_2 behaves identically as a GHG whether emitted at ground level or high in the stratosphere[34]; however, most other emissions from jet engines at high altitudes have a more pronounced effect on the atmosphere than at ground level.[35]

Synthetic aviation fuel, produced by Fischer-Tropsch chemistry, for example, can be made to have a very narrow band of hydrocarbons (mainly shorter-chain ones) and negligible amounts of aromatics. As a 2011 study states "Particle emissions are decreased in particle size, number density, and total mass when petroleum jet fuel is replaced with the zero aromatic fuels"[36] (and see also Saffaripour et al.[37]). Furthermore, synthetic fuel contains no sulfur, an impurity in natural fuel that also contributes to particulate formation; finally synthetic fuel burns more efficiently, producing greater economy per liter.[38]

So, more than a decade after the 2011 study, where is all this wonderful synthetic aviation fuel? Its large-scale production has not been helped by the overwhelming lack of taxation on fossil aviation fuel (as of May 2022). Only a few countries in the European Union apply small taxes. The cost of €2 per liter (excluding excise duty) for synthetic diesel (see Chapter 4) also applies to synthetic aviation fuel (being very similar in chemistry and manufacture). The price of fossil kerosene[39] has typically hovered around a quarter of the price of ordinary car diesel at the fuel pump and is more than five times as cheap as the equivalent synthetic fuel. In the meantime, Russia's war in the Ukraine has done more than taxation to raise the price. Concerned (if not angry) European citizens seem to be changing politicians' minds about this scandalous situation via a petition to end tax exemption of aviation fuel.[40]

However, while politics takes an agonizingly long time to do the "right" thing, many ordinary people have become enchanted by the idea of electric passenger planes. Politicians seems to be moving much faster in the realm of battery-powered X, Y, Z than anything else (likely a result of the attractively simple conceptual model of electric power)—no organic chemistry or Fischer-Tropsch there—though plenty of electrochemistry and economy of mining and manufacture that politicians need to understand. In 2019, I was sitting in a taxi in Paris discussing sustainable road transport with the driver, who was considering an EV as his next car. "*Ah oui, la future, c'est electrique!*" (Yes, the future is electric!) he exclaimed, continuing "*Vouz voyerez: bientôt nous aurons les avions de ligne electriques!*" (You'll see: soon we'll have electric passenger planes!). We are, indeed, seeing the development of small electric planes carrying a few passengers for city-hopping of a few hundred kilometers—though trains would be much more efficient and environmentally superior for that purpose. However, large-capacity electric planes for longer distances seem very unlikely in the foreseeable future. The reason is the high energy density of hydrocarbon fuel compared with batteries.

Briefly revisiting cars, to achieve the same range as a comparable gasoline-powered ICE car, for example, the EV weighs much more (e.g., a Nissan Leaf weights around 250 kg more than an equivalent VW Golf; see Chapter 5, Calculation 4, online at https://doi.org/10.1093/oso/9780197664834.001.0001). A Leaf with a comparable range to a Golf would weigh more than 2 tonnes (Calculation 4): an impressive weight for a small car. So, if range is a primary consideration—*potentially* for a car user and *certainly* for a long-haul passenger plane—then we have a large challenge with electric propulsion. Despite the material burden, a car can lug around such a weight of propulsive energy; however, rudimentary math shows that this isn't an option for a long-haul passenger or freight plane. I use "rough" estimates and equivalences and some assumptions here, but I am only calculating an *estimate*: Let's take a Boeing 737 (the most numerous passenger jet plane ever built) in its latest incarnation (737 MAX). To a first approximation, the 737 MAX in my calculation (see Chapter 5, Calculation 5, online at https://doi.org/10.1093/oso/9780197664834.001.0001) would need a Li-ion battery weighing 373,800 kg (almost 374 tonnes), which is 4.8 times the maximum laden weight of the kerosene-powered plane. Adding the weight of the plane itself, we have between 420 and 430 tonnes (double the take-off weight of a typical jumbo jet, but with tiny wings in comparison). Even if we did get this behemoth airborne, we would never be able to land it safely because Boeing specifies a maximum landing weight of 69.3 tonnes (i.e., six times lighter than our battery-powered thought-experiment airplane). An incidental benefit of kerosene power in this example is that because it is consumed during the flight, the aircraft has an increased range: it lands significantly lighter than it takes off.

The bottom line here is that energy density is more important for aviation than for any other transport medium. The figure of 17.8 that I mentioned earlier (the equivalence ratio of hydrocarbon fuel to Li-ion battery at current technology) is interesting to explore a little further. The best Li-ion battery so far developed (albeit in the lab) has an energy density of 1.8 MJ/kg,[41] and most in practical use are around 0.9 MJ/kg (Tesla, for example), which is around 48 times less dense than kerosene (43/0.9). However, I calculated a ratio of 17.8, not 48: most of the rest went into heat of combustion that was lost in the exhaust gases. Thus, 17.8 divided by 48 is 0.37, or 37% efficiency, which is very close to the published average efficiency of jet engines. Improvements past the 40% mark[42] are already happening, and recent research suggests that they may be able to achieve greater than 60%.[43] The jet engine of passenger aircraft is far from finished in terms of development. However, alongside investments in research and development of renewable CO_2-neutral hydrocarbon fuels, we should also aim for substantial reductions in aviation miles per year, per head of the population.

The personal impact of flying is largely overlooked. Aviation-related CO_2 emissions might "only" make up 2.5% of total anthropogenic CO_2 emissions, but individual "user" behavior is (as with cars) key because it is only at the level of our personal behavior that changes can be effected. We should not allow aviation-related emissions to rise. If air travel became considered as much a "given" (or even an unquestionable, affordable liberty) as car travel, we would have a monstrous problem. Seen at the level of individual behavior and with calls for us to use cars less (and instead shift to public transport, bikes, etc.), one return trip Europe–US East Coast (15,000 km round trip) would equate to half a year's worth of average driving in Europe*. A return flight from Frankfurt to San Diego (West Coast US), for example, at 23,000 km round trip distance, would correspond to most of a year's driving in Europe; and a return trip Europe–Australia (37,000 km) would be like 15 months' worth of driving that average compact ICE car European-style. Of course, a bus or train doesn't have zero emissions or impact on the environment by any means either, but they are substantially lower.

For airplanes, speed is the killer of efficiency. The air resistance and drag of a freight plane at 930 km/h (cruising speed of a Boeing 747-400) essentially causes more fuel consumption per kilogram transported than a large ship traveling through water that is roughly 2,700 times as dense as the air at 10,700 meters above sea level (typical cruising altitude). The *Emma Maersk*, one of the largest freight ships in the world (397 m overall length), pushing through the waves at a stately 25 knots cruising speed—roughly 46 km/h—burns 1/30th of the fuel per kilogram freight as a plane. We look at ships in more detail next.

Giants of the sea: The most efficient means of transport per unit of energy consumed

We can view the efficiency of ships from many perspectives, but one that is manifest in everyday objects is the price component due to transport. Take coffee, for example. In a 2019 analysis published in the *Financial Times*,[44] the components making up the price of a morning coffee bought at a British café were presented. Of the £2.50, only 0.3 pence—0.12%—consisted of transport costs: the largest are the costs of ships, their maintenance, and diesel fuel for the Atlantic crossing (which are tiny in comparison with those of trucking coffee beans thousands of kilometers across Brazil). As a rather depressing aside, the grower of the coffee beans (a Brazilian, in this case) receives a mere 1 pence—0.4% of the price to the consumer. Don't get me wrong, I don't think it's good for increasing amounts of coffee to be shipped long distances across the world in response to an insatiable desire for the drink in Europe, for example. However, if we're going to transport it such long distances, a ship is the most economical way.

---

* A return flight from Frankfurt International Airport in Germany to JFK in New York (2 × 7,500 km) consumes around 46 tonnes of kerosene; at 200 passengers per trip, that equates to 230 kg per passenger. Those 230 kg of fuel create a very similar amount of CO_2 as the same mass of gasoline burnt in a car's ICE: around 850 kg CO_2 (3.67 × 230, as explained in Chapter 6). 230 kg of kerosene is roughly equivalent in energy to 230 kg gasoline, and how far would that drive a car? With a compact car (say our basic Golf), with average consumption of 5 L per 100 km (3.8 kg / 100 km), we could drive 6,000 km (230 kg / 3.8 kg × 100 km), that is, half a year's worth of average European driving.

A large ship will transport goods at roughly 30 times as little fuel per kilogram as a freight plane (see Chapter 5, Calculation 6, online at https://doi.org/10.1093/oso/9780197664834.001.0001). But, as with cars and any other transportation form, we also need to factor the energetic construction costs of the ship into the efficiency equation. This is extremely difficult to do because data for the whole chain from mining of the iron ore, through to rolling of the steel sheets and welding of the hull, deck, infrastructure etc. is very hard to obtain. Little literature exists in this area, but, in 2019, a Finnish thesis author did a model calculation of the CO_2 footprint of shipbuilding based on a cruise ship of 150,000 gross tonnes (GT).[45] This is much smaller than a typical large cargo ship, which is around 230,000 GT on average; still, it gives us some idea. The conclusion was that just over 101,000 tonnes equivalent of CO_2 are produced, which is roughly the same amount as produced by the burning of ships' diesel in the engines over 1 year's operation (which has somewhat fewer days than a freight ship's). It is estimated that a PanaMax freight ship (the largest that can pass through the Panama Canal) uses around 248,000 L of heavy fuel oil per day.[46] Subtracting non-sailing days, that is around 84,300,000 L or 70,800 tonnes per year. This produces around 262,000 tonnes of CO_2, a figure that is reasonably in line with the figure calculated by the Finnish author for the smaller ship (101,000 gross tonnes).

At a typical lightweight tonnage of 100,000 tonnes (i.e., weight of just the ship and all its necessary structures, without fuel and cargo), a PanaMax ship can hence be estimated to have construction energy costs of roughly 0.708 tonnes of heavy fuel oil equivalent per tonne of finished ship. That works out at around 30 GJ per tonne. If we compare with data used in Chapter 6 for modeling of car manufacture economies,* a freight ship is very similar in construction energy per tonne to a car, but typically has almost double the working life span (25–30 years). Even more strikingly, a car only does useful work for around 2.7% of the time (given an average of 12,000 km per year and average speed of 50 km per hour); a ship is doing useful work more than 80% of the time. A ship's large initial energy costs are very small in comparison with its utility and the amount of material that it carries: even semis (articulated road trucks) come nowhere near. Shipping is clearly a hard-to-abate area because of the practicalities involved (ratio of cargo to overall weight, and very long passages, requiring high density energy sources). I won't discuss battery-powered ships except to remark that there *are* some pilot projects for smaller, shorter-range vessels. Ship operators are under pressure to reduce CO_2 emissions and save money on increasingly expensive fuel. Even sail power is making a small comeback in the form of large solid foils mounted on ship's deck or massive kites flown before the ship. In

---

* A conventional ICEV car: 43.2 GJ, or 12 MWh, per car of weight 1,300 kg, which equals approximately 33 GJ per tonne

combination with combustion engines, these measures might produce up to 20% fuel savings, but they are unlikely to take over the propulsion completely in the foreseeable future.

And so, if you want to use a means of transport that is already going somewhere at very high energy efficiency per weight of cargo, *and you are not in a hurry*, taking a ship would be environmentally much better than flying. Some have done the experiment,[47] but at cost of more than €100 per day (more now, because of increases in fuel prices), this is not likely to catch on in the present culture of ever faster and cheaper personal transport. A ship cruises at around 1/20th the speed of a jet airliner (a figure that is comparable with the 30 times as economical). However, the human craving for faster, better, more of everything, combined with ever greater desire for self-determination, is exceeding Earth's capacity to cope with us. We must make better use of everything in an effort to reduce overall consumption, particularly in hard-to-abate areas. We turn next to road freight.

Giants of the land: An awkward combination of weight, distance, and cost
Large trucks working over distances and use cycles typical of current diesel vehicles are hard to imagine in battery electric form: there is substantial consensus that the ratio of battery weight to payload weight, and the relatively small range compared with diesel semis (articulated lorry), mean that electric semis can't compete on the same terms as diesels. For example, Volvo has started production of battery electric trucks, with gross tonnages up to 44 tonnes and ranges up to 298 km,[48] compared with diesel truck ranges of up to 3,000 km in extreme cases (fuel tank holding 1,100 L). But there is also a lot of talk about changing those terms of comparison to reduce distances traveled by one truck and change use cycles to be consistent with battery charging, possibly also introducing more vehicles to service a particular route. Here, as with BEV private cars, the matter of financial cost is raised as a major obstacle; the environmental cost of battery production is rarely addressed, which is worrying, because this transport sector is still growing fast in response to greater demand for goods per capita and shorter delivery time expectations. Moreover, it is already on an environmentally unsustainable trend. Will cost considerations win? Currently, with a battery weighing 2,750 kg, and range up to 300 km, an electric truck must sacrifice almost 2.7 tonnes of carrying capacity compared with a diesel (110 L of diesel give range of 300 km, and 110 L is 0.83 kg × 110 kg, which = 91.3 kg. 2,750 minus 91.3 kg = 2,659 kg). At around 27 tonnes tractor + trailer weight, a 44-tonne articulated truck can carry 17 tonnes of payload. The electric truck not only has a small range and costs substantially more to purchase, but it can carry almost 16% less load. How would this work economically, particularly if long-haul delivery times become longer and/or more trucks are needed to service a particular route? There's much scope for evolution, and we shouldn't rule it out, but the challenges for battery-electric drives are enormous.

Are we helping hard-to-abate sectors by reducing consumption?

In short: we could, but we aren't. The challenge is enormous. One of the hard-to-abate sectors is billionaires, and one that threatens to become ever more "cosmic" in dimension (as noted in a recent report "Billionaires' Single Space Flight Produces a Lifetime's Worth of Carbon Footprint–Report"[49]). Three-quarters of billionaires in a recent study have a "super yacht," the CO_2 footprint of which is calculated at around 7,000 tonnes per year per yacht.[50] Hence, oft-quoted totals of 8,000 tonnes CO_2 per year are not surprising for many of these humans. The global average per capita, including all aspects of life, is 4–5 tonnes.[50]

On that note, let's get back down to Earth with a more mundane example. As a reminder, if we add together the energy needed for steel and concrete production with renewable electricity, we would need 250 TWh annually to power these sectors of German industry alone—more than all of the renewable electricity currently produced in Germany. Now adding the current energy consumption of the chemical industry in Germany (206 TWh in 2020), we reach 456 TWh, twice the current renewable electricity generation in Germany. And that total is not decreasing. An obvious measure in the face of the very serious technical challenges of reaching net zero CO_2 emissions in these sectors is simply for us immediately to use less. Unfortunately we are doing the opposite, both globally and nationally, regardless of country. In 2020, China produced 1,053 million tonnes of primary steel,[51] up 5.2% on 2019; in 2021, it was 2,500 million tonnes (2.5 billion tonnes) of cement,[52] almost 4% more than in 2020.[53] Steel and cement are excellent examples of problematic economies of energy and material in one because once they become steel-reinforced concrete, they are sequestered for decades and are then energetically costly to disentangle and reuse. Recycled steel requires only 20% of the smelting energy of primary production on average; however, extracting steel-reinforced concrete requires a lot of energy and is far from complete: a maximum of 70% of hot-rolled-steel reinforcement is recycled (in individual cases it can be as low as 50%); the rest goes into landfill together with much of the concrete itself.[54]

The sustainable energy problems associated with hard-to-abate sectors of industry should make us very sensitive to industry's habit of making new things to replace old ones in increasingly shorter cycles. But consumer demand is leading not just to more frequent replacement, but increased annual primary production. Take objects made from steel: car production, for example, is still increasing on average, year on year. Even if more of the steel is recycled, in the larger picture it contributes to greater need for primary production of steel. I am very doubtful of claims that new, more efficient cars (whether ICEV or BEV) are better for the environment in total than older ones. The average usage is greatly neglected in these studies, but even more so the need to encourage less—and not more—use as part of the "solution." On average, I tend to think that it's better to keep the car that one has, drive it responsibly (mainly *slower*!), use it less, and maintain it well.

Tiny steps toward reducing per capita consumption-related CO_2 emissions
The practicalities of reducing CO_2 in our atmosphere by changing our behavior are a Herculean challenge; there are few examples, and they usually miss the larger picture. In 2019, the European Environmental Bureau (EEB) published a study concluding that if every European (European Union) owning a smartphone used the device for one year longer before replacing it with a new one, Europe's net CO_2 emissions would sink by more than 2 million tonnes per year. Add computers, washing machines, and vacuum cleaners to the equation, and the CO_2 savings would amount to around 4 million tonnes.[55] Cars were not included in the analysis, but there was at least an interesting and surprising rationale for including small objects that we take for granted: per mass of product, smartphones are the most damaging of the everyday objects studied. That is largely because of the rare metals (and not so rare ones—e.g., aluminum) that they contain and the large amounts of fossil-fuel energy used to mine, purify, and manufacture components from them. Using phones one year longer would save as much CO_2 as that emitted by all the cars in a small European country, apparently. And when it comes to those cars, one can only imagine the scale of the effect of disposing of a car "too soon." As we will see in Chapter 6—congruent with the French Environment and Energy Management Agency (ADEME) report—the energetic (and environmental) impact of the production of any kind of car, but particularly an electrtic one, is a major start-of-life burden that is greatly underapppreciated. Prematurely disposing of a car and replacing it with a brand new one is an extremely bad energy economy.

And so I turn to consumer behavior as one of the bolts that we need to adjust in the larger socioeconomic machine. It has recently been reported that to prevent global warming of more than 1.5°C, humans need to reduce CO_2 emissions to around 1 tonne per year, per person, as noted in a *Nature* paper from 2019[56] and since substantiated by other studies. That would take us back in time to 1955 consumption habits. How far from the magical "1" are we today? Based on International Energy Agency (IEA) data from 2020, the average per capita emission of CO_2 (gross, not net) per year is 4.4 tonnes, and growing at a rate of 1.1% per year (that 1.1% growth is quite reminiscent of the 1% annual increase in coal-burning in China). This is important to know when we consider the contribution that renewable energy is making to reducing CO_2 concentrations: it is still possible, even with increasing percentages of renewable electricity, to be pumping more CO_2 into the atmosphere—we are doing it *now*. The reason is simple: total renewable energy (electrical and primary) that is currently globally produced is around 10% of all energy demand (5% being electricity*), according to calculations from IEA figures.

Annual global primary energy consumption is rising at between 1% and 2% per year,[57]* and the rise in renewable energy is only around 6.5% (estimates from 2019, extrapolated to 2021, indicate a global production of 7,210 TWh in 2019, growing at about 6.5% per annum[58]). For a 1-year period, the 6.5% equates to a growth relative to total primary energy consumption of 0.65% (i.e., 6.5% of 10%). This is much below the current 1–2% per annum growth in global primary energy consumption. This is, of course, why we're very far from "on track" to meet climate goals. The psychology is interesting: companies that report "greenness" by using renewable energy (often passing most of the extra cost on to the consumer) must be viewed in the context that there isn't at present enough "green" energy to go round. And the way we are going, there never will be. Frantically producing all sorts of new things (mainly with non-renewable energy) to enable a "green" technological revolution is doing the opposite. Massive reductions in consumption are the only solution at present.

As with the distribution of global wealth, a relatively small proportion of the human population accounts for most of the CO_2 emissions. Topping the list is Qatar, with a per capita emission of a little under 40 tonnes, if one extrapolates from 2016 figures.[59] The Middle East on average comes in at around 20 tonnes per capita (IEA figures); Canada and the United States around 15 tonnes; Europe on average is somewhere in the region of 6 (Germany is 8, though); China hovers at around 7; South America at a little less than 3; India at 2.2; and, at the very end of the scale, Africa at around 1. So, it seems that whereas other parts of the world have won the competition for economic development, Africa has won the competition for sustainable CO_2 emissions. Unfortunately, it receives no prize, but rather continues to be exploited for its minerals and fossil fuel deposits in very unsustainable ways (both ecologically, environmentally, and socially). A poignant observation is that, if one includes the whole production chain involved in making commodities (of all sorts) that wealthy countries consume, the wealthy countries' per capita CO_2 footprint increases. Externalization of pollution and CO_2 production is something at which Europe, for example, is good: remember, the European Commission considers emissions only from the non-existent tailpipe of electric cars.

---

* IEA sources 2022 (variety of IEA data):
 – Total global energy consumption 2021: 580 million terajoules, which = 580×10^{18} J, which = 5.8×10^{20} J
 – Global renewable electricity capacity estimated at 8,250 TWh in 2021, which = 2.99×10^{19} J, which = 5.1%

Unfortunately, most recent data indicate that we are in the process of overshooting the 1.5°C mark by 2030 unless we reduce GHG emissions by half until then.[60] I believe that a neck-breaking acceleration to solve industry's dilemma by promoting supposedly more environmentally friendly technological solutions will be counterproductive (especially if those solutions are supposed to guarantee equal earnings as presently): it will lead to a short- to medium-term *increase*—rather than *decrease*—in CO_2 emissions, and that rise will probably push global warming into a state from which we cannot easily return (over the tipping point). Development of new technologies can never be as successful in reducing environmental impact as radically reducing consumption; worse still, if consumption continues to increase, no amount of technological innovation will be able to keep up with it sustainably. The writing is already on the wall, and it says "technology itself cannot be sustainable, rather the behavior accompanying it." Unfortunately, messaging of a different kind often appears on advertising boards: one prominent European car manufacturer promoted its newest EV with the hashtag "NOWYOUCAN," in an attempt to assuage pent up consumer guilt. As I mentioned in an editorial,[61] small lexicographical changes produce "NO YOU CAN'T," a message that is arguably closer to the truth. Just for completeness, in most cases, converting functional classic cars to electric (quite a fad at present) is environmentally detrimental compared with simply driving them less.

If one can't reduce, one can "compensate," but how impactful can that be?
Though the car industry would like us to think that buying a BEV is doing something good for the climate (another manufacturer proclaims "THIS IS FOR YOU, WORLD"!), we turn now to more serious initiatives. These amount to assuaging guilt or improving one's environmental conscience by paying for one's CO_2 footprint. For example, schemes abound for making some kind of donation independently, or in addition to, the price of a flight. In fact, more and more schemes are available simply for compensating for one's "bad" CO_2 emission behavior in general. This is quite bizarre, but also quite consistent with human psychology, particularly in richer countries. Paying fines for all manner of transgressions is regarded by many as simply part of life. In the case of CO_2, it amounts to trading an unwillingness to make life "less comfortable or enjoyable" (or at least *seemingly* so) for the relative ease of arranging a regular direct debit to one's bank account.

Planting trees to soak up CO_2 is popular: What could be more glorious and natural than new tree springing up in the name of sustainability? Ah, that it were so simple. Regrettably, the rate of carbon sequestration into semi-permanent new growth is likely to be no more than moderate on the timescale over which we need radical CO_2 reductions. Some modeling scenarios seem quite hopeful, for example, increasing CO_2 uptake by 188 million tonnes per year (4% of current emissions) via fully stocking all understocked productive forest land in the US.[62] However, raising a sapling until it can fend for itself is not easy: literature since the 1990s indicates that between 34% and 50% of intentionally planted trees do not survive past two years.[63] They require intensive looking after. Better than replacing lost trees is the preservation of existing trees and their naturally regenerative environment. However, the Amazon rain forest is currently losing around 12,000 km² per year to bulldozers and intentional fires; between 2020 and 2021 it was more than 13,000 km² (5,100 square miles)[64]—equivalent to a quarter of Denmark—the fastest annual deforestation for 14 years. The Brazilian national space research institute (INPE) has been monitoring the destruction by satellite; INPE's government funding has been cut back considerably lately.

Deforestation and climate change (leading to massive water losses from existing forests) are actually starting to change the status of the Earth's largest forests as the major carbon-sequestering organs.[65] Some researchers now believe that planting herbaceous species (woodless plants) would absorb more CO_2 than trees.[66] But that is certainly not in defense of deforestation, and it harbors massive risks of altering ecosystems, with associated unpredictable consequences for net CO_2 sequestration. Furthermore, local benefits of trees include significantly reducing noise and pollution in towns and cities if planted in the right places: particulates are partly filtered by the foliage, and the waxy lipids coating the leaves absorb incomplete combustion products from all sources.

Globally, forests may be absorbing as much as 7.6 Gt $CO_{2eq.}$ annually,[67] but the error margin of this estimate is very large, reflecting persisting unknowns and the high sensitivity of many habitats to swings toward net $CO_{2eq.}$ emissions. Forest management is a tricky business: one example shows this more than any other: in the 1990s, in response to the first large effects of climate change and associated tree-pathogen response (e.g., bark beetles in beech trees), many natural forests in Europe became targets for strategic planting of Norway spruce. Thirty years later, these spruces are dying faster than most other forest trees, leaving bare areas, exposing beeches and oaks to "sun burn," and accelerating desiccation of the soil. Norway spruces are now considered one of the species most vulnerable to climate change.[68] By contrast, leaving forests to their own devices and being strategically sparing with tree felling for profit and safety concerns has had notable success.[69] Ensuring that as much felled wood as possible goes into long-lasting products is also a priority, hence keeping the wood's carbon bound for decades; paper and cardboard, on the other hand, are notorious examples of wood's carbon re-entering the atmosphere as CO_2 within a few months.

Wetlands are also becoming targets for compensation schemes. But are moorland and peatland reliable targets for CO_2 sinking? The literature abounds with estimates for the quantity of carbon currently sequestered there, but there is very little consistent information on how much CO_2 is presently being absorbed per unit area per unit time. Wetlands are the most variable CO_2 sinks in terms of short-term dynamics: sometimes they can be net CO_2 emitters, other times net absorbers. As they are drained, they absorb less and emit more CO_2, rapidly becoming net emitters. In fact, so sensitive are they that draining of moors and peatlands currently contributes 5% of anthropogenic CO_2 (1.9 Gt per year) despite them representing a mere 2.8% of land area on Earth.[70,71] The reason: they hold 30% of the soil-bound carbon on Earth.[72] It makes sense if we remember that such wetlands produced all of the coal that we have on Earth: if they remain unchanged for millions of years, they do, indeed, sink prodigious masses of carbon into the Earth's upper crust. Preventing active draining for conversion to farmland is a good step toward CO_2 fixation, but many areas on Earth are—because of global warming—drying out on their own, without direct human intervention. Wetlands are vital for the overall health of the countryside, including neighboring agricultural land, so we should, regardless of CO_2, be preserving them—crucially, with a knowledge of their larger value and the massive amounts of CO_2 that they would *emit* if we allowed them to dry up.

So, is this type of compensation good in principle? Probably not, because it kids us into thinking that the way to deal with CO_2 emissions is to continue to make them and simply pay—a strategy that does not encourage reduction of consumption. Furthermore, it is very hard to equate the three components of the equation to find out whether it truly "works" either in general, or for *you*:

- How much net CO_2 *you* actually produce
- How much you are willing to pay for it
- How much CO_2 is truly removed from the atmosphere for that payment

At steady-state (i.e., probably after 100 years or more), we would see the sense and power in preserving natural, biological, CO_2 sinks. However, we don't have that kind of time, and, at the level of planting trees here and there, the success is extremely variable. Next I describe projects aimed at taking a very easily quantifiable mass of CO_2 out of the atmosphere and, with easily calculable quantities of energy, converting it to a form that is—on the timescale of human civilization—permanently sequestered. If one financially supports these schemes, one does, indeed, "compensate." The equation is much more transparent, but, as we will see next, the chances of success are small.

Carbon-negative: Projects for taking CO_2 out of the atmosphere (semi-permanently)

There are essentially only two options for carbon capture: (1) capturing CO_2 post-combustion and (2) capturing CO_2 pre-combustion. Point 2 sounds a bit strange, but it works by virtue of the advantage of producing the CO_2 in a very concentrated form by reacting the combustant (e.g., coal) with steam at relatively high temperature and pressure. This produces a mixture of hydrogen (H_2), carbon monoxide (CO), and CO_2 (a type of synthesis gas). CO_2 is removed, whereupon residual water reacts with CO to produce more H_2 and CO_2 (the water-gas shift). The H_2 is burned to produce electricity. This is a combined process of CO_2 capture-combustion-synthesis gas formation and can therefore be used in combined energy/material/power-to-x plants. We've yet to see these at more than pilot scale, but the potential exists. That said, the feedstock for this process is mainly coal, and not necessarily high-grade coal, hence it does nothing per se to distance us from fossil fuels. Its advantage over collecting typical flue gas from combustion is that it captures the CO_2 even more immediately, purely, and efficiently. Post-combustion CO_2 capture ranges from flue-gas technology to ambient air capture (direct air capture [DAC]), depending on the feasibility of containing the post-combustion gas (e.g., in the case of cars, it is unlikely to be practicable, but for ships it might well be). An accompanying technology is *oxycombustion*: the burning of fuel in an atmosphere of pure oxygen, thus producing flue gas that contains 100% combustion products, most of which is CO_2 and water.

Increasingly, coal-, oil-, and gas-fired power stations are being coupled to CO_2 sequestration facilities, some of which are combinations of deep rock methods and surface absorption. Surface absorption includes pumping CO_2 into freshwater bodies such as lakes, where algae and certain bacteria will convert it into organic material. However, if a lake is already healthy, adding CO_2 and promoting microorganismal growth will not make it healthier; furthermore, it's anybody's guess how long that CO_2 stays in organic compounds. Experimental plants that purpose-culture algae in large incubators with flue gas and sunlight exist but have not achieved large-scale application. In what follows, I will only discuss the possibilities for long-term sequestering of CO_2, regardless of whether it is already in concentrated form or simply from ambient air.

Deep sequestration of CO_2: The further down, the better, but the greater the energy needed

Into rock, cavities, and pseudo-cavities

The key example is CarbFix technology,[73] developed by the company Climeworks, founded in 2009 as a spin-off from work done by two Swiss academics. The principle is simple: use of natural energy to extract CO_2 from the air and pump it into the local basaltic rock (Figure 5.6). In the pilot project,[74] the geothermal energy on Iceland (Hellisheidi) is used. CO_2 concentration in air is 0.04% by volume and

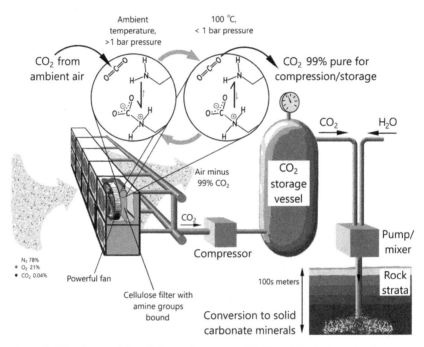

Figure 5.6 Basic principles of direct air capture (DAC) and CarbFix technology developed by Climeworks and applied on Iceland with regenerative energy largely from hydrothermal sources. Technical simplifications have been made for clarity. Ambient air is drawn into a unit containing a cellulose filter that has been doped with an organic amine (carbon compound with at least one NH_2 group). At ambient temperatures and a pressure somewhat above 1 bar, CO_2 binds to the amines to form carbamates. When fully loaded, the pressure in the filter compartment is lowered to somewhat below 1 bar and the temperature raised toward 100 °C. Now the equilibrium favors dissociation of the CO_2, which is pumped off, compressed, and stored. Compressed CO_2 is pumped together with water under high pressure into deep rock strata, where it mineralizes to form permanent deposits.

0.06% by mass, so concentrating it to near to 100% is quite a feat. Climeworks does this via a proprietary filtration technology where CO_2 is first absorbed onto a special surface (organic amines), then released by heating to 100°C, concentrated by compression, and temporarily stored. If destined for "permanent sequestration" (sinking), the CO_2 is then mixed with water and pumped into rock. To maximize the negative CO_2 balance, only renewable energy may be used to drive the fans that pump air through the filters, the heating elements needed to release the CO_2, the pumps that draw the CO_2 off and pressurize it in intermediate storage tanks, and the pumps that force the CO_2/water mixture into the rock. An attractive feature of this technology is that 95% of the dissolved CO_2—as demonstrated by geochemical research—turns into solid carbonate in less than 2 years under ideal conditions; in fact, at temperatures around 260°C, this happens in a matter of months, hence making the location on active volcanic land favorable.

The chemistry is simple: dissolved hydrogencarbonate in the introduced water reacts mainly with calcium ions (possibly magnesium and iron, too) that are leached out of the basaltic rock and present in the natural aquifers. Here the carbonate(s) will stay inert for geological timescales. Only two processes are likely to release it again: (1) uplifting of the rock strata to reveal the carbonate rock to the elements, where the mildly acidic rain will slowly dissolve it; and (2) subduction into deeper layers of the crust, where it ultimately comes into contact with magma and is "decomposed" into its constituent parts. CO_2-rich magma is expelled during volcanic eruptions, and much of the CO_2 within it is then re-released to the atmosphere—a crude explanation, omitting the geochemical transitions between different rock types. We can estimate around 20 million years for a single transition. Hence it is safe to say that we wouldn't see any rock-sunk CO_2 for at least another 20 million years. That would give us plenty of time (we hope!) to get our CO_2-related behavior into order. Pumping CO_2 into rock in an already volcanically active area, such as Iceland, is not the most stable solution seen over geological timescales, but it's certainly adequate to deal with the pressing problem at hand.

Now to metrics: How much of a dent in the CO_2 problem can this technology make? Based on 2019 figures (at that point, Climeworks was sinking 4,000 tonnes annually) I made my own calculation: to remove the annual increase of CO_2 from the atmosphere—hence keeping us exactly at present CO_2 concentrations—one would need 112,000 times that capacity; to keep global warming within 2°C by the end of the century, one would need to reduce the CO_2 concentration by a similar amount again, hence requiring around 225,000 times the 2019 capacity (see Chapter 5, Calculation 7, online at https://doi.org/10.1093/oso/9780197664834.001.0001). In fact, I might have been a bit generous, because even the CEO of Climeworks, Christoph Gebald, reckons that the company would have to build 250,000 more of the current facility. Progress is slow: the first large-scale plant (also on Iceland) was started in September 2021.[75]

Climeworks' plan to suck 1% of global annual CO_2 emissions out of the air by 2025 is not going to happen. The low price of CO_2 in international CO_2 trade has not helped, but the very aim itself embodies problems of a catch-22 nature: enormous industrial production needed to create the materials for these sinking plants should rely *entirely* on CO_2-neutral means. Is that possible over such a short timescale (or even 10–20 years)? And if not, would the use of non-CO_2-neutral energy merely contribute to the temporary additional bubble of CO_2 production inherent in most regenerative energy technology? A bubble that some critics say will push the climate over the point of no return, after which any amount of CO_2 extraction from the atmosphere won't prevent substantially more than 2 °C warming by 2100. There are many unknowns here, and expecting any technology to "save" us by compensating for current emissions is dangerous: we must reduce *production*. It would be unwise to try to review such a young technology in detail; suffice it to say that Climeworks is not the only initiative and that the essential idea can be coupled to high-concentration sources of CO_2 such as industrial exhaust gases, hence reducing the energy needed for CO_2 adsorption and drastically reducing emissions immediately at their source. The steel and cement industries are prime targets, but whatever the target, one needs a reliable (preferably very local) supply of CO_2-neutral energy and suitable (preferably very local) geology for CarbFix technology.

There were several forerunners to CarbFix. For example the European Union-funded project CO_2SINK[76] sank 67,271 tonnes of CO_2 into geological gas-bearing rock strata in Ketzin, Germany[77] between 2008 and 2013—that's 13,440 tonnes per year, or more than three times the Climeworks project on Iceland. However, the EU project was only funded between 2004 and 2010, and no further developments are apparent. As Vattenfall (the Swedish energy company) announced its plan to pump waste CO_2 from a gas-fired power station close to Ketzin into brine-containing rock in the region, there was a regional uproar against the plan.[78] Concerns about the stability of the rock and the possibility of the CO_2-pumping causing mixing of long-term geological water with drinking water were raised. Greenpeace was very vociferously against the plan, proclaiming "CO_2 sinking is unsafe and unnecessary" (German "*CO_2-Speicherung ist unsicher und unnötig*").[79] The idea was shelved. Instead of investing in CO_2-sinking into rock in Brandenburg, interested people can invest in an area of moorland in the same region and pay €64 Euro (US$70) for naturally sinking 1 tonne of CO_2 into the land by preventing it being drained for agricultural use.[80] The so-called *moor futures* seem like a great idea, and if we can reduce agriculture's thirst for land by reducing livestock, they could be expandable.

A similar technique to rock sequestration is to pump the CO_2 into cavities left by oil mining or the depressurized cavities from which natural gas has been extracted. Gas often occurs together with oil, and the benefit of both of these types of sequestration is that the storage locations have already undergone the test of geological time—no leaks. If we manage to put a good stopper on the top of the pipe (crudely put), the CO_2 will likely stay down there for geological timespans. The challenge with this method is to identify feasible sites because many oil or gas mining sites are far from the industries that use their fossil fuels: some are even hundreds of kilometers offshore. To use the more awkward ones, we would need a pipeline or transport infrastructure for the CO_2 (in compressed form). Many unmineable coal seams have been considered for CO_2 sequestration as well, but given their relative closeness to the surface, the long-term full containment of CO_2 there is very doubtful.

All techniques for capturing CO_2 from ambient air and sequestering in rock are very energy-intensive. For example, if we wished to capture the CO_2 from an average-sized coal-fired power plant, we would need roughly 5 times the energy of that plant to do so via Climework's CarbFix technology (see Chapter 5, Calculation 8, online at https://doi.org/10.1093/oso/9780197664834.001.0001). If the developers' aim of 40% reduction in energy is achieved, it's still more than twice the plant's energy output. The bottom line is that to capture and sequester CO_2 from fossil energy activities, we need several times that energy, but from renewable sources—quite a dilemma.

Into water, from saline aquifers to oceans

As we saw earlier, CO_2 is very soluble in water compared with many other gases. Several projects have therefore investigated pumping CO_2 into deep saline aquifers—as would have been done in Brandenburg, Germany. Being essentially incompressible, subterranean water can cause cracks in rock to appear or existing ones to extend. It's a very complicated phenomenon, and one that has only recently started to be modeled.[81] It's further complicated by the increasing acidity of the aquifer as a result of dissolved CO_2, which causes minerals in the surrounding rock to be dissolved and redeposited in a pattern that is as yet unpredictable. Still, it is the fractures in the rock that are the most important to study because of the CO_2 risks leaking out through them into the atmosphere. Pilot plants and modeling are ongoing, but large-scale facilities are some years off.

Some oceanographers advocate a faster solution by pumping CO_2 into the world's seas. Why would that be good, given that a biologically undesirable effect of anthropogenic CO_2 is acidification of ocean water? The reason emerges when one understands the oceans as vertical systems with very variable conditions between surface waters and depths of 11 km in some places; the average is around 3.7 km. As we saw in Chapter 3, at a depth of 5 km, where the water temperature is close to 0°C and the pressure is 500 atmospheres (500 bar), CO_2 is 2.5 times as soluble as at the ocean's surface. The overwhelming lack of currents and the great pressure on the seabed at such depths leads concentrated CO_2 to do something quite remarkable: it can form so-called lakes that remain stable for very long periods. If deposited below 2,700 m depth in concentrated form, CO_2 behaves as a liquid that is denser than water and hence will flow to the deepest place nearby that can hold it. When that becomes full, the overflow will flow to even greater depths until it reaches another hollow, and so on. Essentially, this physical property prevents the CO_2 from mixing with, and reducing the pH of, the surface waters containing most of the life.

Seawater does cycle between surface layers and the deep, but slowly: depending on depth and location, this can be between 300 and 1,000 years. If sequestered at 3,000 m depth, modeling indicates that 80% of the CO_2 would remain after 200 years. The deeper it is, the better the performance of the site.[82] So, should we do this on a large scale? Probably not yet because it is a major intervention in a part of the planet about which we still understand very little, and negative impacts on biology surrounding a given CO_2 lake are already foreseen. And yet even at small scales it is very hard to do experiments: in 2002, the first of its kind—to pipe 60 tonnes of liquid CO_2 down to 800 meters just off the Hawaii coastline—met with such local resistance that it was canceled. Furthermore, putting CO_2 into the ocean—wherever we put it—will lead to some drop in pH, however small: it's a tradeoff against allowing less CO_2 to dissolve into surface waters by pumping that CO_2 into deeper waters. This might at least buy us time, but only if we simultaneously reduce CO_2 emissions.

Pilot projects are promising, but progress is far too slow
The story of Climeworks is typical of this industry sector in general: massive investments need to be made and these are principally coming from venture capitalists and private people wishing to compensate for their CO_2 footprint. Getting enough renewable energy is expensive and very difficult in the quantities needed to make a real difference. Basically, to impact CO_2 concentrations noticeably, carbon sequestration projects need to go into a furious frenzy of building and sinking that will require a horrendous amount of renewable energy. We don't have it. A glimpse at the latest figures[83] is depressing: 27 plants in operation, 4 in construction, 58 in advanced development, 44 in early development, 2 suspended. The numbers are too small to produce a statistically significant trend, but basically, since 2009, when Climeworks CarbFix technology was published, 133 plants have arisen. That gives us 133 divided by 12 years = 11 plants per year. At that rate, to reach the 250,000 Climework-similar plants that we need to start to diminish CO_2 concentrations in our atmosphere, we'd need more than 22,700 years. We'd better speed up.

Conclusion: Recognizing the absurdity of "decarbonizing" whole economies and the scale of necessary reductions in CO_2 emissions

I hope to have given sufficient examples of the place of carbon compounds in various sectors of our economy for it to be clear that decarbonizing quickly and at a large scale is impossible; the very idea is absurd. The concept of decarbonizing wholesale has worrying parallels to human developments that started long before CO_2 concentrations in the atmosphere turned into the major challenge for humanity—developments that would lead to the current environmental crises because of the way that we humans "think." It started with the "cognitive revolution," in the sense coined by Yuval Noah Harari in his book *Sapiens*[84] (i.e., the rapid advance in human thinking, organization, and collaboration occurring 70,000–30,000 years ago). This led to the growth of modern societies that could achieve more than could small groups of nomads. Arguably this was the beginning of humanity's divergence from non-human nature. It started the process that led us, and continues to lead us, to believe that we are "above" nature or at least separate from it; that our solutions to problems need not have anything to do with ways in which biology or the natural world works; that we can invent completely "artificial" ways of shaping the planet and its life—and similarly artificial solutions to the problems that we cause. We have surely caused a problem via unsustainable use of fossil fuels, but that does not mean that we should replace it with the use of technologies that take us even further away from the biological realm. If we do so, we distance ourselves yet further from the mechanism of the natural balance that keeps the whole place habitable. The forest, discussed in this chapter, is a microcosmic example for the larger Earth: more and more experts are realizing that attempts to engineer forests to compensate for human-made environmental insults (climate change included) do not, on the whole, work. They lack insights as to how nature with its enormous organismal and genetic diversity, and given the chance, would respond to the challenge.

The great GM debates of the 1990s and early 2000s showed how many people could be led to believe that GM, and even genes themselves, are something unnatural despite biology having used them for 3.5 billion years. Some were even persuaded that natural fruit and vegetables are "gene-free": we were on the road to "degeneticizing" public perception, aided by highly emotional arguments and a lack of scientific input. The reality was that we had, since the Agricultural Revolution, bought into this optimization of crops via indirect, and then direct, methods because it worked so well. Similarly, carbon-based energy sources and materials have worked well for humanity in many respects because they are superbly versatile, interconvertible, and easy to apply; our mistake was to take them from the Earth's crust in quantities that we knew could not be regenerated quickly enough. We have within our grasp models from biology for creating cyclical carbon economies that additionally embody self-regulation, hence keeping us within sensible limits.

Faced with the consequences of fossil-fuel–based economies that could never have been sustainable, we must also acknowledge that no energy source will be sustainable in the context of our incessantly growing demand. However, we now have a choice with regard to the principles of the "biological" Earth: to try to get our carbon-based economies into cyclical pathways or abandon carbon to a large extent and try something different. The task is enormous because we have, wittingly or unwittingly, integrated the products of fossil oil, coal, and gas into every realm of life. We have hence become blind to the extent of the challenge of replacing their fossil source. We might fantasize about electric passenger planes, ships, and even heavy-goods vehicles; however, simple calculations show how unlikely these are to emerge at any scale without alternative energy carriers to hydrocarbon fuels that have similarly high energy densities. We may wish—in order to break the horrifying task down to more manageable chunks—to regard energy as a separate problem from material, but a theme that runs throughout this book is that the two are inextricably linked. To break the link by inventing a completely mineral-based large-scale energy economy does not seem wise to me (i.e., digging up the Earth's crust, and [possibly in the near future] scraping off the bottom of the oceans or making inorganic mineral-based energy-storage devices and attempting to recycle them in ways that consume enormous amounts of energy but that don't produce any energy in the product).

And yet we risk mixing the two economies in uncontrollable ways if we are not properly organized: this is already happening with refuse, much of which is plastic. This makes waste-material burning a significant source of anthropogenic CO_2 that is, ultimately, from fossil reserves at present. However, if full CO_2 recycling (via flue gas capture) is implemented, why not use waste as fuel and give it a second life? On the other hand, for many synthetics, recycling is increasingly possible—for an increasing range of polymers if we make them conveniently separable. With the drive to break free from fossil feedstocks for plastics and the chemistry industry in general, recycling must increase. However, we will surely be left with a demand that needs satisfying from primary production: that is where recycling CO_2 from the atmosphere, reduction to CO, and combination with H_2 to make syngas comes in. Here is a very low-level point of contact between energy and material economies because synthetic methanol and methane from syngas are as much fuels as they are feedstocks.

Priority targets for CO_2 capture at source are the stationary hard-to-abate sectors, the iron and steel industry being one of the hardest. Hydrogen is the "dream" fuel for these sectors, offering heat energy and chemical reducing power in one, but the infrastructural challenges are enormous. Progress, albeit slow, will probably be made with a combination of strategies, including hydrogen, methane, and further improvements in efficiency; however, nothing will work as well as reducing our thirst for primary iron and steel. The cement industry also continues to grow in output and is increasingly finding a very attractive business model in taking refuse (much of it plastic) off the hands of authorities to burn in its kilns. Is this an acceptable use? Perhaps in the short-term at least, rather than land-filling it. But the challenge of these hard-to-abate sectors leads to an inescapable medium-term conclusion: we must ramp up our carbon capture capabilities faster and take advantage of the high-concentration CO_2 in flue gases. Carbon capture directly from ambient air (DAC) via "thermo-swing" filters, followed by deep rock sequestration, is currently most famous for its application with Climeworks. However, the growth of this sector appears to be stagnating, perhaps in part because it is extremely expensive and produces nothing that is of direct value to the investor. Hence conscience-abating schemes are unlikely to be the solution to reducing ambient CO_2 to any tangible extent: more "involuntary" measures at the levels of taxation, international politics, trade, and legislation are unavoidable. Would it help also to use this CO_2 for making valuable CO_2-neutral fuel and storing it (e.g., "e-fuel futures")?

At present, a tiny proportion of the world's population continues to increase its already massive CO_2 footprint while the rest emits hundreds or thousands of times less. One could be forgiven for thinking that there can be no environmental justice before social justice is implemented. But this is not an easy task for capitalist economies that have raised economic growth, freedom to become wealthy, and freedom of choice to the status of holy cows. In the next chapter, we will see how one holy cow in particular has become ensnared in a quandary partly of its own making but partly of political indecision, well-intentioned activism, and major legal challenges. Still, bespattered with its own feces on one side while shining with clean, green promise on the other, it remains a holy cow. More than that, regardless of whether it is truly improving in environmental terms, the produce from this cow remains one of the most common objects of the human desire for newer, bigger, faster, better: the car.

6
Decarbonizing the car
Trading off CO_2 against larger environmental problems?

I am devoting a whole chapter to the car because it's emblematic of a seriously incomplete analysis of environmental impact in general and because it has been distorted into a massive and unbalanced public "attractor." Most "over-hyping" relates to car usage (in contrast to car manufacture). However, car use is connected with a much smaller part of the overall carbon dioxide (CO_2) emissions problem than industry as a whole and agriculture, for example, sectors that less frequently enter the public consciousness via the media. A narrative has been created in which internal combustion engine vehicles (ICEVs) are, in general, greatly inferior to battery electric vehicles (BEVs) in terms of their energy consumption and environmental impact, and e-fuels are seen as no more than a desperate, very costly, attempt to save outdated, filthy technology. However, closer and much fuller analysis shows that the truth is very different: ICEVs running on e-fuels do not generally turn out to be energetically or environmentally worse and more polluting than BEVs at all. My and others' analyses reveal that, in environmental impact, BEVs (and certain other EVs) are, in larger areas of application, even worse than ICEVs running on current fossil fuel. Most of the difference originates from manufacture of batteries and drive-train components. Battery technology and recycling are bound to improve, but, for average global driving habits, they must improve 4-fold in environmental impact and between 10- and 20-fold in human health impact in order to beat ICEVs running on e-fuel—and they must do so very quickly. I've heard the following argument from those who understand the problems, but still intend to buy an EV: "Yes, but if we don't use this technology, it's never going to improve." My counter argument: "The priority NOW is to reduce harm to the climate and environment, not to support a particular technology." To be blunt, no cars are good for the environment, but some are worse than others. Simply driving less would be a good start to reducing emissions, but most people agree that that won't happen voluntarily.

Sustainable personal transport: How can we reduce the current impact?

Globally, passenger car CO_2 emissions are 3 gigatonnes (Gt: i.e., 3,000,000,000 tonnes), or 8.3%, of the current 36.3 Gt fossil-fuel related CO_2 emissions.[1] The power industry accounts for 36% of the emissions,[2] and agriculture is a similarly striking case, if one includes methane and other greenhouse gases (GHGs) as CO_2-equivalents ($CO_{2eq.}$): the Food and Agriculture Organization of the United Nations (FAO) figures from 2018 give 9.3 Gt,[3] or almost 26%—more than half of which is methane and nitrous oxide—an underreported fact. I need to write a lot about the car to explore what has become a very disproportionately large focus of attention—distracting us from even more important things. In fact, this distraction is so successful that one might wonder whether it is intentional. What better way of thinking that you have done your bit for the environment than buying an EV—as some car manufacturers are expressing? Even more importantly, I conclude that we're heading in the wrong direction for sustainability with a personal mobility plan based almost entirely on BEVs replacing all ICEVs. I'm increasingly convinced that we must embrace synthetic hydrogen-based and hydrocarbon fuels and reduce consumption.

"Inconvenient" results lead curious minds to investigate: EVs are not nearly as sustainable as we are led to believe

My trigger for writing this chapter is the following: in 2019, the renowned Fraunhofer Institut in Germany published a research report on the sustainability of road vehicles that use different drive-trains/fuels in terms of GHG emissions.[4] One might think that this report is only about BEVs: it does a good job of convincing readers of their virtues. However, more diligent journalists noticed a tiny part of the report where it was concluded that

> At present, the use of completely renewable electricity in the production of synthetic fuels is far from a commercially viable application. For that reason, these [fuels] will only be able to play a relevant role after 2030. If one assumes a PtL-scenario [power-to-liquid, essentially e-fuels] with 100% renewable electricity, then such vehicles [i.e., powered by internal combustion engines running on synthetic fuel] perform better than electric vehicles because of their lower greenhouse gas emissions in production (see also Ludwig et al. 2018).* [Ludwig et al.[5]]

---

* Original in German: "Derzeit ist die Verwendung von ausschließlich erneuerbarem Strom bei der Herstellung von synthetischen Kraftstoffen noch weit von einer wirtschaftlichen Anwendung entfernt. Deshalb werden diese wohl erst nach 2030 eine relevante Rolle spielen können. Wenn

Two salient observations: the first sentence witnesses the current price differential between synthetic and fossil fuels, but clearly that is (1) getting smaller (rather suddenly as a result of Russia's invasion of the Ukraine) and (2) subject to state-side modulation via taxation. The Fraunhofer report is, in my informed opinion, a little too pessimistic about the speed of progress in e-fuels (synthetic kerosene production is growing fast, for example). Second, as pointed out by a previous study,[5] if both types of car drive with renewable energy (CO_2-neutral electricity for the BEV and CO_2-neutral e-fuel for the ICEV), but still use fossil energy for their manufacture, the GHG emissions from battery production outweigh the advantage that the BEV has over the ICEV during use.

Unfortunately, few people outside German-speaking countries will even get as far as the well-hidden paragraph in the Fraunhofer study: the report is only available in German. Other published analyses that cast doubt on the superiority of the larger environmental balance of BEVs are continually reaching the public media: one recent magazine article carried the title "Is It Ethical To Purchase an EV Lithium Battery Powered Vehicle?"[6]; others note a growing aversion, accompanied by large protests from environmentalist groups across the world, against lithium mining.[7–9] In the peer-reviewed literature, there are many research papers showing how, depending on energy source, location, and use scenario, the perceived superiority of the BEV over the ICEV either shrinks to zero, or even reverses in favor of the ICEV.[10–12] Some of these studies analyze material fluxes in great detail in the context of "lightweighting" (decreasing the weight of the total car), and, as we will see, material flux—and the associated energy requirement—is a very large environmental disadvantage of the BEV at present. Note that these studies do not even take the possibility of CO_2-neutral e-fuels into consideration, whereupon the disadvantages of BEVs versus ICEVs would grow even larger. Such work has regularly been "rubbished" by a variety of people, often with the conclusion that BEVs are simply better in all situations than ICEVs. This is patently not true. Why are the environmental concerns voiced in many studies so vehemently played down? Surely we should be highly sensitive to them. The observation about GHG emissions during production implies great energy requirements, and those usually go hand-in-hand with other large, environmental impacts. That is what I ultimately wanted to investigate further.

An earlier study (2016) that, similarly, could not demonstrate the general environmental superiority of EVs over conventional cars came from a French

---

man ein PtL-Szenario auf Basis 100% EE unterstellt, dann schneiden diese Fahrzeuge aufgrund der geringen THG-Emissionen in der Herstellung besser ab als BEV [Strom aus erneuerbaren Quellen] (siehe auch Ludwig et al. 2018)."

consortium under the direction of the French Environment and Energy Management Agency (ADEME: www.ademe.fr, an independent public agency for the development of policies to support ecological transition).[13] The study was the second coordinated by ADEME with participation from a wide range of stakeholders, including environmental protection groups and industry. During the first study (concluded in 2011), the coordinator, Michel Dubromel, President of France Nature Environment, noted the unease of the car manufacturer Renault at study findings that disadvantaged the EV. According to Dubromel, Renault asked for them to be removed, though that was not done. Finally, after an inexplicable delay, the report was published. The follow-up report (2016) essentially reached the same conclusions about environmental balance, noting, for example that "over the lifecycle of the vehicle, these [negative environmental] impacts are of the same order of magnitude for an EV as for an internal combustion vehicle."[13] As we will see, other analyses suggest that EVs can be significantly worse than ICEVs on balance.

Naturally, buying a BEV, or any new car, has, per se, little to do with sustainable behavior. Even if it did, it depends on the specifications of the car and how one uses it. The current fashion of tax breaks, manufacturer discounts, and massive government subsidies for EVs of any sort (some of which are plug-in hybrids) is surely blinding many people to a big problem: the larger environmental consequences of the personal financially motivated decision to switch to electric. Many people today think that an EV is, per se, a more sustainable option than a conventional car. Clearly, we face massive environmental challenges in battery production and recycling, and massively expanding electricity production and distribution, but that's not all. An equally important challenge is to ensure that our *use* of EV technology is environmentally sustainable. That is not simple because our use of technology is conditioned by consumer behavior that is fundamentally at odds with environmental sustainability—be it driven by emotional factors, incomplete/incorrect understanding of facts, media reporting, or political or commercial persuasion.

Many studies report EV owners' pride (at owning something so seemingly "environmentally friendly" at great personal cost) and guilt (at having driven something so seemingly "environmentally unfriendly") rather than analysis and understanding of true environmental impacts.[14] EVs' positive environmental impact, however, is small and local compared with the negative impact of manufacturing-associated mining, waste, and CO_2, which is remote and hence hidden from most people. A large proportion of respondents in a recent survey of EV owners[15] noted driving more in their EV than in their ICEV, particularly for short trips into town. Here a few survey quotes: "It's easy to cruise into town for a forgotten item or even go for a drive for pleasure. . . . Short trips I used to avoid in my ICE if they were not completely necessary, or could wait until

I had several things to do. Now, I will happily go out for that one thing in my EV, knowing that I am not using petrol.... I'm happy to go into town and do just one thing now whereas when I had my ICE I planned multiple things because of the cost of petrol."

The worrying emergent message is that having a (B)EV uncouples people from the responsibility of saving energy—indeed, it seems to make many completely unaware that they are still using energy, the source of which is a wide mixture (currently largely fossil fuels) and will always have environmental and human health impacts, even when we break free from fossil fuels. A few acquaintances of mine think that ICEVs simply "befoul" the air, but they don't think twice about flying in a jet airliner, which has no particulate filter or catalytic converter; or they consider the environmentally problematic nature of mineral mining for batteries to be a political problem of countries like Bolivia, Chile, or the Republic of Congo, which haven't yet "sanitized" the processes. These are devastating impacts, as we will see in a later section on the full economy of driving *and* manufacture.

Implications for my personal car: Consider driving much less and sticking with an economical ICEV

Many people are faced with a decision on which car to buy next. Nothing beats talking to real people, so I quizzed a couple of friends: one had bought a brand new medium-sized cross-over vehicle (with switchable 2-wheel/4-wheel drive) and reported average diesel consumption of 4.5 l/100 km (2021 Dacia Duster diesel, in 2-wheel-drive mode), which, incidentally works out at a maximum range of 1,100 km (684 miles). The other had purchased a brand new medium-sized BEV SUV and reported average electricity consumption—read from the dashboard instruments—of 19.5 kWh per 100 km (2021 Hyundai Kona Electric with 64 kWh battery, for which I found very similar real-world consumption values in consumer advice reports). The plot in Figure 6.1 represents the scenario where a car purchase was a "must" (e.g., because the person had no car and "needed" one). For that reason, the total energy of car manufacture is not incorporated, merely the differential in manufacture between the BEV and the ICEV, which I calculated with general values from the research literature.

Because we are living in an era where CO_2 emissions take center stage (we are still heavily reliant on fossil fuel) I calculated the break-even points for CO_2 emission. The bottom lines (see plots in Figures 6.1 and 6.2) are

Case 1 (Figure 6.1): If the friends absolutely *needed* to buy a car:

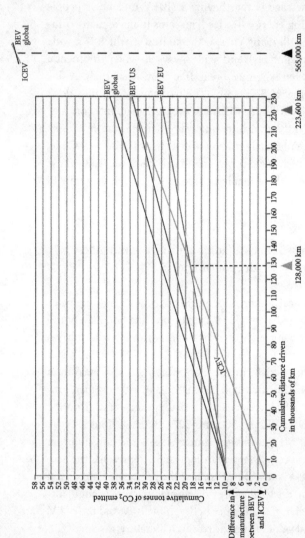

Figure 6.1 Plot of cumulative difference in CO_2 emissions from a single car, either battery electric vehicle (BEV) or internal combustion engine vehicle (ICEV) (see Chapter 6, Calculation 1, online at https://doi.org/10.1093/oso/9780197664834.001.0001). Only the difference in manufacturing CO_2 emissions are considered at 0 km (i.e. the extra emissions that BEV manufacture makes in comparison with comparable ICEV manufacture). Battery/ancillaries for BEV made with combination of fossil fuels and global electricity, of which 66% is fossil (electricity). Running on a global electricity mix (66% fossil), the BEV breaks even with the ICEV at 564,854, rounded to 565,000 km (off the graph); on US electricity mix (61% fossil), the break-even is at 223,588, rounded to 224,000 km; on EU electricity mix (37% fossil), the break-even is at 128,103, rounded to 128,000 km. Values used: BEV with 64 kWh Li-ion battery (current technology status) and realistic (considering all calculable energy losses) consumption of 23.7 kWh/100 km (present value); ICEV consuming 4.5 L diesel/100 km. BEV battery and drive-train ancillaries produced with energy including a global electricity mix of 66% fossil. Well-to-tank/well-to-plant emissions of an additional 23% are included in all calculations. Note that if battery manufacture is in China, with a fossil electricity component of 71%,[16] all break-even points become somewhat greater.

- If they were "global average people," they'd need to drive more than 500,000 km before the CO_2 burden of the BEV broke even with that of the ICEV.
- If they were in the United States, the figure is more than 220,000 km (137,000 miles).
- If they were in Europe, it would be almost 130,000 km.

Case 2 (Figure 6.2): If the friends had wanted to replace a well-functioning and economical ICEV with a BEV of comparable size and utility (i.e., they were responsible for the production of a car that otherwise would not have been produced):

- If they were "global average people," they'd need to drive over 800,000 km before the CO_2 burden of the BEV broke even with that of the ICEV.
- If they were in the United States, the figure is near 330,000 km (205,000 miles).
- If they were in Europe, it would be near 190,000 km.

The effects on the *environment* and *human health* will be comparable in scale between the two mobility options as presented as whole economies in Figures 6.4 and 6.6, which appear later in this chapter; that is, one would have to drive considerably more than an average number of kilometers or miles per year in order for the BEV to be better than the ICEV (yes, even running on fossil fuel). Furthermore, the relationship between the two cars remains remarkably constant when moving from lighter comparisons through to heavier cars (e.g., Audi Q7 versus Audi e-tron). So the break-evens in Figures 6.1 and 6.2 are likely to be very similar for most pairs of comparable cars.

Some people reckon that they can make the BEV substantially better by charging only with "green" electricity; however, that neglects the fact that they are still drawing from a national grid that must, in total, feed all requirements: what *you* take from the "green" percentage is not available for others, including industry. You might expect the straight lines for BEV driving in Figures 6.1 and 6.2 to be curves (i.e., getting "better" into the future), indicating gradually less fossil energy consumed per kilometer driven; however, that is very unlikely because, on the timescales of current developments, electricity demand is increasing greatly, and the infrastructure needed to provide sufficient "green" energy is creating its own, massive CO_2 emissions. This is obvious because global CO_2 emissions are continuing to increase despite growing renewable electricity production.

To make matters worse, the production of cars continues to rise year on year, by between 1.5% and 3.5%, and some industry analysts predict a boom because of the EV "revolution." Already now, the production increases are outstripping the growth of regenerative energy as a percentage of total primary energy.

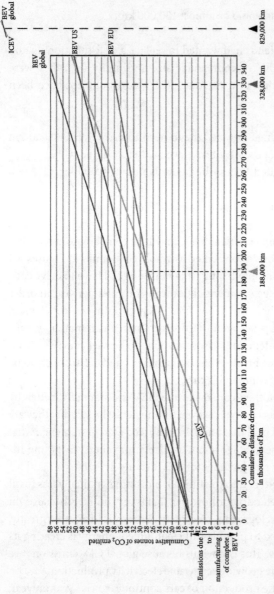

Figure 6.2 Plot of the cumulative difference in CO_2 emissions that result from the scenario of replacing a well-functioning and relatively economical ICEV with a comparable new BEV. The BEV value at 0 km corresponds to the total BEV car (i.e., ready to charge and drive) made with a combination of fossil fuels, and global electricity, of which 66% is fossil (electricity). The break-even points (determined similarly to Calculation 1 for Figure 6.1) are: for global electricity mix, 829,298, rounded to 829,000 km; for US electricity mix, 328,264, rounded to 328,000 km; for EU electricity mix, 188,077, rounded to 188,000 km. Electricity mixes and other assumptions are the same as in Figure 6.1. The differential at 0 km between BEV and ICEV is the *total* CO_2 emissions from the production of the BEV, taking the ICEV as zero, because it already exists in this particular scenario and is being replaced with a BEV. Well-to-tank/well-to-plant emissions of an additional 23% are included in all figures. Note: if battery manufacture is in China, with a fossil electricity component of 71%,[16] all break-even points become somewhat greater.

Again, even if a car manufacturer reports using large percentages of "green" electricity, one must see this in the larger economy, where many other sectors need "greening." According to my analysis, on a global scale, the car economy is not helping anything to decarbonize or improve environmentally.

Approaching 10 years ownership with a BEV is much more worrisome (personally and environmentally) than with an ICEV

At a global average of 12,000 km driven per average car per year—somewhat above the European average—the CO_2 emissions break-even point between new ICEV and new BEV (Figure 6.1 scenario) would be reached at more than 40 years (when neither car will be around anymore); in the United States, with an average annual mileage of 14,260 miles/23,377 km (2022), break-even is at around 9.6 years. In Europe, it would be 11.3 years. The average ownership of a new car in Northern hemisphere countries is somewhere between 8 and 9 years (in the United States, it's 8.4 years[17] and likely to be comparable in Europe[18,19]). Even if it's 10 years, there's a problem because, at current technology status, a BEV at 8–10 years (and outside battery manufacturer warranty) will, on average, not have broken even with a comparable ICEV; and it may well not be so easy to sell with its original battery. If the owner wants to keep it substantially longer with good usability, the probability of a necessary battery replacement will rise quite steeply, at a large financial—and environmental—cost. An average EV at 10 years may well need a new battery, as suggested by one of the few thorough small EV analyses published in 2019[20]—either to maintain a reasonable range or to persuade someone to buy it as a second-hand vehicle. End-2022 opinion from the United States National Renewable Energy Laboratory suggest 12 – 15 years battery life in moderate climates. The global average is likely to be lower.

Battery deterioration is well-studied, and, at 80% maximum charge-holding capacity, an EV battery is generally defined as unsuitable for a car.[21] Its aging accelerates further with time, raising the question of how attractive it is for long-term second-use applications. Economic analysts reckon that people buying a hybrid car wish to see cost-savings (over an ICEV) in 2–5 years, but for a *pure* BEV it's 8–10 years or more. Hence, before reaching break-even and providing benefit, a large number of BEVs are likely to become substantially less desirable for their owners and for resale. For the scenario of replacing a well-functioning and relatively economical ICEV with a comparable BEV (Figure 6.2), the disadvantage to the owner is even greater, both financially and in terms of environmental conscience. Yes, a used ICEV can be bought by someone else, but in Europe second-hand cars—particularly the now-despised but climate superior diesels—are accumulating in unheard-of numbers at dealers: the second-hand market is overflowing.

In conclusion, it is likely that current EVs have shorter life cycles than ICEVs, a development that is not environmentally advantageous. As we will see from another perspective later (according to my calculations), overall it is without a shadow of doubt a bad thing for global warming, the environment, and global human health to replace a good, economical ICEV with a comparable BEV at the current state of technology and manufacturing. It is highly questionable whether most BEVs will have "beaten" comparable ICEVs in environmental impact over their lifetimes: it is likely that most will have caused significantly more impact than an ICEV.

The nagging "driver" of what seems to be turning into EV madness is reduction in CO_2 emissions, but even that equation already seems not to be working in their favor (Figures 6.1 and 6.2) for most normal applications. Then come political problems, leading Germany, for example, to start burning more coal to generate the electricity that is, among other things, used by its EVs (both in production and driving). Coal has the worst CO_2 balance per kilowatt-hour of generated electricity. Any CO_2 savings that might be envisaged in the current EV economy are highly sensitive to situations (as noted by many other authors), and they are often theoretical. For example, though I chose 66% as the fossil electricity for battery manufacture in the previous plots, 80% of global Li-ion battery manufacture is in China,[22] where 71% of electricity is produced with fossil fuel (trend stable).[16] Is it sensible to trade theoretical advantages of BEVs for the certitude of serious impacts on the environment and human health in the present battery-driven vehicle economy? We are currently damaging the climate, environment, and human health with a transition technology (current BEV) instead of greatly reducing consumption with existing technology (ICEV running on fossil fuel) and more carefully exploring a greater diversity of alternatives.

Doing the "right" thing when buying a new car: Modest ICEVs used little are probably the best option at present

In Europe as a whole, the average kilometers driven by private cars per year across all types of use is falling, and was around 11,300 km before COVID hit.[23] It then dropped considerably because of COVID, and it may well remain below 11,000 given that many companies are introducing more home office–based work. Basically, the message emerging from Figures 6.1 and 6.2 with regard to a vehicle for routine use and assuming a car-keeping period of 10 years, is this: If one doesn't use a car much, but still needs one (e.g., one lives far away from public transport hubs and car-sharing initiatives), a modest (even better, second-hand) ICEV is going to be better for the climate than an EV up to annual use of 13,000 km (7,930 miles) per year in Europe and up to 22,000 km (13,670 miles) per year in the United States (the global scenario break-even is 50,000 km [31,070 miles] per year). Because of the substantially lower mineral mining impact of the ICEV's manufacture chain, the ICEV also starts with a large environmental advantage over the BEV. In conclusion, the less one drives, the stronger is the argument for an ICEV over a BEV of equivalent utility. If one generally drives short distances that equate to a very low annual total, a modest ICEV is the very clear winner: it's obvious from Figures 6.1 and 6.2. And a *second-hand* car obviously comes with the advantage of having amortized its production impact to an extent dependent on its age.

Routine maintenance is often cited as an area in which BEVs beat ICEVs hands down. However, data indicate somewhere around 21% greater total wear on BEV tires compared with ICEV tires (as calculated from mean values from table 1 in Prenner et al.[24] for example). Simple math shows that this equates to a quantity of energy and material comparable with the material needed for the oil/filter change of the ICEV (Chapter 6, Calculation 5, online at https://doi.org/10.1093/oso/9780197664834.001.0001). Both old tires and old oil are recycled to some extent (much oil is recycled to engine oil again or to fuel). However, rubber that is lost as fine particulates on the road and in the air cannot be recovered: a recent study places microparticulates from tire and road wear at the top of the list of environmental microplastic in terms of mass per capita in Germany,[25] where it's estimated that tire wear particulates total at least 150,000 tonnes per year.[26] Old tires have, furthermore, long since been identified as a notorious waste problem. Just a couple of thoughts.

Clearly, both the oil and the energy for both tire and engine oil manufacture can increasingly come from renewable non-fossil sources so that proportion can be neglected, and, in the end, it basically comes down to energy balance and resource cycling: a BEV is not likely to be any cheaper energetically or materially than the ICEV in terms of *routine* maintenance over its lifetime. Large repairs can be necessary, but latest figures, don't indicate any major difference in the incidence of those between ICEVs and BEVs as measured from reliability metrics (e.g., see Forbes WHEELS[27]); moreover, it seems that many EVs are significantly less reliable than their ICEV counterparts.[28] As EVs grow in popularity, we will also get a clearer picture of major events. If there is serious battery damage/failure in a BEV, for example, it ends up being several times the financial *and* environmental cost to put right (replacement battery) compared with a serious engine event in an ICEV.

Painting a fuller picture of the impact of global car manufacture and driving

In this section, I present modeling and calculations, the upshot of which is very consistent with the studies that cast serious doubt on the wisdom of the EV revolution in general. One of these, the ADEME study,[13] explicitly models a type of economy where the energy of manufacture and driving are considered as part of a multidimensional application space reflecting extent and frequency of driving.[13] I go further in presenting likely break-even points between two whole economies of manufacture and driving. It is, as I increasingly realized, crucial to view the topic in this way: I have modeled a scenario for 2030 and its eventual consequences. The most important reason for doing this is the great difficulty in *fully* comparing individual vehicles with such radically different energy sources and drive-trains—essentially, a battery and a tank of fossil fuel. A *direct* comparison is impossible, but it *is* possible to compare whole economies (manufacture and driving) moving through time and looking at their yearly impact. Importantly, I imagine a world where all energy is CO_2-neutral; clearly, this will not be the case by 2030 (perhaps by 2050?). However, 2030 gives much more reliable figures for extent of vehicle manufacture and usage, and we must try to foresee eventualities before they happen. Whether the scenario is for 2030, 2040, or 2050 matters little; rather that it is a model that we can use in planning for an acceptable future. If you want the bottom lines, they are in Figures 6.3, 6.4, and 6.6, but the next subsection, at least, is important to digest because it's there that I reveal the true cost of battery manufacture.

Energy impact: It's much more for EVs than usually reported

Full production chain energies for battery manufacture appear to be greatly underestimated

Frustrated by the differences in methodology, assumptions, basic data etc. between numerous recent studies, I finally bit the bullet and did some calculations and modeling of my own. I tried my best to get to the bottom of all the energy and environmental impacts involved in car production and driving using the comparison between mid-sized BEVs and ICEVs and making some reasonable assumptions. I have the necessary background in chemistry and physical chemistry, and I was bothered by a nagging thought that the figures I was reading for the total energy of manufacture (starting with mining of raw ingredients) of lithium-ion batteries were too low—much too low. I was right. Finally, it was confirmed in an article where the author noted in the abstract that the figure of between 50 and 65 kWh per kWh battery capacity referred only to the processes going on in the *assembly* factory (which I had increasingly assumed). I investigated every part of the chain of manufacture, drawing figures from numerous peer-reviewed research articles—the most recent that I could find on the topic. I tried hard to correct for various assumptions and differences in methodology, and finally, there it was: a figure of five times the often quoted "approximately 60 kWh." And, furthermore, my figure—around 292 kWh—was a mid-range estimate (Chapter 6, Calculation 2, online at https://doi.org/10.1093/oso/9780197664834.001.0001). This figure is also very consistent with another recent review that concludes that 328 kWh are consumed in producing 1 kWh of battery.[29] So, round about 300 kWh per kWh battery capacity sounded likely.

An average BEV without battery and large parts of drive-train and electronics is at least as energetically costly to produce as an ICEV ready to drive

Extra components of the BEV that contribute to the consumption of more total energy of manufacture than for the ICEV are the charging electronics/converters, temperature regulation electronics, high- to low-voltage converters, parts of the drive motor, cooling pumps, and all of the copper wire that goes into and between them (on average 80 kg, compared with 20 kg in an ICEV). We're talking about a very complicated material and "embodied energy" composite: relatively large quantities of numerous mineral-derived and non-mineral components with numerous stages of manufacture, all of which consume energy. For *any* high-voltage/high-current EV battery, regardless of what kind (supercapacitor, redox flow, solid state) one needs all of these accessory electronics/drive-train ancillaries. Hence there is a sizable component of energetic and environmental burden that is intrinsic to the EV *concept*. How large that is, proportionally, depends on exactly how an EV is to be powered, and that, to be honest, is still a bit of an experiment.

According to the energy breakdown from the Argonne National Laboratory (ANL; a world-renowned think tank for sustainable technology development),[30] the energy needed to make a BEV's battery and ancillary technology results in a situation that is surprising for many people: a whole ICEV (including everything that one needs to step in and drive away) has very similar energy costs of manufacture to a comparable BEV that is missing its battery and large parts of the ancillary electronics and drive-train technology. In fact, some researchers have calculated that the energetic costs of the battery production are even greater[31]—much greater—than the figures that I have found. Claims that seem extraordinary deserve a pretty good explanation. Here it is, in two parts.

1. The manufacturing energies (and hence CO_2 emissions) of the battery/drive-train components are dominated by metals that have much higher

primary production energies—from ores/mineral deposits—compared with the steel and plastic of the rest of the car: steel requires a mere 6 MWh/tonne; in contrast, copper requires 8 MWh/tonne; aluminum, 17 MWh/tonne; cobalt, 28 MWh/tonne; nickel, 80 MWh/tonne; and lithium hydroxide, 73 MWh/tonne (unweighted average between brine route [32 MWh/tonne] and spodumene—rock ore—route [114 MWh/t]).

2. The enormous energy and CO_2 footprint of electronic appliances.[32,33] The seemingly bizarre relationship between large, but electronically poor devices (e.g., a fridge), and small but electronically rich devices (e.g., a laptop computer), has long been appreciated by researchers.[34] It is reported in Williams,[34] for example, that the yearly life cycle energy cost of a computer, at 3,000 MJ, is 50% greater than that of a refrigerator (2,000 MJ), despite the fridge's much greater mass of material (2004 figures).

The International Energy Agency (IEA) acknowledges the disproportionate energy consumption during manufacture of electronic devices compared with their energy consumption during use.[35] The explanation is the very high energy-intensiveness of extracting and purifying the unusual metals and semi-conductors increasingly needed in electronics and the particularly energy-intensive manufacturing processes. As noted in Irimia-Vladu[33]: "The manufacturing process of a significant amount of a high quality inorganic semi-conductor or other nanomaterial of any modern electronic gadget requires up to six orders of magnitude or more energy than the energy required for processing a plastic or a metal component." EV batteries are highly reliant on nanomaterials (not to mention energetically costly lithium, aluminum, and, still in many batteries, nickel and cobalt); the ancillary electronics and traction motor contain substantial quantities of rare earths and semi-conductors. The reduction of between 400 and 800 volts from the traction battery to between 12 and 48 volts needed for on-board electronics and electrical appliances requires a relatively large onboard transformer with associated electronics. Furthermore, recycling of electronic devices is highly energy-intensive and has a bad record of completeness compared with many other areas[36,37]: major improvements are needed here.

The low-tech glider (chassis-frame for the battery) is quite a massive structure, and I consider it part of the rest of the car to compare with the ICEV. The extra drive-train technology for the BEV adds (*at least*) an extra 37% to the energetic cost of total drive technology manufacture (as calculated from ANL figures). The "ancillary technology" for an EV scales largely with the size of the battery, so I add that extra quantity to our previous figure. Hence, if we take 292 kWh as a realistic energy for full-chain of battery manufacture, we need to add a further 108 kWh. That gives us 400 kWh (i.e., 400 kWh per kWh or battery capacity *for the technology that differentiates a BEV from an ICEV*).

A crucial aspect of the energy burden of Li-ion battery manufacture is that most of it is upstream of the assembly factory. Many studies note that larger factories are more efficient than small ones. Improvements (of any sort) at the factory-level shift the burden upstream, namely to the metals/mining industries and the chemical processing plants.[38] Battery factories are relatively young and still have potential for improvement; the industries of metal/mineral mining, purification, and chemical treatment are, however, mature, and further efficiency gains are much smaller in comparison. Most improvements arise from the primary energy source, but they are hard to achieve because we are talking about heat/chemical energy. Next we will look at the effect that energy consumption has on the simulation of future BEV economies.

A model for comparing the incomparable: Electrons in a BEV's battery with e-fuel in an ICEV's tank

First, I wanted to "level the playing field" concerning the concept of "fuel" between BEVs and ICEVs. Here there are some very large obstacles to rational thinking. The fuel for an ICEV is the liquid that we pour into the fuel tank. Outside that tank, it still has exactly the same potential as fuel and energy content as inside the tank. For the BEV, the electrons *and* the battery are essentially the fuel (a compound of two "things" that are hard to compare with gasoline in a tank). One cannot abstract the electrons from the container in which they sit, and that container is a very large, heavy, and energetically costly piece of technology to build. I therefore set about constructing scenarios where I factored in the energy for producing the battery as part of the total "energy economy" of the vehicle culture at a snapshot time interval of 1 year in the future. That year was 2030—no coincidence, because in the Fraunhofer study (and others), scenarios for sufficient e-fuel for ICEVs in 2030 had been discussed (theoretically).

I used published predictions of growth in BEV production to envisage a scenario where, globally, in 2030, we have 140 million BEVs worldwide on the roads, and we produce 35 million new BEVs in that year (figures typical of several reports). I also imagined that we have enough regenerative energy for BEVs to drive "green." It does not matter whether we are on a trend upward or downward or even standing still: we could do the simulation for previous and subsequent years with comparable results (a bit smaller and a bit larger, respectively), but for that year, 2030, we have a glimpse of what's going on. Here it is.

Assumptions for the 2030 scenario:

- 140 million BEVs with lithium-ion batteries on the roads, globally (e.g., International Energy Agency [IEA][39]).
- 35 million new BEVs manufactured.[40]

- Some reduction in full-chain production energies for batteries and ancillary technology as a result of improvements, thus reducing the 400 to 349 kWh per kWh battery capacity (around 15% reduction). I could not factor in reductions in weight because trends do not yet indicate any significant reduction in weight for BEVs.
- An average battery capacity per car of 50 kWh (this will likely be higher because my figure is a prediction for 2022, and it is widely noted that average battery capacity is growing).[30] In the United States, it is 75.5 kWh (Gohlke et al.[30] battery capacity for BEVs produced in 2020 = 81.1 GWh [table 2]; number of BEVs sold in 2020 = 239,000 [table 5]); in Europe, the average battery capacity is around 40 kWh at present according to the IEF,[39] where it notes that "battery electric cars in most countries are in the 50–70 kWh range." The 50 kWh average by 2030 is, therefore, likely to be an underestimate; still, I will keep the 50 kWh for my calculation.

You might be thinking "but where is the factor that incorporates the number of years for which an average car is *in service*"? Answer: the beauty of this simulation is that we don't need that factor: the predicted annual production figure already incorporates the demand caused by cars rotating out of use (for whatever reason). We are moving through part of the life span of existing cars with a 1-year window that accounts for their replacement. The calculations that produce the results in the following sections are given online.

The BEV and ICEV economies break even on energy consumption within the range of 6,700 to 9,800 km per year

To reveal the break-even points, we first need to calculate the annual energy consumption from BEV "culture" (manufacturing and driving) at one value that is in a "reasonable" range (I chose 12,000 km because that's likely to be the global average at present). I count the starting point as the differential manufacturing energy between a BEV and an ICEV represented by the battery pack and a substantial part of motor, electronics, and ancillary systems (as describe in the section "An average BEV without battery and large parts of drive-train and electronics is at least as energetically costly to produce as an ICEV ready to drive"). Then I turned to the ICEV. Here, the starting point for energy is zero because we are only considering the *difference* between the energies of manufacture. Similarly to the BEV, I then calculated the energy needed to power the cars with e-fuel. In both economies, both electricity and e-fuel are made with completely renewable (CO_2-neutral) methods. Calculation 3 (Chapter 6, Calculation 3, online at https://doi.org/10.1093/oso/9780197664834.001.0001) gives the detailed methodology and figures, which were then used to plot Figure 6.3. It's very important to realize that the break-evens here are not comparable with those in

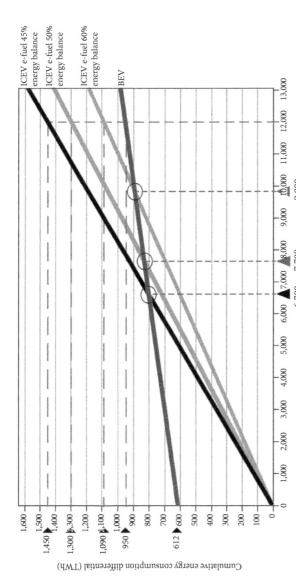

Figure 6.3 Plot of total cumulative energy consumption of light duty vehicle (LDV) economies of battery electric vehicles (BEVs) and internal combustion engine vehicles (ICEVs) (2030 scenario) against average yearly kilometers driven per vehicle (Chapter 6, Calculation 3, online at https://doi.org/10.1093/oso/9780197664834.001.0001). N.B. the data zero (where the lines cross the 0 km point, i.e., at vehicle manufacture) represents only the *differential* between the manufacturing energy between the average BEVs and average ICEVs envisaged in the simulation. In a scenario with 45% full process energy balance of e-fuel production (from CO_2 capture and H_2 production through to finished fuel), at roughly 6,700 km per year (6,652 km as per Calculation 3), the BEV economy breaks even with the ICEV economy in terms of energy consumption. Above 6,700 km per year, the BEV economy is energetically more efficient than the ICEV economy. If, as widely quoted, the energy balance of large-scale e-fuel production transpires to be 50%, the break-even point between the two economies is at roughly 7,700 km per year. Scenarios incorporating 60% full system energy balance for e-fuel produce an intersection at 9,800 km per year per car.

Figures 6.1 or 6.2 because now we are looking at a whole socioeconomic situation in which all energies are renewable and there are no net CO_2 emissions.

It's often reported in the media that an ICEV is up to 8 times less efficient (or even worse) than a comparable BEV. We can now work out the likely relationship based on the whole economy of manufacture and driving (Chapter 6, Calculation 2b, online at https://doi.org/10.1093/oso/9780197664834.001.0001): at 12,000 km per year, the ICEV economy consumes 37% more energy than the BEV economy: it's nowhere near eight-fold (let alone ten-fold). The reason is precisely the one noted earlier: we must consider the battery as a component of the fuel supply and view the whole of the automotive economy as one large unit that has energetic costs that are cumulative once it has got enough momentum (which it certainly will by 2030).

Figure 6.3 still looks OK for BEVs, at least if, in this economy, the average one is driven more than 6,700 km per year. In scenarios considered realistic for future large-scale e-fuel production with optimized process and energy recuperation, energy balance figures of at least 60% emerge. In such a scenario, we would see a break-even point between BEV and ICEV economies close to 10,000 km per car per year (Figure 6.3). However, the truly worrying problems lay in the area of environmental and human health impact of battery manufacture (and disposal/recycling).

Environmental impact: The BEV economy is several times as bad as the ICEV economy at average mileages

Here I used a published system that is applied by many researchers to calculate environmental impacts of technologies by identifying the materials involved in making an object and the materials involved in generating the energy to do so: impact of manufacture, cycling, and use of substances of all kinds, from natural resources through to highly processed synthetics. The framework is called ReCiPe 2016 v1.1 (the most recent version at time of writing), developed by the National Institute for Public Health and the Environment of the Netherlands.[41] Impact is judged in three categories: resource scarcity (how much a given technology reduces available resources), environmental impact (via reduction in ecosystem diversity), and human health (via reduction in average healthy years of life). Basically, one takes a given mass (kg) of finished product, breaks it down into its constituents, and calculates the impact of procuring them, processing them, and manufacturing them into the object in question (which also includes energy use); further impacts can be calculated during use by knowing the quantity and nature of the energy in question. Most people with access to data and a calculator can do this.

There are 18 environmental impact indicators in ReCiPe 2016:

- Particulate matter emission during production, use and recycling
- Tropospheric ozone formation—effects on humans
- Production of ionizing radiation
- Stratospheric ozone depletion
- Human toxicity, carcinogenic
- Human toxicity, non-carcinogenic
- Global warming (GWP in $CO_{2eq.}$)
- Water use
- Freshwater ecotoxicity (potential of human-derived stressors to affect inland aquatic ecosystems)
- Freshwater eutrophication (overloading with minerals and nutrients, causing dense algal and cyanobacterial growth that depletes the oxygen supply and prevents other life existing)
- Tropospheric ozone formation—effects on ecosystems
- Terrestrial ecotoxicity (potential of human-derived stressors to affect terrestrial ecosystems)
- Land use/transformations
- Marine ecotoxicity (potential of human-derived stressors to affect marine ecosystems)
- Marine eutrophication (overloading with minerals and nutrients, causing dense algal and cyanobacterial growth that depletes the oxygen supply and prevents other life existing)
- Mineral resources (extraction costs—leading to damage to resource availability)
- Fossil resources (energy costs—leading to damage to resource availability).

The first stage of impact assessment is calculating "midpoints" (i.e., equivalent weights of substances that have been well studied in their effects on living organisms). For example, the midpoint of freshwater eutrophication (depletion of life-giving oxygen) is in units of equivalence to kilograms of phosphorus. The endpoint values are then derived from the midpoints by ReCiPe's published conversion factors. This gives, for example, the rate at which species can be expected to go extinct as a result of the particular impact. I chose the 100-year (hierarchic) scenario for my model (Chapter 6, Calculation 3, online at https://doi.org/10.1093/oso/9780197664834.001.0001) because this is the period over which humans can be expected to "think" of the next generation. The plot that I produced shows a break-even point between my hypothetical BEV and ICEV economies of roughly 48,000 km driven per car *per year* (Figure 6.4). Below that distance, the BEV economy is more damaging than the ICEV.

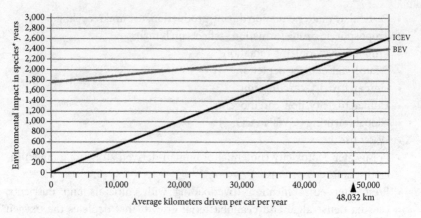

Figure 6.4 Plot of total cumulative environmental impact of light duty vehicle (LDV) economies of battery electric vehicles (BEVs) and internal combustion engine vehicles (ICEVs) (2030 scenario with 35 million manufactured BEVs or ICEVs and 140,000,000 BEVs or ICEVs on the roads, all energies being renewable) against average yearly kilometers driven per vehicle. N.B. the data zero (where the lines cross the 0 km point, i.e., at vehicle manufacture) represents only the *differential* in species*year impact due to manufacturing between the average BEVs and average ICEVs envisaged in the simulation. The intersection is at 48,035 km per year, or four times the assumed global average (average km driven per year in Europe = 12,000) (Chapter 6, Calculation 4, online at https://doi.org/10.1093/oso/9780197664834.001.0001).

Human health impact: The BEV economy is 10–20 times as bad as the ICEV economy at average mileages

This is the most frightening revelation from my modeling and calculations, and it is unarguably strongly influenced by the high value attributed to the human toxicity potential of battery manufacture, which has long been known (see table 2 in Ellingsen et al.[42]) and likely underestimated. Here, the human toxicity midpoint is given as 15,900 kg 1,4-DCB-eq for a 26.6 kWh battery. That figure is several-fold higher than any of the environmental midpoint scores. Though not easy to find, other publications also note the relatively very high human health impact of battery vehicle technology, even when compared with fossil diesel or petrol used by modern ICEVs: that is, the types of fuel that produce higher quantities of harmful emissions compared with synthetic fuels (e-fuels/power-to-x fuels/Fischer Tropsch fuels).[43] Karasu[43] provides a graphic from which one can assess the relative *midpoint* impacts, importantly over a 100-year time window (Figure 6.5), which is the same as the ReCiPe time span that I use for my simulations

This reveals a sad thing about the large-scale EV battery economy: it is comparatively very damaging to human health, but people in wealthy countries that

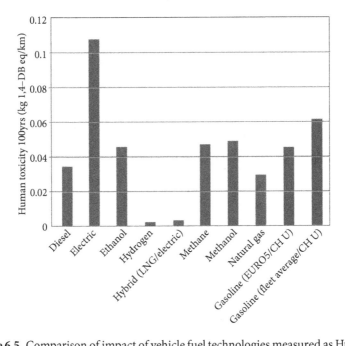

Figure 6.5 Comparison of impact of vehicle fuel technologies measured as Human Toxicity Potential over 100 years in terms of kg of 1,4-DB-eq. (equivalents of 1,4-dichlorobenzene) per kilometer driven. Redrawn from Karasu.[43] Note that figures for internal combustion engine vehicles (ICEVs) are based on the EURO 5 norm, which is less strict on emissions than the current EURO 6 norm or soon to be announced EURO 7 norm.

only consider zero tailpipe emissions will suffer less of the damage than people in the poorer countries of this Earth. It is there that most of the mining, ore extraction, and metal production happen, at present largely with fossil energy, which has additional detrimental health impacts. To attempt a fair comparison between BEV and ICEV economies in the 2030 scenario, I should try to estimate the human health impact of e-fuel production and combustion during driving—as opposed to fossil fuel production and consumption. Used in engines, it is known that e-fuels are definitely substantially "cleaner" than fossil fuels in all respects. However, concrete figures still await further research.

It is possible, from a handful of publications, to calculate the human toxicity endpoints (HTP) from *fossil* fuels, the mining and distillation of which are assumed to be substantially more damaging for environment and human health than e-fuels. I use values from Karasu[43] (Chapter 6, Calculation 3, online at https://doi.org/10.1093/oso/9780197664834.001.0001). The resulting comparison of human health impact between BEV and ICEV economies is presented in Figure 6.6. It suggests a break-even point between the two somewhere between

238   THE DECARBONIZATION DELUSION

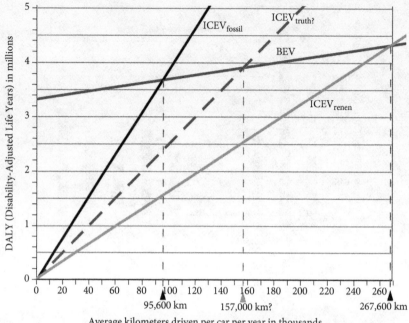

Figure 6.6 Plot of total cumulative human health impact as Disability-Adjusted Life Years (DALY) of light duty vehicle (LDV) economies of battery electric vehicles (BEVs) and internal combustion engine vehicles (ICEVs) (2030 scenario with 35 million manufactured BEVs or ICEVs and 140,000,000 BEVs or ICEVs on the roads, all energies being renewable) against average yearly kilometers driven per vehicle. N.B. the data zero (where the lines cross the 0 km point, i.e., at vehicle manufacture) represents only the *differential* in DALY due to manufacturing between the average BEVs and average ICEVs envisaged in the simulation. The two continuous lines for ICEV refer to (1) an extreme case, where the human health impacts of mining, refining, and burning fossil fuels are incorporated into the simulation (labeled ICEV$_{fossil}$), and (2) an extreme case, where only the energy of making regenerative fuels with regenerative energy (from hydroelectric, as in Figure 6.1) is counted as the contributor to the DALY score from driving the ICEV (labeled ICEV$_{renen}$ to signify, ICEV running on renewable energy). The reason for plotting these two extremes is that currently there are no robust data on the human health impacts of manufacture and combustion of e-fuels. However, it can, with reasonable confidence, be assumed that the human health impacts would be markedly lower than those resulting from fossil oil mining because of the following advantages of e-fuels: (1) no oil mining with associated environmental degradation and human health risks, including production of mildly radioactive bore-hole sludge; and (2) greatly cleaner combustion associated with absence of sulfur in e-fuels and their greater chemical homogeneity. For the sake of simulation, a mid-range plot is therefore suggested (broken line, labeled ICEV$_{truth?}$) to describe the

DECARBONIZING THE CAR 239

rough region in which the human health impacts of the ICEV economy running fully on synthetic fuels made with regenerative energy potentially lies. The break-even points of the simulations are, respectively, 95,600 km per year for ICEV$_{fossil}$ versus BEV; 267,600 km per year for the ICEV$_{renen}$ versus BEV where only the energy of fuel manufacture is considered and not the manufacture and combustion themselves; and a speculative value of around 157,000 km per year for the ICEV running on e-fuel made with regenerative energy and incorporating a speculative value for the human health impact of manufacture, supply, and combustion of that e-fuel (i.e., 13 times more than the assumed 12,000 km global average). At values below the intercepts, the BEV economy is more damaging than the ICEV; above them, the converse is true (Chapter 6, Calculation 4, online at https://doi.org/10.1093/oso/9780197664834.001.0001).

the extremes of 95,600 and 267,000 km per year per car. Below those values, the BEV economy produces greater harm than the respective ICEV one (i.e., using fossil fuel or e-fuel). Just to be clear, these kilometer values are not lifetime values for average cars: they are the average yearly kilometers covered by an average car as a component of the scenario. Obviously they are many times more than 12,000 km per year, the average that is reasonable to assume as a global value.

Matching the ADEME and Fraunhofer conclusions: ICEVs running on e-fuel have lower full-cycle impacts than BEVs

My modeling, from a complementary perspective, shows the basic phenomenon that led the ADEME (2016) and Fraunhofer (2019) studies to come to the following conclusion: in a scenario where ICEVs run on e-fuels, they have less total negative impact than BEVs. ICEVs still seem to be considerably worse than hydrogen-powered vehicles, but techno-economic modeling of large-scale practical hydrogen-based mobility also has large unknowns related to the establishment of massive new distribution infrastructures and storage facilities (static and mobile): hydrogen seems to lie much further in the future than e-fuels. To summarize, according to my calculations based mainly on published, peer-reviewed data and methods: in a scenario where, in 2030, we could use regenerative electricity to support a BEV economy or an ICEV economy of car production and mobility, the BEV would break even with the ICEV in terms of energy consumption (production plus use) of between 6,700 and 9,800 km per year per car; in terms of environmental impact, the figure would be 48,000 km per year; and with respect to human health impact, we're likely to be looking at somewhere around 157,000 km per year.

Remember, these are *not* total distances, but are instead yearly average distances driven by an average car as a component of the larger economy. These figures are very far removed from the figures that we mostly read of and hear in the media and on websites claiming to present information for the public, but not on all: "While [EVs] are 'zero emission' when being driven, the mining, manufacturing and disposal process for batteries could become an environmental disaster for the industry as the technology goes mainstream," the *Financial Times* noted in 2021,[44] because of the rapid growth in primary metal mining for EV batteries and drive-train technology (copper and aluminum—mined and processed with fossil energy—are reaching new records of tonnes mined almost every year). A typical EV contains around four times as much weight of copper as a comparable ICEV[45] (ball-park: 80 kg compared with 20 kg). Passenger EVs represent almost three-quarters of the challenge. It seems that the energetic advantages of the BEV (which are considerably smaller than often reported) are blinding most people (including politicians) to serious environmental and human health impacts.

Here is a current example: the European Union has decided to source more of its battery-relevant minerals (notably lithium) from Europe and manufacture more batteries in a shared European facility. It might be coincidence, but the opening of the Tesla factory in Germany's Brandenburg is soon to be accompanied by a large lithium hydroxide production facility that takes its raw material—the rock-based lithium ore "spodumene"—from the mountain range between Germany and the Czech Republic: the "Erzgebirge" (English = Ore-Mountains). Spodumene contains lithium in much lower concentrations than brines (as mined in Bolivia, Chile, or Argentina) and requires substantially more energy per kilogram of usable lithium extracted. Moreover, rock-mined lithium takes considerably more fresh water from local sources than does brine-mined lithium to be converted to hydrated lithium hydroxide ($LiOH \cdot H_2O$) for battery manufacture.[46] To give an idea of the range, a recent report from German Institute for Geosciences and Natural Resources[47] contains figures from which it can be deduced that roughly 100 tonnes of strongly contaminated water per tonne of $LiOH \cdot H_2O$ are produced, constituting 20% of the total water use presented in that paper; a 2022 research paper[48] contains figures from which over 600 tonnes of water per tonne of $LiOH \cdot H_2O$ can be calculated. In reading and assessing a variety of recent sources, I come to the conclusion that a realistic figure could well be in the region of 400 tonnes of fresh water per tonne of $LiOH \cdot H_2O$ from rock ore if one includes all identifiable parts of the production chain. The intended output of 24,000 tonnes $LiOH \cdot H_2O$ from the Brandenburg factory[49] multiplied by 400 gives 9,600,000 tonnes of necessary water per year, or the equivalent water consumption of a town of 220,000 population; if we include the likely water consumption of the gigafactory for battery *assembly*, we're talking about an equivalent water consumption of a town with over 280,000 inhabitants. Germany, one of the most threatened countries on Earth in terms of falling water tables, is currently losing 2.5 gigatonnes of water per year.[50] Ships on German rivers are already finding it almost impossible to transport cargo because of unprecedented low water levels.

The break-even plots between BEV and ICEV economies that I presented earlier are sensitive to the source of the lithium for traction batteries. Currently, most lithium hydroxide is sourced from brines; however, as demand continues to grow so rapidly, hard-rock ores (e.g., spodumene) are providing ever more of the world's lithium. Current literature gives values of around 2.3 times as much energy being required to produce LiOH from hard rock as compared with brines[46]—and, at current energy mixes, 2.2 times as much CO_2 emission.[46] It is, hence, not unlikely that the energetic and environmental burden of lithium-ion battery production will increase in the foreseeable future. This would shift the intercepts of the plots (Figures 6.3, 6.4, and 6.6) further to the right (i.e., requiring more driven kilometers for BEVs to break even with ICEVs running on e-fuels), or it might cancel out battery manufacturing improvements, holding the break-even points static.

Car life spans and mileages: Consumer habits and culture do not bode well for the sustainability of EVs

The mean age of cars (across all types) in the European Union at present is around 11 years[51] (in Bulgaria and Poland, e.g., it is 20 years). On average, a car built today survives for around 20 years across Europe before being scrapped or "decommissioned." These averages are—at present—made up mainly of ICEVs, and our century-long experience of ICEs is that, if regularly and correctly serviced, their fuel efficiency and range do not noticeably decline with age unless some major mechanical "event" or process of unusual wear occurs. EVs, incidentally, are similarly not "maintenance-free": to promote maximum battery life, it is essential that the battery capacity and performance be checked at intervals and software updates carried out (EVs communicate with servers more than ICEVs, hence contributing somewhat more to the growing CO_2 footprint of the Internet of Things [IoT]); brakes and braking system need servicing; battery cooling systems need checking and topping up. Customers are already being surprised by the first few years of EV service costs, which, if the study is correct, considerably exceed those of ICEVs.[52]

An ICE can be left for relatively long periods without suffering noticeable degradation if stored properly. Changing oil and a few minor components will bring it back to 100% working condition. Batteries, in contrast, age and deteriorate irreversibly; to minimize aging, they need to be kept at a constant charge and temperature of around 10°C. With present battery technology, it is highly questionable whether many completely electric cars will still be using their original battery after 10 years from new, let alone 20 years. That means that if we wish to preserve individual EVs for lengths of time that are comparable to the ICE cars currently on our streets, they will probably need battery changes. At present, at least, the energetic costs of producing an EV battery (not to mention the other environmental impacts and the difficulty of recycling their constituents economically) are substantially greater than those of producing the drive-train of an equivalent ICE car (Chapter 6, Calculation 1, online at https://doi.org/10.1093/oso/9780197664834.001.0001, data from Argonne National Laboratory,[30] and see the section "An average BEV without battery and large parts of drive-train and electronics is as energetically costly to produce as an ICEV ready to drive"). Many studies make forward-looking CO_2 emissions estimates that increasingly favor EVs over ICEs, but they usually don't factor in the future use of e-fuels in the ICE cars used for the comparison; many fail to use correct energy values for battery manufacture; nor do they consider other environmental damage apart from CO_2 emission.

The superiority of BEVs over ICEVs running on renewable CO_2-neutral fuel is also not clear when we consider all factors in the current culture of car use. Let's first take a "limiting case scenario" (an extreme-case experiment that scientists often do). If we apply this to a small car, at present status of technology and sources of energy for industrial manufacture, buying an *electric* car and not using it at all (i.e., the extreme scenario of annual km = 0) is more environmentally damaging than buying an *ICE* car and not using it. At somewhere between 0 and a number of thousands of kilometers total distance covered, the environmental damage of the two cars equals out (fossil fuel impacts "catch up" with the greater manufacturing impacts of the EV); but that cross-over point depends greatly on the source of the electricity for charging. Essentially, it is not true that an EV in the form of a rarely used means of transport is better than an ICE car of equivalent characteristics: in fact, it might well be worse. That is not to say that EVs should not be developed or used; rather that—as with all cars—the type and the use scenario should be a good match. Even if they are charged with 100% renewable electricity (which most of them are not really, because they're simply part of a mixed total grid), EVs are certainly not without emissions and serious environmental impact during manufacture and recycling. Here electricity use from renewable sources *seems* to be increasing in many parts of the world, also in conventional car manufacture, but not nearly fast enough.

Manufacture and recycling: ICEVs have many more "moving" parts than BEVs, but that argument is weak and deceptive

Proponents of BEVs often cite the enormous number of components needed to build an ICEV's drive-train in comparison. More parts, more to go wrong—possibly even more trouble to recycle (though my reading doesn't support the last claim). Figures of 2,000 moving parts in an ICEV's drive-train compared with a BEV's 20 abound on the Internet. It's hard to find out where they come from: there is so much uncertainty as to what truly constitutes a moving part and how many of them there are in an average ICEV drive-train. For sustainability, whether a part moves is unimportant; what *is* important is how much energy and environmental impact are involved in its production and recycling. The Nissan Leaf S (basic model, 2019) has 192 individual cells in its battery; the SV model has 288.[53] Those cells need to be produced individually and assembled into the final battery: one reason why EV batteries are more energy-intensive to produce than the drive-trains of most ICE cars—even if that difference is becoming smaller as technology improves.

With Li-ion technology—or any other electrolyte battery technology in application today—there is no way around multicomponent batteries. The reason: minor faults, damage, or aging in one part of the battery would quickly spread and "ruin" the rest of the battery if it were made as one block. Still, 288 cells doesn't come close to the 2,000 components of an ICEV drive-train, you're thinking; however, a Tesla Model S has more than 7,100 individual cells in its battery[54]; some versions even have 8,756.[55] Such individually insulated and physically isolated units are quite a task to recycle. In fact, we don't yet know how much of a task at "steady state" because we haven't started doing it seriously. The overwhelming majority of the numerous components of an ICEV are, by contrast, easily separable and hence relatively easy to recycle (a massive literature on this topic already exists).

EV batteries also go hand in hand with additional electronic components such as a power regulator, a voltage chopper to make alternating current (in the case of induction motors), and, in many EVs, a cooling system and even a battery-heating system consisting of numerous other parts. The present rate of increase in lithium-ion battery production is outstripping our understanding of its environmental consequences; spent batteries are starting to roll out of the back end of the car industry in numbers that far outpace recycling capacity and technology.

It is estimated that no more than 5% of all lithium-ion batteries are recycled globally (in some places it's as low as 2%); most of them (mainly the smaller ones) are tipped into landfill sites.[56] Stockpiling (with its attendant risks of auto-ignition and chemical leakage) is rapidly becoming the intermediate solution for retired EV batteries. Recent research notes, for example "The main concern that emerges from our forecast above . . . is the possibility that by 2025, the UK's dynamic stockpile could exceed 100,000 redundant battery packs, or 42,000 t of lithium-ion battery waste, for which—as we will discuss below—there is there is no readily available sustainable solution."[57]

Recycling of components from ICEVs is a well-established process that—despite the *claimed* larger number of individual components—mechanically breaks up sorts the scrap into a limited number of metallic and non-metallic components that are easy to smelt back or dispose of. For recycling BEV batteries we're talking about a whole different ball game. Many recycling plants for Li-ion batteries already exist and report being able to produce, at high efficiency, "black mass"—a mixture of various thermo-stable battery components, including relatively high proportions of lithium, manganese, cobalt, and nickel. The big challenges arise at this point, because (1) the black mass requires elaborate and energy-intensive treatment to separate the metals, and (2) there are several types of black mass, depending on the exact battery chemistry and design. A 2021 review of battery component recycling concludes that a "transnational network of dismantling and sorting locations, and flexible and high sophisticated recycling processes with case-wise higher safety standards than today" is required, and that, for low-cost batteries, it may be impossible to make recycling economically viable.[58]

Such recycling is much more energy-intensive, difficult, dangerous, and incomplete than recycling of any battery that we have hitherto attempted, and we are lagging very far behind *primary* production rates. A 73-page 2021 report from the US Environmental Protection Agency (EPA), for example, outlines the problems of Li-ion battery disposal, noting 245 fires in waste management/storage facilities between 2013 and 2020 caused by Li-ion batteries.[59] Until late 2020, the European Union, for example, still listed Li-ion batteries under "other waste batteries and accumulators," requiring only 50% recycling by mass.[60] The directive was then updated to require 65% by mass recycling by 2025.[61] However, the devil is in the detail, because within that 65%, lithium need only be recycled to 35%—and, indeed, high-percentage lithium recycling is very hard. Furthermore, the target for *collection* of Li-ion batteries by 2025 is 65% by mass, so multiplying 65% collected batteries by 35% recovered lithium, gives a requirement for 23% lithium recovery: not much. Rapid improvements are needed. Where and how batteries are made and recycled is a major unsolved problem—one that creates pollution hypocrisy, as I'll discuss next.

European-style zero emissions in exchange for environmental degradation elsewhere

In any scenario that one chooses, and even readjusting for up to several-fold errors in estimating environmental and human-health impact (see Figures 6.4 and 6.6), I conclude that it is very likely that almost all BEVs will not turn out to be better than comparable modern ICEVs. But much of the negative impact is, unfortunately, being exported out of the sight of the consumer, as we will see next.

High-level politics is supporting an unsustainable consumption culture, conscious of the false underlying logic

Electric supercars (sports cars and massive SUVs) are also touted as unproblematic: after all, if a small EV is portrayed as having a neutral CO_2 balance, and a larger one is simply more of the small one, then surely the large one must also have a neutral CO_2 balance. Who is painting this picture? In Europe, prominently the European Commission (EC). In my many correspondences with politicians and political parties in 2019, I also quizzed heads of directorates of the European Commission about their intentions with policy and their view of a sustainable future. One such exchange with a civil servant in Brussels was quite interesting. I must add that I was always impressed by the European Commission's responsiveness (all my inquiries were replied to in a timely manner)—all the more by this particular person, who engaged in a to-and-fro with me. My starting question concerned the ultimate wisdom of considering BEVs unquestionably superior to ICEVs in terms of CO_2 balance. The EC official at Unit C4 "Road Transport" at DG Climate Action replied, among other things, that "the co-legislators have decided to set targets that apply for the tailpipe emissions of CO_2, which means that battery electric vehicles are considered as zero-emission vehicles."

Upon my remarking that "Measuring tailpipe emissions is no good for assessing the environmental impact of a car," and that, in the case of EVs, "we're exporting our CO_2 conscience somewhere else just to 'get rid' of it," the official replied, "the Commission will evaluate the possibility of developing a methodology for assessing and reporting the full life-cycle CO_2 emissions of cars and vans. A preparatory study on this topic is currently ongoing."

At the time of writing (May 2022), the European Union's study on "real-world emissions" (as it calls them) had not been started. Curious readers can keep up to date with the progress at https://ec.europa.eu/clima/policies/transport/vehicles/regulation_en, which stated (past tense) in 2020 that "The Regulation also includes a mechanism . . . to incentivize the uptake of zero- and low-emission vehicles, in a technology-neutral way." That is good because it means that different drive-trains and fuels will be assessed purely on their environmental merit. However, the latest version of this information (accessed March 9, 2022) omits the "in a technology-neutral way": Does that mean that the European Commission is becoming less open to technology diversity? In April 2023, it announced a ban on sales of ICEVs in Europe by 2035; however, the ban is now tempered by the Commission's agreement to make an exception for ICEVs running on CO_2-neutral e-fuels. The Commission's website originally contained the following: "By 2023, the Commission shall evaluate the possibility of developing a common methodology for the assessment and reporting of the full life-cycle CO_2 emissions of cars and vans." That is probably too late, given the incredible speed at which the car industry is being forced to convert its production almost entirely to EVs; but now the text doesn't even exist on that page anymore. It strikes me as alarming that a body as important as the European Commission can request advice from experts, swiftly set ambitious targets, and then be so indifferent as to how they can be reached by scientifically justifiable means. Using which measures, and how, exactly, do politicians intend to compare environmental impacts of ICEV car manufacture and use with BEV manufacture and use as they plan into the future? My conclusion after reviewing the literature is that nobody has a good answer at present.

And not heeding studies that show how important it is to recognize environmental impacts beyond CO_2 emissions

In most studies and debate, the question of CO_2-neutrality of manufacturing energy distracts from the many other environmental impacts of car production: increasing renewable energy will help to bring the GHG footprint of BEV manufacture closer to that of ICEVs. However, in terms of *total* environmental impact, the BEV economy is encumbered by some very hefty problems, according to my own calculations and several recent studies that apply the ReCiPe 2016 framework.[41]

I summarize here a study that uses the 18 ReCiPe environmental impact indicators (see "Environmental impact: the BEV economy is four times as bad as the ICEV economy at average mileages") for comparing a BEV with an ICEV.[62] The study is highly complementary to the results of my scenario-modeling discussed earlier. Furthermore, it relied on analyzing an ICEV and comparable BEV cradle-to-grave in order to identify and quantify all material and energetic costs. The study is also interesting because it incorporates the same kind of broad environmental impacts as the French study from ADEME (described in the earlier section "'Inconvenient' results lead curious minds to investigate").[13]

Finally, it cannot be criticized for being done with money from the car industry or some other car industry–related body: rather the researchers work at the University of Applied Sciences Trier, an overwhelmingly publicly funded institution, and received no external funding. Here are two high-level conclusions of relevance to current technology:

- At current levels of renewable energy production, industrial processes, and fuel consumption figures, a medium-sized European BEV with 51.8 kWh battery made in China (for comparison, the Tesla Model S battery can be between 70 and 100 kWh) would break even with an equivalent diesel car (running on fossil fuel) at somewhere between 200,000 and 300,000 km.
- If we factor in second-use of the battery—which is still a largely unexplored technology—the figure drops to close to 175,000 km. Battery manufacture considerations also make a large difference.*

The most recent figure available for kilometer traveled per car per year is 11,300 km[23]; the average age of scrapping for cars in Western Europe is 18.1 years.[63] Hence, the average Western European car will have covered around 205,000 km in its life. Distances traveled by car per year are clearly falling across the European Union,[23] hence that figure might even fall in future. So with figures from the Helmers et al. study,[62] we might see a break-even between BEV and ICEV in a few cases over a car's lifetime; however, if that lifetime is 18.1 years on average, it's likely that an average BEV will need a new battery and will hence immediately fall behind the ICEV again in environmental impact.

A 2022 study[64] that considered only driving-associated CO_2 emissions from current energy mixes concludes that "with an annual mileage of 3,000 km, the purchase of a new electric car [to replace an ICEV] results in higher cumulative CO_2 emissions throughout the analyzed period, whereas for an annual mileage of 7,500 or 15,000 km, replacing the car with an electric car 'pays back' in terms of cumulative CO_2 emissions after 8.5 or 4 years, respectively." However, the authors explicitly exclude from the model the CO_2 emissions from car manufacture—which, we have seen, are much higher for EVs than ICEVs; they also exclude the possibility of running ICEVs with CO_2-neutral e-fuels. If these two factors are incorporated into the modeling, it is highly unlikely that the EV would break even with the ICEV regardless of the time span considered, a conclusion that also emerges from my modeling.

Looking across the EV sector as a whole, taking into account locations of manufacture, source of energy for manufacture, and lifetime usage styles, the present reality of break-even between the environmental impacts of a BEV and ICEV probably lays somewhere between the extremes. About 190,000 km strikes me as likely (a figure not far removed from other estimates). The much more attractive figures emerge when scenarios are stretched out to the year 2050, at which point the Helmers et al. study[62] suggests the following scenario: the car could be produced with the most environmentally conscious renewable energy, driven with that same energy, donate its battery to second-use, and then be recycled as completely and economically as possible, producing a break-even with the ICEV of around 20,000 km.[62] However—and this is a *big* qualifier—the study explicitly states "Fossil fuel choice" in the fuel comparison table: it did not factor in completely renewable hydrocarbon fuel (e-fuel) for the ICEV in 2050. Had it factored in e-fuel, it would have produced a very different picture—similar to the Fraunhofer study mentioned earlier in this chapter and my own modeling: that is, that in a future where all energy for production *and* driving of ICEVs is completely renewable and fossil-free, the ICEV has the edge over the BEV in terms of environmental balance from the beginning (including that due to energy economies). But something tells me that had the car industry and human behavior been more modest in ambitions, we might never have gotten into such muddy waters in the first place.

Car makers under pressure, the failure of the world's most economical car, and the road to madness

The heat is on for automobile manufacturers because the temperature is rising

The car industry, particularly in Europe, is under enormous pressure: it is widely known that the European Commission has set legally binding CO_2 emissions targets for 2030[65] that translate into a very, very low consumption of diesel or petrol fuel in conventional ICEVs: 2.2 L per 100 km. Currently, a small, modern diesel car with a medium-sized motor has manufacturer-side consumption figures around 3.5 l/100 km. But there once was a car called "Lupo 3L" by Volkswagen,[66] so named because it was designed to consume no more than 3 L of diesel per 100 km if driven reasonably. It had a three-cylinder turbo diesel engine, various drive-train innovations to reduce consumption, many components made of light alloys, and an almost complete absence of "knick-knacks" for comfort or entertainment.

The Lupo 3L weighed in at only 850 kg curb weight, could carry four people (plus a little luggage), and was, in fact, capable of even better diesel consumption than its eponymous 3 L: in 2001, a Japanese researcher, Dr. Shigeru Miyano, set out, in a properly scrutinized experiment, to prove just how economically one could get around in a small car. He drove a Lupo 3L around Britain (5,600 km) with a final average diesel consumption of just 2.36 L per 100 km, setting a new Guinness Book of Records record.[67] Few people know about the Lupo 3L, so what happened to this "heroic" little car? Lupo 3L production started in 1999 and ran until 2005; nevertheless, only 5.5% (27,000) of the almost 490,000 Lupos produced during this time were 3Ls. The wide success of the 3L was never to be. Ironically, one could imagine that if Volkswagen had been serious about expanding production of the 3L to meet consumer demand, it might not have become so ensnared in cheating to cover up the emissions shortcomings of its larger diesel engines a decade or so later in the celebrated "Dieselgate" scandal[68]; also, it would have contributed to a lowering of average vehicle emissions in the meantime.

Cars are becoming bigger and heavier despite the need to reduce environmental burden

Unfortunately, the demand to which the car industry seems to be pandering is one for size and power. In 2019, in the United States, a record 67% of the new car registrations were sport utility vehicles (SUVs), crossover utility vehicles (CUVs), or pickups,[69] and, even in Europe, the figure was an all-time high of 40% SUVs.[70] In 2020, the even larger pickups outsold all other types of car for the first time.[71] Worldwide in 2020, 42% of new car sales were SUVs or similarly large vehicles.[72] In all areas, the SUV trend is growing year on year. A greater percentage than ever before of these cars was electric or hybrid, and some industry experts see the BEV revolution essentially killing off the small to medium-size non-luxury car market. Larger ranges are much easier to achieve with larger BEVs, which also make much larger percentage profits than smaller cars. A recent German auto club (ADAC) analysis clearly shows a strong correlation between size and range[73]: having the highest range, the BMW iX xDrive 50, a large SUV weighing in at 2.44 tonnes, with a nominal consumption (before distribution and charging losses) of up to 21.1 kWh/100 km (manufacturer figure)[74]; at the bottom of the range table is the Smart Forfour EQ passion, with one-sixth of the range, 17.6 kWh/100 km consumption, and curb weight of 1,200 kg.[75]

Electric sports cars are also starting to accumulate record orders, for example the Tesla Model S P90D, with more than 500 horsepower and a curb weight of 2.25 tonnes (more than two-and-a-half times as heavy as a VW Lupo, and almost twice the weight of a comparable Porsche 911 sports car). Meanwhile, Porsche unveiled its electric supercar in 2019: the Taycan "Turbo" (yes, that's really what it's called) with 680 horses under the hood and a curb weight of 2.3 tonnes. That's an impressive amount of energy-intensive materials, largely (at present) produced with fossil fuel energy. About 30,000 orders were placed for the Porsche in a matter of weeks after its announcement, and, at the time of writing, and if the 2022 first-quarter sales of the Taycan are indicative, almost 40,000 of them will be sold in 2022.[76]

Audi's latest e-SUV, the e-tron, weighs in at between 2.4 and 2.7 tonnes empty,[77] has 408 brake horsepower (bhp) as standard, and can accelerate from 0 to 100 km/h in 5.7 seconds. Order books are full, years into the future, and figures showed sales of the e-tron in the United States surging to record levels in 2021.[78] A super sports version, the e-tron GT RS, is offered with up to 646 bhp and can accelerate 0 to 100 km/h in 3 seconds (leaving quite a lot of its tires behind as microplastic in the process). The trends are clear, as an alarming graphic from the Argonne National Laboratory shows: from 2011 to 2020, the average motor size for a newly bought BEV increased from 80 to 290 kW, and the acceleration times decreased from 10.3 to 4.8 seconds.[30] Are such trends consistent with environmental sustainability?

And range increases are largely made simply by increasing battery mass

All of this is a far cry from the culture of parsimony and modesty that would have made the Lupo 3L a success! So what went wrong? The car makers' strategy is clearly fueled by a growing political conviction that BEVs have finally solved all environmental problems of the motor car. Maximum ranges are apparently also increasing modestly, but under typically "optimistic" test methodologies. Government test regimens for estimating a BEV's range, particularly the US EPA, are very "gentle" in terms of acceleration and maximum speeds (averages: city driving 20 mph/30 km/h; highway driving 48 mph/79 km/h). At a speed of 79 km/h, one would even be holding up heavy-duty vehicles. At these speeds, maximum ranges for higher-end models of around 300 miles or 490 km are reported, notably from a full battery charge (not the 80% recommended for maximum life span) and discharge to the point at which the vehicle can no longer hold a constant speed (likely below the recommended 20% residual charge).

The EPA applies a "correction factor" according to real-life data: minus 30%, hence producing a customer-facing figure of 210 miles or just over 343 km range. That windscreen-label range, particularly at higher highway speeds than used by the EPA, can suffer a further reduction of more than a quarter in real-life use,[79] producing 157 miles or 257 km. Naturally, the diesel or petrol consumption of ICE cars is also traditionally estimated based on unrealistic tests and questionable adjustments: only the end user knows the truth, and that truth, of course, also depends on the type and style of driving; however, many modern diesel cars can easily achieve ranges of 1,000 km or more between tank fillings. A final truth is the cost to the customer of a new battery—be it after 10 years, as a result of normal aging, or as a result of an outside-guarantee problem or accident: as a recent article in *Current Automotive* revealed, the cost to the customer (in this outside-guarantee case) was nearly US$16,000 (€13,400), consisting of US$13,500 for the battery pack; the rest being the 13 hours of labor needed to remove the floor of the car, remove old and install new battery (480 kg), fill cooling system with new fluid, etc.[80]

Because we should be concerned about the sustainability of this economy, we should ask how many people would wish to invest US$16,000 in their car at 10 years, particularly given that, by that time, many other things will probably need replacing. BEV proponents continue to proclaim massive cost reductions as batteries become ever more efficiently made. But my estimate is that massive cost reductions will not happen if we embrace more environmentally conscious manufacturing right from the beginning (mineral mining) and majority-recycling of battery components (as mentioned earlier): the true price for environmentally responsible production will be much higher. But the true price is currently obscured by the many financial advantages that EV research, development, and purchase receive and that are arguably disadvantaging research in other areas.

The electric mobility revolution has been financially supported enough: Now it's the turn of e-fuels

Taxpayers in countries that offer large subsidies for buying an EV and either scrapping or trading in one's existing car are never asked whether they agree with such a policy. In Germany, for example, until the end of 2022, one could receive up to €9,000, plus road-tax exemption (and in some cases free electricity for a number of years). Government support in Germany for the development of e-fuels to economically viable scales is relatively small, but with the latest government that might well change. Some e-fuel companies, such as Sunfire (which started life in Germany but has changed its focus to Norway), have already realized the scientific and technological wisdom of using e-fuels in many large-vehicle applications (heavy goods vehicles, ships, airplanes).

Scale-up to levels that will make a significant dent in the use of fossil-derived fuels for the large-vehicle transport sectors is under way (e.g., Sunfire and Norsk e-Fuel AS) and looks very likely to continue, helped in Norway by a high percentage of hydroelectricity in the national electricity mix. If e-fuels made with renewable electricity are produced in increasingly larger quantities, the bulk costs decrease, hence making them financially ever more attractive to other transport sectors, including ordinary cars. Furthermore, it is true that facilities for producing e-fuel require enormous amounts of energy to build and run at steady state; however, the same is true of battery factories, and we haven't even started to factor in the energy economy of large-scale recycling of batteries, which is currently only in development. As mentioned earlier, as little as 5% of the constituents of batteries is currently recycled, and that percentage is not rising because production of batteries is accelerating so quickly in reaction to demand.

The financial costs of recycling compared with those of primary mining are still rather sketchy; the environmental ones even more so. What of the energy requirements of large-scale recycling? The honest answer is that nobody really knows. Hence we must make many assumptions in comparing regenerative e-fuels with battery technology at the moment because there is a crucial part of the battery full-life-cycle analysis missing. What is more, that part is missing because the technology of battery recycling is lagging far behind where it should be. Calculations of break-even points for BEVs against ICEVs give us a glimpse into questions of the true environmental cost of an average full life cycle from another perspective.

The argument that e-fuels are ridiculously energetically costly to consider does not "work" when viewed in a larger economy of manufacture and driving. A similar argument is that clean energy generation technology needed to supply the energy to make e-fuels would create environmental and climatic impacts that are unacceptable. My calculations suggest that they would not be greater than those of the EV economy. But, importantly, if we immediately cut car driving by 50% (preferably permanently), we could use part of the "saved" 50% to invest in clean energy generation for e-fuels. That solution would, in my analysis, have less environmental and human health impact in the medium to long term than EVs. We should be doing this experiment *now*. More than ever before, we need openness to a variety of technologies, because so much is at stake.

How e-fuel technology could help us reach net-zero CO_2 as fast as possible

Driving industry harder with newer technologies: A good way to reach net zero CO_2?

There is a serious paradox implicit in the developments surrounding mobility: we need to reach net zero emissions of CO_2 as quickly as possible and stay there. We also need to be acutely aware of other serious environmental side effects of technology development and industrial activity—as embodied in the 18 factors of environmental impact developed by the Netherlands National Institute for Public Health and the Environment (ReCiPe 2016)—see the list in the section "And not heeding studies that show how important it is to recognize environmental impacts." Concentration on battery-powered mobility currently neither respects the environment as a whole, nor does it *immediately* address net zero CO_2 emissions. We seem endlessly to be kicking the proverbial can further down the road instead of thinking long term. And there's a lot that we can do better in the here and now, I believe. We are overlooking more immediate measures to reach zero emissions.

There are very insidious forces driving this thinking: the need for industry to make profit within politically set legislative bounds (emissions limits and CO_2 quotas) and the willingness of humans to take solutions made for them by some organization and avoid thinking too much about the problems themselves. Could these also be the reasons why the concentration on personal mobility via ordinary cars (which generate 7–8% of GHG emissions worldwide, compared with agriculture, which generates more than 20%) is rather disproportionate? Often-quoted figures of between 14% and 17% refers to all road transport, by the way.[81,82] Personal road mobility (cars, motorcycles, buses, and taxis) accounts for around 7.3% of global CO2 emissions.[83] CO_2 emissions from power generation and fuel combustion in industry in total account for almost 60%,[2] but they are much harder to tackle from a political perspective. How much easier to focus on an aspect of everyday consumption and stimulate more purchases in the name of the environment!

Could greatly reducing the use of what we already have be a better immediate solution?

There are many hundreds of millions of private cars, many of them very new and with plenty of life left in them, running on diesel, gasoline, and gas. One automotive digital marketing company gives the 2022 global figure for all motorized road vehicles as 1.45 billion, the highest per capita ownership being in the United States (0.89 vehicles per capita).[83] We know of the massive environmental cost of the production of new cars and hence the massive consequences for the environment of trying to replace—as quickly as possible—all conventional cars with different ones. That simply feeds the earnings of the car industry, and keeps politicians happy in countries with large auto manufacturing sectors (e.g., in Germany, 800,000 people work in car and car component manufacturing,[84] almost 2% of the working population).[85]

Driving of *private* cars accounts for even less than the 7.3% of CO_2 mentioned above—which fell to 3.7% in 2020 because of COVID—but far from the consciousness of the private consumer are the emissions caused by the car industry, tucked away in the almost 60% of CO_2 emissions made by power generation and industry in general in 2020.[2] We know that steel production accounts for around 7.2% CO_2, and manufacturing industries accounts for around 10.6%,[81] but where, exactly, is the car industry in all of that? Research by Greenpeace in 2019 concluded that 9% of global emissions are from the automotive manufacturing sector.[86] So, seen from an aggregate production perspective, rather than at the level of an individual car, it appears that car manufacturing emits very similar quantities of CO_2 as car driving: we're not talking about several times less, as the analysis of a single car's life cycle would suggest. This is a growth industry, and it partners with a consumer culture that buys more and more, increasingly heavier, cars. Regardless of whether it's a BEV or ICEV, the production of *any* kind of car cannot possibly help to reduce CO_2 emissions quickly, but the much greater per-car manufacturing energy costs of BEVs add a further problem on top.

Choosing areas in which to move quickly with least environmental impact: Driving less and adding e-fuels to the mix

The "quickly as possible" that we should be concentrating on at least as much as new drive-train development is (1) driving less and (2) replacing the fuel in the existing cars with CO_2-neutral fuel—and doing it with as little environmental impact as possible. In Chapter 4, I presented a model based on reducing fossil fuel consumption immediately and using half of the saving to build e-fuel infrastructures. We need alternative solutions to oil palm plantations and similar cheap bio-fuel schemes that simply steal land from nature and degrade it or compete with human food production. Instead we must bite into the sour apple that is the true cost of environmentally conscious fuel production: it is expensive! Why politicians see fit to provide massive subsidies for battery development and little for e-fuel development is a question that should bother all voters: politics seems quite capable of artificially lowering the price of EVs to make them attractive to buy while not wishing to support in equal measure the development of e-fuels as a component of energy solutions for mobility. That is the long and short of it. Moreover, it should be clear by now that no technology will save the planet unless it is accompanied by responsible behavior: for example, we might be well-advised to consider limitations on the amount of gasoline or diesel that we are allowed to take from the pump each month—forcing us to drive less. There is, in my opinion, no such thing as "sustainable technology," but there is certainly sustainable behavior.

The fact that is often neglected in political pronouncements and media reports of research on environmentally sustainable development is that technology development in a variety of areas is progressing simultaneously. A non-expert observer could be forgiven for thinking that particular research has only started when reported on the radio or recognized as important by a government minister. As for alternatives for powering road-based transport, research has, in fact, been ongoing for almost 100 years in (petro)chemical laboratories and production facilities, mainly in Europe. Oil in Europe is not as abundant as in certain other parts of the world, and economic development and security in times of war are therefore dependent on other regions. These realizations propelled Europe to pioneer research into synthetic alternatives. In World War II, this led to the ingenuity of using wood-burners fitted to cars to produce carbon monoxide (CO; via incomplete combustion of wood) that would burn in their ICEs. Back then, synthetic fuels were not truly synthetic because at least one component had to be made from an organic source (often natural gas); they were also hard to produce in quantity and extremely expensive. Prospects have since changed greatly.

Cost of e-fuel to the customer is now "right"

Science and technology in this area have advanced considerably in the meantime, and truly synthetic hydrocarbon fuels can now be made in relatively large quantities[87]: a principal route is the Fischer-Tropsch process, as described in Chapter 4. In the latest synthesis pathways, CO_2 from the atmosphere is reduced by sunlight to CO; hydrogen is obtained by electrolysis of water, again using energy from sunlight. Diesel fuel from Fischer-Tropsch, for example, is not available at your local filling station because until recently it was rather more expensive than fossil diesel. Russia's war in the Ukraine has brought fossil fuel prices near to parity with synthetics. And it could get even better for synthetic fuels: in 2019, one company press release forecast the price of synthetic diesel falling to between €1.20 and €1.40 by 2030 (excluding excise duty).[88] Inevitably, the price of synthetic fuel will fall with further progress and upscaling, and, given that the final price of fuel at the pump is largely government tax, the state can do a lot to encourage or inhibit such developments.

Indeed, with the introduction of CO_2 emission taxes, fossil diesel, gasoline, and gas are becoming more expensive anyway. The €2 per liter that Europeans paid for automotive diesel in Spring of 2022 would be the cost of e-fuel if it were exempt from the majority of current fuel taxation. That would, in fact, be environmentally sound politics: if made and distributed with renewable electricity, e-fuel would, at steady state, produce no CO_2 emissions. Even methanol—which is an early (and therefore energetically economical) product of Fischer-Tropsch chemistry—can be used for ICEVs and has, for decades, been powering many racing cars, as described in Chapter 4 (see also Wikipedia[89]). Methanol has already proved itself in many trials to be a serious option: it is only politics and industrial considerations that are holding it up.

And we desperately need to develop more geopolitical independence in energy

The reasons why e-fuel schemes *in general* have not (yet) taken off stem partly from political and legislative problems and the resulting lack of funding and reassurance to the fledgling industry that it will be taken seriously; but there is the ever-present argument that we don't have enough regenerative electricity to share with fuel production. Reflecting on the massive amounts of energy needed to make new cars, and new electric ones at that, I cannot accept that argument. However, politicians and legislators might have to rethink, and not just for reasons connected with climate change and the superior combustion properties of e-fuels. For geopolitical reasons, nations dependent on imports of oil, as well as strategic metals that are used in large quantities in EVs, are well-advised to become more autonomous. This will unlikely be achieved by replacing most of the transport sector with machines that don't use hydrocarbon fuels, but rather by embracing CO_2-neutral hydrocarbon fuels (Figure 6.7). The production systems already exist and require scale-up; distribution infrastructures are those that we have already been using for a century. Peak energy from renewable sources that cannot currently be stored can be made into fuel. We must not make the mistake of branding all fuels that come from a gas pump "dirty in principle": the fossil oil industry *is* dirty, but we cannot assume that the e-fuel industry would be dirtier than the industry of battery production, for example. In my analysis, it would likely be significantly less dirty.

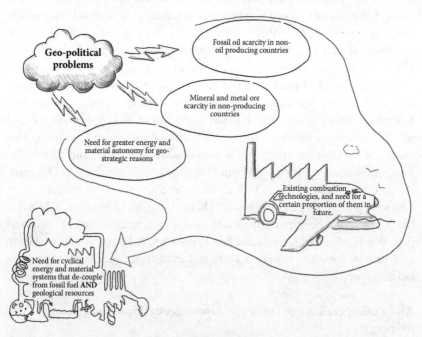

Figure 6.7 How geopolitical instability can rapidly favor e-fuels in countries where oil and minerals for battery/electronics manufacture are scarce or hard to mine.

Conclusion: Understanding full environmental impact, true recycling, and the need for openness to technology diversity

There is, I concede, a lively exchange of views on the matter of the energy-intensiveness of EV battery/ancillaries manufacture compared with ICE manufacture. "Views" is an appropriate word, because depending on what one "looks at," from which angle, and how fairly, one can obtain very different results. At the very least, we must have the honesty to admit the following: the energetic costs and full environmental impact of ICE car manufacture are hard to calculate; those of Li-ion battery manufacture are *even* harder to calculate. The mere observation that recycling of spent batteries is not contributing much to the global production of new batteries (which are still more than 90% produced from de novo mined minerals)[90] is worrying: mining new minerals for new batteries is less expensive than recycling. Does that also tell us something about the true *energetic* costs of a recycling-based battery economy? Smaller Li-ion batteries still go into landfill sites in large proportions the world over: larger batteries are mostly stockpiled pending further developments. It *is* possible to recycle an EV traction battery, but there is no doubt that it is more complicated and energy-intensive than recycling the engine of an ICEV. With batteries, we are talking about taking apart the complex composite consisting of hundreds or thousands of individual cells, mechanically breaking them into small pieces, and (usually electrolytically) extracting the metals into a mixture (with many other substances) called "black mass," separating and purifying them.

The significant improvements in costs—both energetic and financial—of EV battery production that are celebrated in report after report these days are based almost entirely on freshly mined lithium, nickel, and cobalt. They do not consider the costs of a battery economy based increasingly on recycled battery constituents. It is unlikely that batteries from majority-recycled minerals will be cheap. If Li-ion batteries remain the mainstay of EV power, and production from primary sources continues largely unchanged, the following problems are obvious: (1) the increasing need to mine lithium from sources other than subterranean brines (i.e., from rocks containing percentage-wise much less lithium, which is greatly more energy-intensive and water-consuming to extract than from brines) and (2) the imperative of recycling lithium from spent batteries, which is also more expensive than taking it from brines. Put simply: the rapid fall in the price of EV batteries—and hence EVs themselves—is, to a degree, occurring at the cost of environmental sustainability. That aspect is completely ignored in the optimistic reports about EV development that I have read.

As noted in an editorial in *Nature Energy* 2019 (which cites several reports of the rapidly increasing challenges of battery recycling and the environmental urgency of addressing them), we are very far away from a sustainable EV battery economy, and we must do better. In a 2021 news item, the journal *Science*, notes continuing serious problems with EV battery recycling: in particular, the fact that they are still not designed and manufactured with recycling in mind and that it is still cheaper to build them from de novo mined materials.[91] In another sense, research and development *is* continuing to do better: different types of battery (e.g., based on solid-state technology instead of electrolyte chemistry) are being developed. A survey and comparison of these technologies is outside the scope of this book, but the acute problem centers around Li-ion batteries because they are still, by a wide margin, the mainstay of the BEV industry and are seen as the fastest way out of a dilemma. Their production is increasing rapidly and will probably continue to do so for years to come. This whole story is arguably less about environmental sustainability and more about an emergency plan to sustain the car industry and personal mobility. According to my calculations, ironically, at current technology and energy mixes, we are probably producing more net CO_2 globally per vehicle via the full EV economy than via the full ICEV economy running on fossil fuels.

Before ending this chapter, I wish to make a plea for technological openness at a time where diversity of thinking appears to be shrinking dramatically, guided not only by incomplete arguments and predictions, but also by three (mis)conceptions or prejudices:

1. The ICE is an old-fashioned technology that should now make way for something more "modern": untrue, because (a) engine development—particularly in conjunction with energy recovery technologies—is far from finished and (b) CO_2-neutral portable fuels are a technological development that inevitably accompanies the ICE, and these fuels still have great development potential.
2. The ICE is noisy: this one gets amusing. Undeniably, EVs and hybrid cars are almost silent at low speeds and are definitely less "noticeable" to everyone, whether pedestrian or cyclist. The European Commission therefore issued legislation requiring all EVs and hybrids manufactured from July 1, 2019 onward to be fitted with a noise-generator ((EU) 2017/1576).[92] At the time of writing, it is reported that a sound of between 56 and 75 dB should be emitted up to a speed of 20 km/h (i.e., in the range of noise emitted by an average ICE car).[93] Interestingly, 20 km/h is very close to the speed at which road noise overwhelms engine noise in most modern ICE cars driven reasonably. Imagine, then, a town seething with EVs as opposed to seething with ICE cars. With such repeated stopping, starting, and generally relatively low-speed driving, this town would be exposed to very similar levels of traffic noise as one containing the same number of ICE cars. There is essentially no difference in the road noise produced by

a BEV in comparison with an ICEV. Finally, picture highways full of cars "hissing" past at over 100 km/h: it matters naught in terms of noise pollution whether those cars are powered by electric motors or ICEs: road and air noise massively outweigh.

3. The ICE is "filthy": untrue, because modern ICE technology has reduced all types of emission in line with legislative limits, the proof of which is, for example, the fact that the major contributors to town air pollution (especially particulates and nitric oxide (NO_x) in most European cities is not the combustion engine but other sources, notably domestic heating and industrial processes. Technology is still making major advances in exhaust gas treatment (see Chapter 4, the section "Much achieved, but still great scope for improving the ICE"), and synthetic fuels burn much more cleanly than fossil ones. Persistently labeling the ICE as a "filthy" technology risks creating another problem: the ever-diminishing, local environmental impact of ICE *use* risks being eclipsed by much greater environmental impact (of many kinds) elsewhere on our planet (i.e., environmental degradation linked to the "renewable" energy and EV industries). In due course, the consequences will likely reach back to us in our nice "clean" towns in a variety of forms.

In March 2021, 27 leading German professors from the areas of mobility and automotive technology study sent an urgent message to Brussels: they beseeched the European Union not to be so dogmatic about the switch from the combustion engine to electric drive-trains.[94] They note the lack of wisdom inherent in a plan to get rid of technologies that still have significant improvement potential and have not been definitively proved inferior to EV technology, thereby also squandering billions of Euros of research and development funds: to adapt an adage, the perceived ideologically perfect course (a one-technology solution to net zero CO_2) is the enemy of a good, future-proof, developmental strategy. It's almost about a cultural wind of change, threatening to blow away good strategy—a strategy that I believe can, and should, be applied to all realms of the anthropogenic carbon cycle and energy balances of manufacture, use, and recycling. There is nothing worse than staking all one's chances on a single card at a time when so many uncertainties exist and calamity looms large.

In this chapter, it has certainly not been my aim to denigrate EVs out of some principle-driven dislike of them: they definitely should be part of the research and development for mobility of the future, as should a variety of technologies, simply because there are so many uncertainties related to their ultimate impact on the environment. The car industry, beset by scandals and the inevitable retribution of brutally hard legislation, understandably feels pushed into a corner where it must guarantee its survival based on developing a technology that is preferred by policymakers: splitting its efforts between even two technologies (BEV and ICE) seems too much to handle. However, the reality seen from an environmental perspective is very different, and there I conclude that the maxim of "one size fits all" is seriously flawed. As discussed in Chapter 4, synthetic CO_2-neutral fuels are very feasible replacements for fossil fuels; as I hope to have presented in this chapter, the oft-raised argument that synthetic fuels are energetically massively too costly to consider feasible (in comparison with electric mobility) is fundamentally flawed if one analyzes the whole economy of car manufacture and driving. Furthermore, e-fuels have very much lower environmental impacts and are not associated with massive quantities of metal and mineral mining, recycling, and waste production.

The car is a relatively modest part of our CO_2 and energy problems, but it has become a major focus of public debate, political actions, and industry reactions. The danger that I see is that the small problem that the car currently represents will become a larger problem because of the speed and force with which the so-called mobility revolution is being pushed through: we risk producing not only a CO_2 bubble via more energy-intensive and more numerous new car production (EVs); additionally—as many studies have shown—if we take into account the broader environmental impact of car manufacture, particularly EV battery manufacture, we risk creating an environmental degradation bubble, too. This is accompanied by highly emotional consumer decisions and behavior, rather than rational and responsible analysis. Could, for example, the dream of supposedly cleaner mobility lead to the nightmare of towns crammed full of cars—possibly even more than at present? Psychology is a complicated matter, but I conclude that a much more mundane and common-sense question should be asked by potential buyers: Is it not, in the short term, simply better to keep one's existing car and use it sparingly?

7
A carbonaceous, biology-inspired recipe for sensible and environmentally conscious energy economies

A technological revolution is afoot, and it's driven by the knowledge that hundreds or even thousands of times more energy reach the Earth from the sun and are circulating in wind and water on this planet than our present energy demands. So what are we worried about? It should be child's play to satisfy those demands and even allow humans to use much, much, more energy than at present, right? In the long-term perhaps, but there are two shorter-term problems that have long been recognized: first, rapidly creating renewable energy installations also generates environmental impacts (particularly if most of the necessary construction energy comes from fossil reserves); second, and of greater relevance to this chapter, finding environmentally conscious ways of storing energy is as important as developing technology to generate that energy: if we can store energy, we can keep generation facilities to a minimum, hence minimizing their environmental impact. In my assessment, in the storage department we are lagging very far behind because we are not thinking holistically. This is complicated because one has to "get one's head" around many factors at the same time. I describe this larger system as the "economy of energy and material production, storage and use," and it has many interesting parallels with biology—in particular the metabolism of carbon-based intermediates for energy transport and storage, and for synthesis of other materials. Crucially, in the short term, we must reduce our consumption in order not to overload the environment's capacity to buffer. I conclude that the following measures are absolute requirements if we are to reduce the worst impacts that modern energy and material economies have on the environment:

1. Reduction in consumption and use
2. Striving for true recyclability (no net harmful emissions of any kind from the human system)
3. Integration and economy of "metabolism": that is, *cycling of energy and material*

4. Opportunism: that is, *making use of material and energy flows that would otherwise go to "waste" and cause environmental damage*
5. Economy of scale and decentralized production in appropriate measures
6. Diversity of technology
7. Experimentation, close monitoring, and adaptation
8. Preparedness to pay more for almost everything
9. A new materialism

I will now expand on each, explicitly discussing the development of energy economies relying largely on carbon-based energy-carriers:

Reduction in consumption of almost everything is mandatory

The principle and practice are very simple: we must all—mainly in the minority of countries (relatively wealthy ones) that cause the majority of environmental impacts—reduce our consumption and futile cycling of energy and materials: from the fuels we use, through the objects we buy, to the food and drink that we consume. For example, given that the reduction in travel in 2020, due to COVID-19, produced a net 7% reduction in carbon dioxide (CO_2) emissions, simply reducing our driving to half its pre-COVID levels would be a major and immediate improvement, without the need to buy a supposedly more efficient car. In concrete terms, in Europe, that would mean on average driving only 6,000 km per year instead of 12,000. Another obvious area is meat and dairy consumption: reducing that by half would have a comparable effect.[1] Particularly in wealthy countries, we should transition to more plant-based diets.

Reduction in home heating impact would mean weighing up the environmental factors influencing when to change to a more efficient heating system and/or improve home insulation and putting on a pullover instead of turning up the boiler. Consumption of all types of objects, from cars through electronics to clothes, should be greatly reduced. We should simply make them last longer, look after them well, mend them where possible, and resist the temptation to buy new (remembering the scale of the energy—and at present CO_2—savings that would result[2]). And the damage done by scrapping cars too soon (as one consequence of, for example, subsidies and tax-breaks for buying more economical cars, primarily electric vehicles [EVs]) is many, many-fold worse than that.

The transition to an energy economy that is free of fossil fuels and that makes no net CO_2 emissions is vital: there should be no compromise on that. However, we must also drastically reduce the cycling of materials of all sorts because not only does this consume unnecessarily large amounts of energy, but it results in serious environmental degradation. Battery EVs (BEVs) are still substantially more expensive than internal combustion engine vehicles (ICEVs), and my analysis suggests that they should remain so. Though having to pay more for all manner of things will certainly reduce consumption somewhat, I think that we need more directly tangible connections between people's behavior and its environmental impact—ones based on a knowledge of the environmental cost more than the cost to one's pocketbook. That is because paying more is a type of exoneration strategy—particularly for wealthier people who tend (on average) to have more negative environmental impact than poorer people simply because they can afford it.

How much should we reduce? Based on my research for this book, and the oft-quoted observation that we're currently using 2.5–3 Earths, I'd say half to two-thirds, in as many areas as reasonably feasible. Of course, I don't envisage us all eating half as much as we do today, but cutting meat consumption in half, for example, would have a highly positive impact on ecosystems (as shown by ReCiPe criteria), water use, and CO_2 and methane emissions. Weighting by current environmental impact would therefore be very important for identifying priority areas, and that would imply certain "don'ts" as well. So, here are a few basic suggestions, some of which we can do immediately.

Reduce

- Meat consumption by 50% (or better still, switch to a vegetarian diet)
- Car-driving (km per year) by 50–70%
- Flying (km per year) by 50–70%
- Heating by an amount that you can compensate by simply putting on another layer of clothing

Don't

- Buy that next brand new car because it looks great and attracts you with environmental arguments and a fat price reduction funded by taxpayers; instead, make your existing car last longer, and, if you really *need* a replacement car, buy second-hand.
- Buy a new electrical or electronic device sooner than necessary; try to mend (or have mended) instead of replacing.
- Shy from talking openly about your adjustments in the above areas, but don't come across as "superior"; mention them when appropriate and watch for the reaction.

Independent organizations consisting of natural scientists, technologists, land-use experts, and social scientists should produce simple information on environmental impact of all commonly bought products. It's not enough, for example, to slap a sticker on a fridge giving its operational energy efficiency as A+ if the making of the fridge is done in an environmentally very unsound way. I leave to your imagination what extra information one would need to provide next to a new car's windshield sticker proudly announcing an energy efficiency of A+.

We must strive for true recyclability

The principle is simple: we must develop procedures of production and disposal that form conceptually closed cycles (cradle-to-cradle[3] instead of cradle-to-grave). For example mixing of anthropogenic CO_2 with "natural" CO_2, and other substances with natural pools *is* allowed as long as they do not degrade the environment; but no net outgoings or incomings are permitted. The practice, by contrast, is hard, but certainly not impossible. Major scientific and technological strides have been made in monitoring the outgoings and incomings of human society since the 1980s, and the substance about which we arguably know most is CO_2 and carbon in general.

Expanding on that example, we are starting to see the emergence of an industry of synthetic hydrocarbon fuel (e-fuel/power-to-x) production. Crudely put, this takes CO_2 from the atmosphere, mixes it with hydrogen, and turns it into fuel, the burning of which returns exactly the same amount of CO_2 to the atmosphere. Another example is the recycling of synthetic polymers; either by reforming them into useful materials/objects, or—less desirable—by burning them cleanly to produce energy while preventing the CO_2 from leaving the economic system (more on that later). Capture of outgoings is a crucial component of this system because unnecessary energy consumption can be prevented by capturing matter to be recycled before it mixes with the environment. Plastic waste is a regrettable case in point. A large part of the success of recyclability rests on the next point: economy of metabolism, crucially embodying material capture.

Integration and economy of metabolism are crucial for making the most of natural energy and energy carriers

This is a long section, so I've further subdivided it. But it's interesting, so please bear with me, because here I explain how we can learn from biology.

The basic concept

Economy of metabolism means capturing and/or reusing the different forms of energy and materials that are generated or involved in making a specific product (e.g., heat generated in various industrial and domestic processes). Heat and pressure are applied to make e-fuels via Fischer-Tropsch chemistry, and heat is produced during combustion of these fuels in ICEs or other heat engines (as covered in Chapters 4 and 6). The concept of "grade" of energy is important: for example, in many districts, low-grade energy from power stations in the form of low-pressure steam at temperatures of around 200°C is supplied in well-insulated pipes to residential areas. There it provides for heating and warm water. What of the CO_2 emissions from these power stations? If we foresee e-fuels as an important part of our energy economy, we must capture CO_2 at high concentration at source, wherever possible (from the power station chimney). It would then be transported to power-to-x production facilities, or—arguably better—the fuel production facilities would be built on-site with power stations (more on that later).

Economy of metabolism also applies at very small scales: for example, the burning of e-fuels in ICEs must be coupled to mechanisms that generate useful work from the waste heat of combustion: some applications of this principle are already in use (as covered in Chapter 6), and many others are being developed. Hydrogen production via water electrolysis produces oxygen, which must be captured and used in relevant industrial processes; it can, of course, also be used in power-generating reaction processes from fuel cells through to heat engines: oxycombustion of C-H fuels (Chapter 6) is an example. All such technologies require as much investment from governments and support from regulators as the development of new fuels.

We must distinguish between fuels for different uses

A clear division must be made between fuels for mobile uses or for specialist purposes (e.g., iron/steel production) and for the generation of primary electricity. As much primary electricity as reasonably possible must be produced from non-fossil natural energy sources: water, wind, solar, geothermal. These are the only long-term sustainable option for *primary* production. However, as I discuss later, carbon-based energy carriers will probably be very important for *storage* of primary energy and generation of *secondary* electricity (defined as electricity that is regenerated and supplied to the grid after one round of storage in a suitable medium).

We must be careful in considering CO_2 emitted from industrial/power generation sources as a convenient feedstock for e-fuel generation; after all, in the long term, we should be aiming to have industries and power generation facilities that produce no net CO_2 emissions. However, in the meantime, it is better to capture the CO_2 than to let it escape immediately into the atmosphere: if it is turned into e-fuel and used in sectors of the economy, it at least prevents the equivalent volume of fossil fuel production and the environmental degradation associated therewith. Furthermore, it saves the CO_2 emissions from that fossil fuel. It is clearly less energetically costly to harvest CO_2 at typically 100,000 ppm from flue gas for this purpose than filtering it out of normal air at 420 ppm (almost 250 times less concentrated). The long-term future of e-fuel production (i.e., looking forward to the 2050 target of zero net CO_2 emissions globally) will likely depend on a combination of exhaust gas CO_2 capture and filtration from ambient air.

Natural energy must be stored sensibly when abundant

One of the most important facets of economy of metabolism—and this is also seen in biology—is the ability to store energy when it is abundant and use it when it is scarce. Already today, hundreds of terawatt hours (TWh) of water, wind, and photovoltaic electricity generating potential are wasted globally each year because the electricity cannot be stored. Just 1 TWh would satisfy the electricity demand of more than 36,000,000 average American homes for 1 day. A major hurdle to using any renewable energy source at our disposal efficiently is its variability: natural water columns are not constant, the sun only shines in the day—sometimes not brightly—and the wind is a law unto itself. Provided a constant supply of fuel (from coal to nuclear) a conventional power station can keep going for very long periods with little or no necessary variation in output. This is the energy to which we're used. Regenerative energy, by contrast, needs to be stored when in excess (of grid capacity or end use) and be released or regenerated when there is too little wind, water, sun. Next we will see how much we are talking about as we consider how to capture it.

Should we mainly use batteries for renewable electricity storage?
There are many net carbon-free ways, potentially, to store energy and feed it back into the grid as electricity: pumping water into high reservoirs and letting it flow back via turbines; pumping air into large receptacles underwater, or compression tanks, and letting it flow out via gas turbines; conserving energy in massive flywheels that can then be re-engaged with a motor working in reverse as a generator; storing it as heat in massive thermal reservoirs (blocks of rock or other mineral); salt gradients; chemical reactions that release heat upon

reversal; supercapacitors; batteries of various kinds (static electrolyte and redox-flow, for example); and power-to-x (including e-fuels). All have advantages, disadvantages, and preferred applications; none is ideal or universally suitable. I review them in more detail later, but here I focus on batteries because we are very familiar with them in domestic applications, and there is talk of using them for much grander applications. The likely very large environmental impacts of using battery technology to serve most of our energy storage and portability problems, however, lead me to think that "megabatteries" are not the best solution. They do not score well in terms of full life cycle analysis of impacts, including sizable environmental ones (as judged by ReCiPe criteria).[4] Sodium ion batteries seem better than lithium ion ones in that regard, but they have other problems.

How "mega" would these batteries be? How much wasted energy are we talking about? The United Kingdom estimates that from wind power alone, 3.6 TWh were wasted in 2020, enough to power 1,000,000 homes for 1 year. The ability to store 40 GWh of energy during peak production would allow wind farms to put all of their potential into service.[5] 40 GWh is 670,000 Tesla Model S 100 kWh batteries (the high-power, long-range ones), or almost 420,000 tonnes of battery equivalents (Chapter 7, Calculation 1, online at https://doi.org/10.1093/oso/9780197664834.001.0001). Seen yet another way, it's 1,300,000 mid-size BEV batteries. And we're only talking about one renewable energy source in one relatively small country. Competition with the car industry is a major consideration, particularly given the fact that the limits of easily minable lithium are already foreseen. It's still extremely hard to recycle lithium from batteries economically.[6,7]

My calculations (Chapter 7, Calculation 2, online at https://doi.org/10.1093/oso/9780197664834.001.0001), based on a 20-year period in which the first battery is from primary resources (e.g., freshly mined/processed lithium carbonate) and the second battery is made more efficiently and with some recycled material (but without factoring in installation/maintenance/repair) suggest a round-trip efficiency of 36% as an absolute maximum (i.e., you get 36% of the energy out compared with the energy you put in). That is lower than you might expect because I have factored in the energetic costs of producing the batteries: I justify that because hydrocarbon energy carriers, with which I compare batteries, are both the *substance* and the *energy*. Thus, 36% is also no better than the present-day realistic estimate of round-trip methane storage economy for peak energy, which I calculate at between 34.5% and almost 40% (Chapter 7, Calculation 3, online at https://doi.org/10.1093/oso/9780197664834.001.0001). If I restrict the modeling period to 10 years instead (because in 20 years much more technological "improvement" can happen), then the apparent efficiency of batteries is 33.2% (slightly less than e-methane; i.e., the batteries would re-release 33.2% of the energy stored in them, when taking their construction energy into account).

Incidentally, you might also wonder whether the much-vaunted idea of using BEVs as storage capacity for excess grid electricity would be a good idea. Here are just a few observations that might make you think again: (1) variations in renewable energy can only be statistically predicted, so at some point there will be an "unexpected" and much longer drop in renewable supply (cases have already been recorded); (2) the batteries of BEVs in this system will go through many more cycles of charge/discharge than normal, hence shortening their life spans more quickly; and (3) imagine that you want to use your BEV, but, at the same time there has been a drop in grid electricity, and it has "spent" too much of its battery on feeding electricity into the grid. Carbon-based fuels have none of these disadvantages, though, admittedly, they have others.

Another point that is often neglected is the realistic energy loss in the grid, house wiring, and—most importantly—the voltage converting electronics for charging and discharging. Here are some real-life estimates: globally 8% loss through the grid + transformers + house wiring to the socket[8-10]; 16% loss from wall socket to battery charge.[11] Multiplying 92% by 84% gives 77% total efficiency for these steps. Now, on feeding the electricity back into the grid (and to someone else's socket (!), we have the same figures, but in reverse. Multiplying 77% by 84%, and then by 92% gives a little under 60% of the original energy (round-trip efficiency). If, for the moment, we neglect the megabatteries' manufacturing energy, an average efficiency of 75% (storage/release cycles for excess grid electricity)[12] could be expected, so 60% from the BEV-storage/release isn't great. Furthermore, it is very likely to cause premature aging of the EV battery, hence reducing the benefit of spreading the battery manufacture impact between two uses (domestic and mobility). It could look even less good depending on where you live because the loss from power plant to socket is as large as 10% in some areas,[10] leading to a round-trip efficiency for BEVs as a storage medium of only 57% in extreme cases. Basically, building storage capacities (be they batteries or e-fuel facilities) as close to power plants as possible (cutting out most of the transmission and distribution network) seems a better idea.

A final thought about megabatteries for peak energy storage: cars have flattened batteries not only because they fit better under the floor, but also because that shape dissipate heat more effectively: flattening increases the surface area to volume ratio compared with a cube, and greater surface area sheds heat better. The cooling challenge increases with the square of the volume because, for a constant three-dimensional shape, the surface area decreases by the square of the volume: double the volume, and the relative surface area drops by 50%. The bigger the battery, the flatter you need to make it: not very practical. Alternatively you can build increasingly large cooling facilities into the battery, but those consume increasing amounts of energy and have consequences for manufacture, maintenance, and recycling—all of which also consume energy and produce other environmental impacts. By the way, I factored none of that into my earlier calculations. Such scaling problems do not exist with power-to-x fuels as storage media.

A variety of options for energy storage: From physical to chemical, from stationary to mobile

Having discussed batteries for storing grid electricity, I owe it to other technologies to review them briefly:

- *Pumped hydroelectric storage* (PHS) uses the potential energy of water that is pumped into a container or reservoir that is substantially higher than the source, and letting it run "downhill" again through turbines. Advantages include the use of existing hydroelectric facilities—where an additional water-holding facility can be built downstream and water pumped back into the reservoir using the turbines working in reverse; the technology is very mature and very long-lasting (ca. 100 years). Round-trip efficiency: 55–75%*; use: stationary applications only.
- *Compressed air energy storage* (CAES) in its current large-scale application involves compressing air with the surplus energy of a power plant (so far a few nuclear and gas-fired plants have been piloted) and storing it in a large subterranean cavern. In one variant, water from a nearby lake is given access (via a large pipe) to the bottom of the cavern, hence stabilizing the pressure. During times requiring peak energy generation, the compressed air is then directed to a gas turbine (burning natural gas) and superheated to generate more energy than it would otherwise do if it simply drove an air turbine. The technology is largely limited by the need to find large enough caverns (so far, this has been hard) and has the disadvantages that it uses, at current fuel mixes, fossil energy, and has relatively moderate efficiencies. Round-trip efficiency: 45–60%*; use: stationary applications only.
- *Flywheels* are very old technology, but they have been revived in a number of forms that essentially store energy via a rotor of relatively high density rotating very fast (between 20,000 and 60,000 revolutions per minute) in a high vacuum and on a low-friction bearing (often magnetic). They tend to be rather heavy for the amount of energy that they can store (hence low MJ/kg), requiring an extremely robust shell to protect against the consequences of the rotor breaking free or breaking up during use (a 10 kWh flywheel can weigh between 200 and 2,000 kg). Hence mobile applications are challenging; that said, they have been applied in small measure in road vehicles (where special mountings are necessary because their gyroscopic effect can influence the directional stability of the vehicle). They are suitable for short-term energy storage, having energy-loss rates of 3–20% per hour.

Round-trip efficiency over short time intervals: 70–95%*; use: mainly stationary applications for short-term bridging of energy troughs or as an emergency energy source for direct electricity generation or starting a diesel generator.
- *Supercapacitors* (also termed "supercaps") consist of two flat electrodes having a very large surface area separated by a very thin dielectric material (insulator), typically rolled into a cylinder, though often also as a block of flat plates. They store electrical energy by virtue of the static electrical effect: one electrode simply becomes more positively charged as electrons are "pumped," using the electricity to be stored, to the other electrode, which consequentially becomes more negatively charged. Supercaps can manifest roughly 10% of the energy density of a good Li-ion battery, but they can be charged and discharged much more quickly (10–100 times) with larger amounts of energy, hence making them interesting accessory storage devices for recuperating braking energy in road vehicles—this they do much more efficiently than Li-ion batteries and can survive greatly more charge-discharge cycles. They are also relatively cheap to produce. Greatest progress is being made in the coating of the electrodes to increase their surface area—notably by using activated carbon. Round-trip efficiency: 80–95%*; use: application as bridging storage for electronic devices and in certain road vehicles in regenerative braking.
- *Superconducting coils* (also known as superconducting magnetic energy storage systems [SMES]) store energy by virtue of having almost no resistance because they are cryogenically cooled to just below their superconducting temperature. Once charged with electrical energy, they keep the electrons circulating almost indefinitely, which also produces a magnetic field, the ultimate medium of energy storage, that is responsible for the extremely high energy efficiency and energy density of SMES (typically 100× more than Li-ion batteries). Theoretically SMES have unlimited charge-discharge cycles because no chemical or thermal degradation occurs. Disadvantages are the very high costs because of the cryogenics, bulkiness, and the need for very robust magnetic shielding to minimize disturbance with electronic devices. Round-trip efficiency: 90–97%*—from which substantial energy demands of the cryogenic system must be subtracted; use: stationary applications only, large-scale.
- *Batteries* (typically Li-ion) store electricity in the context of an electrochemical imbalance (roughly the same way as an old-fashioned lead acid accumulator, but with a crucial difference). Essentially, when charged, lithium ions (originally from the lithium hydroxide used to make the battery) are embedded in a nano-porous layer on the electrostatically neutral anode. Upon discharge, the anode releases lithium ions, thereby leaving negative

charge behind in the form of electrons, which flow through the load device (e.g., the traction motor of an EV). The Li⁺ ions travel via the water-containing electrolyte to the cathode. This is typically made of cobalt, nickel, and manganese, but increasingly iron phosphate instead—albeit permitting a significantly lower energy density. At the cathode, the Li⁺ ions do not change chemically (e.g., from lithium ion to lithium solid), rather they charge-compensate for a reduction reaction (gain of electrons) by the material of the cathode that is taking the electrons returning to the battery from the other side of the load device. Hence cobalt IV is reduced to cobalt III: $Co^{4+} \rightarrow Co^{3+}$, for example. In the case of iron phosphate cathodes, iron III is reduced to iron II. The reduction is reversed during charging. In contrast to lead-acid batteries, the crucial element (lithium) is not converted chemically during charge/discharge. A major advantage of Li-ion batteries is their relatively high energy density compared with other batteries or capacitors. Disadvantages include their high energy of production, environmental impact, and energy-intensive and laborious recycling. Round-trip efficiency: 75–90%*; use: in a variety of applications from hand-held devices through to vehicles and houses as storage for photovoltaic electricity, possibly also for a small proportion of peak grid electricity storage.

- *Flow batteries* (also termed "redox-flow batteries") have two relatively large containers holding the respective electrolytes (the reduced chemical and the oxidized chemical). Electric current is generated by pumping the respective solutions at equal and constant rate through a device possessing a membrane that insulates against short-circuiting but allows the redox reaction between the two chemical solutions to happen. This is basically like a conventional battery, but, instead of having the bulk electrolytes constantly almost in contact with each other, it holds them back, letting them react only at the electricity-generation interface, hence allowing greater power at increased safety; major advantages include easy recyclability (because the electrolyte solution is simply removed and replaced), long life (15–20 years), reduced production cost (relatively cheap technology), and further economy because the same electrolyte can be used in both halves of the cell. Round-trip efficiency: 60–85%*; use: overwhelmingly stationary applications because of weight and bulk.
- *Hydrogen* may be produced directly by alkaline electrolysis using electricity, or, more recently, directly within specialized photovoltaic cells; it can also be produced from organic matter via steam pyrolysis. As the "ultimate" reducing agent, so to speak, it is theoretically usable universally because it can, for example, even be used to generate electricity directly in a hydrogen fuel cell. However, the thermal efficiency of such cells is at present very modest. Major challenges are storage and distribution because H_2 is an

extremely small molecule and therefore easily escapes from containers and fuel lines designed for natural gas. It requires cooling and enormous energy to compress it to liquid form, hence making it a very energy-intensive and economically challenging energy carrier in its gravimetrically most energy-dense form. Expensive storage vessels are required to assure high-pressure leak-free containment. Round-trip efficiency: 60–75%*; use: diverse applications, upward in size from small road vehicles.
- *Thermal storage* relies on pumping heat energy (e.g., from electrical heating elements or heat pumps) into a substance to change its properties: these can be very basic (e.g., raising the temperature of a mass of water or rock), more subtle (e.g., using the latent heat of a phase change such as ice to water or salt hydrates to dehydrated salts), or thermo-chemical reactions (e.g., conversion of a gaseous mixture of N_2 and H_2, to liquid ammonia (NH_2). Round-trip efficiency: 50–90%*; use: mainly large, non-mobile installations.

A conclusion from all of this is that for mobile energy uses that would benefit considerably from a high-density energy source, and for requirements to store large amounts of energy in relatively simple forms (e.g., liquid, at or close to ambient temperature/pressure) and distribute them with simple (largely existing) infrastructures, hydrocarbon fuels are very advantageous.

I now explore the potential of methane. An obvious advantage is that methane would allow us to use existing devices, storage, and distribution infrastructures for powering part of our domestic energy needs, such as heating houses. Methane synthesis (one of many power-to-x technologies) has recently advanced greatly in thermal efficiency,[14] employing recycling of excess heat of reaction instead of using primary electricity to generate the temperatures needed for synthesis. Individual power-to-x methodologies have arguably not received as much research and development funding as solutions to energy storage based on Li-ion batteries; however, both technology areas are credible competitors, and they are still developing. Using existing well-tested methane-based energy technology, we could currently achieve between 36% (from direct air capture) and 40% (from flue gas capture) thermal efficiency in producing liquid methane for storage on a large scale (Chapter 7, Calculation 3, online at https://doi.org/10.1093/oso/9780197664834.001.0001). But one of the most promising directions for fuel generation as a storage means—the reversible solid-oxide cell (RSOC, covered in more detail later)—may reach a round-trip process efficiency of 47.5% for methane (a cycle of production from CO_2 and H_2O using surplus electricity/reaction of methane with O_2 in a fuel cell to produce electricity, CO_2 and H_2O).[15] If CO_2 and water produced in the fuel cell mode are recycled (an easy step), and we factor in intermittent addition (minor topping-up) of CO_2, we would likely arrive at a realistic round-trip efficiency of 47% from this so-called power-to-x-to-power system.

That is much more than the 33.2% round-trip efficiency that megabatteries would likely afford over 10 years. A major advantage of methane-based energy storage is its flexibility. First, the technology can be scaled up to generate and store more methane very easily and at relatively small costs. Extending existing capacity is very easy and both energetically and financially cheap: just add a new storage tank. Second, the technology can feed the methane into other facilities, from the chemical industry, through cement factories, to home heating, using existing infrastructure. Third, the plant can easily switch product according to demand—just like biological metabolism—producing and storing, variously, methane, methanol, hydrogen, syngas, or ammonia. Indeed, methanol could also constitute a very important storage medium, given its advantage over methane of being liquid at ambient temperatures and pressures and not contributing to global warming via small percentages of leakage (methane slip). Even if battery manufacture becomes energetically much less costly, I believe that the flexibility of carbon-based energy storage would still pay dividends: it would contribute to the creation of integrated industrial metabolism of energy and materials. Moreover, it would have a sustainability that promises to be relatively high because it closely mimics the recycling that occurs in cells, through organisms, up to whole ecosystems.

A hybrid system consisting of a mixture of batteries (for small-scale, short-term storage) and carbon-based fuel cycling (for longer periods and industrial demand) could well provide the best of both worlds. The similarity to the hybrid system in living cells is clear: the mitochondrion's "electric" electron transport chain (ETC) and the larger metabolism of carbohydrates and fats/oils. Ironically, instead of first developing small-scale battery solutions for quick peak absorption/release of grid electricity, in combination with power-to-x for longer term storage, we have leaped far ahead by applying small batteries in EVs. We clearly don't have the clean energy and resources to do both simultaneously. Arguably the better long-term solution would be small batteries for absorbing a small, but quickly releasable, proportion of peak electricity, and power-to-x for generating longer-term, massively scalable, renewable energy storage and mobility fuel. This might also be better for curbing the very environmentally damaging, runaway car manufacture culture.

Methane and methanol can form technically feasible cyclical systems of synthesis and use

The power-to-x-to-power concept works by incorporating methane and primary electricity from regenerative sources (water, wind, photoelectric) in a cyclical alternating system: regenerative electricity (from the methane fuel cell mode) is fed into the grid as much as possible and stored in the form of methane when in excess. The methane would, at present, power existing natural gas power plants—which would then simply be called methane power plants. These would burn it to produce electricity and heat for domestic heating or industrial processes; but, increasingly, the methane would be generated in RSOC plants that synthesize and "burn" methane in alternation. The flue gas from these plants, rich in CO_2, would be collected (an easy task) and recycled within the plant itself, using peak excess renewable electricity. The "burning" would occur inside the reversible unit (details of which follow later), and a one-way efficiency (electricity generation only) of around 66%[16] could be reached (i.e., in the range of the best combined heat and power generation plants that we have today).

The numbers that I have presented so far assume fairly constant and regular cycling between excess electricity "absorption" and electricity regeneration. Now comes the real challenge for megabatteries: What if there is a large *and* long decrease in wind energy? This recently happened in Germany, for example, where the first few months of 2021 were much less windy than usual.[17] Around one-third less power than "average" was generated from wind turbines, and the shortfall was—alarmingly—made good by coal-burning (largely the filthy brown variety) power stations. In massive contrast, generating and storing power-to-x carbon-based fuels to even out the fluctuations in availability of natural energy is a very flexible system: it is simple to expand production and storage. That is what biology also does when organisms stock up on food when plentiful, evolutionarily programmed to expect a short supply next. And, most importantly, in terms of environmentally sympathetic recyclability, it leaves no troublesome residue. Yes, there are containers that need servicing and replacing, but ordinary steel cylinders (scaled to demand), can easily last 70 years.[18] The yearly residual environmental load of such a system is very small compared with that of Li-ion batteries (or any batteries).

According to the International Energy Agency (IEA), 6.6 million EVs were produced worldwide in 2021.[19] If, on average, each will produce 0.25 tonnes of lithium ion battery (LIB) waste per vehicle (according to figures in Harper et al.[20]), that's 1,650,000 tonnes in total that needs to be dealt with at end of life. Already at present, hundreds of thousands of tonnes of LIB battery waste materializes each year, and figures of several million tonnes per year of LIBs needing recycling as of 2030 are highly credible. At current growth rates, the battery recycling industry seems unlikely to keep up. Moreover, current recycling methods are, besides being very energy-intensive, acknowledged as not being sustainable[21]; there is patchy information on the exact methodology used for metal extraction[7] and the extent of recovery of different components—though for lithium, at least, it is known to be very small. Furthermore, there is limited information on the economic viability of LIB recycling.[22] As noted in a recent review paper, "most research in batteries is entirely focused on performance while the sustainability of all battery components making up the cell, as well as the battery chemistry itself are much overlooked."[23] Does all of this constitute a good basis on which to expand LIB applications so fast?

True, carbon-based energy solutions also require much investment, development, and energy input, but, in reviewing the larger field, I conclude that they are based on a sounder principle for up-scaling. Power-to-x-to-power systems, such as the methane and methanol examples, are just the beginning of a low-residue (waste) and highly flexible metabolism of energy and materials. I will describe latest progress in the larger area of power-to-x-to-power at the end of this section.

We must critically compare storage media, their advantages, disadvantages, and acceptable tradeoffs

How can we compare different means of storage and portability of energy from primary production (renewable electricity)? Can we compare them at all? I've tried very hard to do so between batteries and e-fuels, but I can only make partial comparisons. This is because they are very different energy carriers in their materials, quantitative measures, qualitative features, and their principles of energy storage and release. That observation, however, is the key because it is the strongest justification for pursuing both technologies (experimenting, trialing, adapting, etc.) and integrating them in ways that maximize their efficiency and minimize wider environmental impact (not just CO_2 emissions!). Table 7.1 presents a rundown of the advantages and disadvantages.

From Table 7.1 you will see that methane is not at present an ideal fuel for ICEs: much work is still needed to optimize precombustion fuel supply, the combustion cycle itself, and engine design. Methanol, by contrast, is recognized as very well-suited to ICEs (with minor modifications), right up to ships' engines. Methanol is interesting for another reason: it is extremely versatile as a major feedstock for the chemical industry (for CO_2 emissions of synthetic methanol from different sources, see International Renewable Energy Agency[24]). Economies of scale of methanol production from renewable (CO_2-neutral) sources can be reached with the right political support, hence serving both transport and industry. Using existing large-scale technology, we can currently make e-methanol—via direct air capture of CO_2—at a thermal efficiency of 30% (Chapter 7, Calculation 4, online at https://doi.org/10.1093/oso/9780197664834.001.0001) without energy recovery (i.e., this is a minimum value). RSOC technology would achieve around 41%,[16] but potentially as much as 45%[25] with efficient heat capture/reuse (and see Chapter 7, Qualification 1, online at https://doi.org/10.1093/oso/9780197664834.001.0001).

Table 7.1 Advantages and disadvantages of Li-ion battery and e-fuel technology

Li-ion battery technology		E-fuel technology (methane/methanol as examples)	
Advantages	Disadvantages	Advantages	Disadvantages
High energetic efficiency because of low heat production.	Hard to recycle, and leaves considerable residue.	Easier to recycle, and leaves no residue.	Substantially lower energetic efficiency because of greater heat production.
Energy transfer produces no by-products that are emitted	Can only be used to store electricity.	Can be flexibly balanced between fuel and chemical feedstock use.	Energy conversion produces by-products that must be captured.
	Requires very energy-intensive and highly specialized production methods.	Can use existing fuel storage and distribution infrastructure.	Requires costly new installations to achieve reversible and efficient energy/material conversions.
	Requires mining of limited resources, causing significant environmental impact.	Is not associated with limitations on mineral supplies.	Risk of methane leaks leading to significant greenhouse effect, particularly from ICEs.
	Low energy density, leading to heavy/bulky installations.	High energy density, leading to relatively light and small installations.	
	Relatively low energy fluxes, because of internal electrical resistance.	High energy fluxes possible.	

Interconnecting energy and material transformations is essential for minimizing waste and facilitating flexibility: How does biology do this?

To capture available energy optimally (i.e., with sensible tradeoffs), we basically need to think on much larger scales and more integratively (considering interconnectedness) across the whole energy and materials economy. We clearly need short-term storage *and* long-term storage. Carbon-based energy carriers such as methane and methanol would, in fact, serve both purposes, and they are also vital for the manufacture of so many things that we need, from pharmaceuticals to everyday objects. The cyclical electricity/e-fuel systems described earlier are also very analogous to biological energy economies, the plant cell—embodying solar-energy-harvesting and reversible biochemistry in one—being the prime example (Figure 7.1). In an important sense, the plant cell is the biological equivalent of the apparatus for collecting solar energy, converting it into chemical energy, feeding it into a synthesis facility that includes an RSOC cell (or similar), and producing energy and material in equilibrium according to needs.

Mitochondria run pretty constantly, using stored energy (hydrocarbons, carbohydrates, and, in emergencies, protein) to produce electricity at a regular rate, whereas the input (food) into the larger organism (plant or animal) varies considerably in availability, both in the short and long term. Mitochondria produce an electrical charge, which they store, separated across their membranes (a biological capacitor). They represent very short-term stationary energy stores, while the carbon compounds with which they work are the mobile energy carriers and the substrates for building up long-term energy reserves. Mitochondria have both forward (energy-producing) and reverse (material-producing) pathways (Figure 7.2), which exist in equilibrium depending on the cell's (and larger organism's) needs.

In the human economy, we can, in principle, store and distribute massive volumes of carbon-based energy carriers (made from renewable electricity) at very low infrastructural cost, hence also reducing CO_2 in the atmosphere. But we're in a catch-22 situation: we need to reduce CO_2 levels quickly, as well as develop technologies (of all kinds, from industrial processes through to transport) that produce less net CO_2. Use of fossil fuel energy to sequester and sink CO_2 into rock, for example, would produce more CO_2 than it eliminates; that must be done with entirely natural, and preferably local, energy sources. However, we do not yet have enough renewable energy to sink CO_2 on a large scale. Could we be doing something useful and also CO_2-neutral with the CO_2 instead? Having reviewed the literature, I conclude that developing a renewable energy economy analogous to biology would be wise: it would benefit from our existing hydrocarbon fuel infrastructures and machinery and would lead to massive savings in

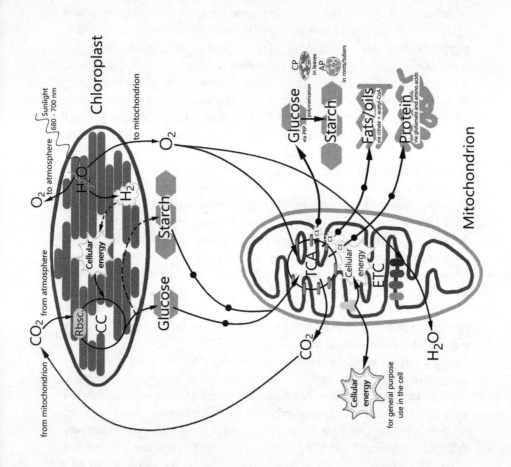

Figure 7.1 Schematic of energy and material transitions in a plant cell—highly simplified and incomplete. Chloroplasts in leaves of green plants use sunlight (wavelengths 680 and 700 nm) to split water into oxygen and hydrogen. Oxygen is partly emitted into the atmosphere and partly used in respiration (burning carbohydrates). The hydrogen is the reducing power that the chloroplast uses, together with CO_2 (partly from the atmosphere and partly from the mitochondrion) to synthesize glucose and starch (polymerized glucose)—the crucial CO_2-fixing enzyme being RuBisCO (Rbsc) in the Calvin cycle (CC). Glucose (also de-polymerized from stored starch) is burned by the mitochondrion, together with oxygen, making CO_2 (from the tricarboxylic acid [TCA] cycle), and H_2O (from the electron transport chain (ETC) and F_1F_0-ATP synthase combination), that is, the equivalent of classical combustion. With the combustion energy, the mitochondrion synthesizes the universal cellular energy-carrier, adenosine triphosphate (ATP), which it partly exports to the rest of the cell for use in anabolic and cell-maintenance processes and partly uses for its own anabolic (substance-building) purposes: it can produce glucose, which can then be polymerized into starch in chloroplasts (CP; in plant leaves) or in amyloplasts (AP; in plant roots and tubers); fatty acids for fats and oils (e.g., for storage in seeds); and amino acids, mainly for protein synthesis. These anabolic reactions depend critically on intermediates from the TCA cycle. All energy and material needs of the plant cell are achieved via chloroplasts and mitochondria as the hubs where energy and material economies converge and interconversions take place. In a dissociated sense (i.e., between plants and other organisms in food-chains) these two organelles are the crucial principle of cyclical material economies, where material and energy carriers are embodied largely by the same substances or close relatives: from an individual plant, through ecosystems with all manner of organisms, to the whole Earth. They work sustainably at the centralized level (within the same cell and same plant) and at the decentralized level, represented by the organismal divide between primary producers (plants) and primary consumers (herbivores). For more details of the biochemical reactions involved, see Chapter 1, Box 1.4 and 1.5, and Figure 7.2 for the mitochondrion.

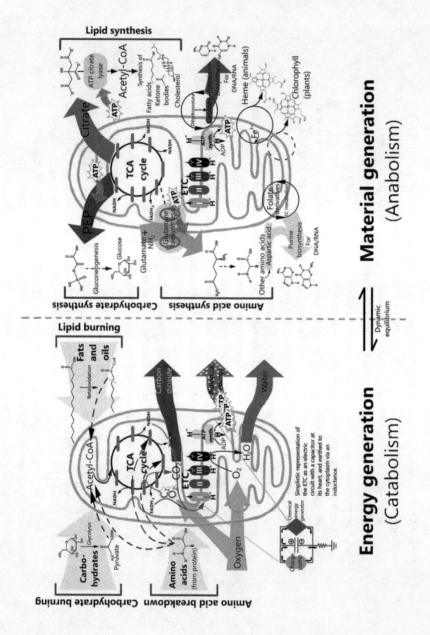

Figure 7.2 The mitochondrion (Mt) as an interchangeable hub of energy and material economies in the cell, much simplified to aid visualization. Note that the urea cycle (largely involved in degrading excess protein-derived amino acids to NH_2 and then urea) is omitted for clarity.

Left: Energy-containing biomolecules are catabolized into energy, ultimately in the form of adenosine triphosphate (ATP), which partly leaves the Mt to be used in endothermic reactions in the cell's cytoplasm and partly remains inside the Mt to be used in biosynthesis reactions there. Carbohydrate metabolism in the cytoplasm (glycolysis) produces pyruvate, which enters the Mt, is converted to acetyl-CoA, which enters the tricarboxylic acid (TCA) cycle. Fatty acids (FA; from fats and oils) are broken down via beta oxidation, similarly producing acetyl-CoA, which enters the TCA cycle. The TCA cycle produces reducing equivalents in the form of biological hydrogen carriers (nicotinamide adenine dinucleotide [NADH] and flavin adenine dinucleotide [$FADH_2$]). These donate electrons to the electron transport chain (ETC), which uses their energy to build up a charge across the inner Mt membrane, very similarly to a capacitor, by pumping protons (H^+ ions) across the membrane. The charge then flows back (as H^+ ions), this time through the F_1F_0-ATP synthase, a molecular machine that makes the biological energy carrier, ATP. Catabolism produces CO_2, from the TCA cycle when the two-carbon units (in the form of acetyl-CoA) from sugars and/or fats are oxidized and, in return, produce reducing equivalents (NADH and $FADH_2$); H_2O, from the ETC, which is the result of the now "worn-out" electrons being accepted by oxygen in the presence of H^+ ions. The production of CO_2 and H_2O by the mitochondrion in catabolizing biomolecules is exactly chemically equivalent to the combustion of organic molecules with which we are familiar. Amino acids are catabolized quite similarly, via the acetyl-CoA route, and also directly entering (as high-energy carboxylic acids) the TCA cycle at two points (their NH_2 groups are converted to ammonia and then urea via the urea cycle, not shown). Right: Energy in the form of ATP is used by the mitochondrion to create the high-energy intermediate molecules that feed the biosynthesis of numerous crucial molecules. Gluconeogenesis (remaking glucose) starts with an intermediate from the TCA cycle, which is converted (via energy from ATP) to the high-energy phosphoenolpyruvate (PEP), which leaves the mitochondrion and is the substrate for glucose biosynthesis in the cytoplasm. Citrate, another intermediate of the TCA cycle, is the initial substrate for FA, leaving the mitochondrion to be converted (via energy from ATP) to acetyl-CoA in the cytoplasm, where it is used to synthesize FA, cholesterol, and ketones. Amino acids are synthesized starting with energy from ATP plus glutamate (an NH_2-containing carboxylic acid) (likely both inside and outside the mitochondrion). The resulting glutamine is the substrate for further amino acids, made in the cytoplasm. One of these, aspartic acid, is used, in a reaction with folate derivatives exported from the mitochondrion, to make the two-ring purines (adenine and guanine) found in DNA and RNA. A different pathway (relying on conversion of dihydroorotate to orotate in the mitochondrion) is responsible for synthesis of pyrimidines (cytosine, thymine, and uracil [uracil not shown]) also for DNA and RNA. Roughly half of the heme biosynthesis pathway (producing heme for hemoglobin in blood) occurs in the mitochondrion, notably the addition of the iron atom (Fe), which occurs in the space between the two mitochondrial membranes—a very rare situation. Ten of the more than 200 reactions in chlorophyll biosynthesis happen in the mitochondrion.

In all catabolic and anabolic reactions, the TCA cycle is the most important central metabolic hub in the interconversions of energy-containing biomolecules and energy in the forms of reducing equivalents (H-carriers).

energy, material, and environmental impact compared with establishing completely new energy storage means.

Toward an industrial metabolism with high similarity to biology: From theory to practice

Ideas for practical facilities are emerging

In October 2020, a paper appeared that struck me as key to creating a sustainable energy and material economy.[26] It was the latest of many based on the RSOC (Figure. 7.3) an energy/material converter that can work in one direction as an electrolyzer and in the other as an electricity-producing fuel cell. The device, which would form the basis of a fuel synthesis plant, consists of a ceramic unit with several layers and a honeycomb pore structure (to increase surface area). These are coated with various metal catalysts and work best at temperatures between 680°C and 800°C. The cell contains an anode and a cathode, which electrolyze steam into hydrogen and oxygen when current is applied (i.e., when excess grid electricity is at stake); in reverse mode, they work as a fuel cell for hydrogen or hydrogen-carriers (e.g., methane) or even the syngas mixture (H_2 and CO)[16]: in this mode, we essentially get much of the energy out again that we had put in to make the fuels. All reaction products of the fuel cell mode are captured and stored for the reverse pathway: interestingly, the apparatus can also "burn" ammonia or methanol to release electricity and give, respectively, nitrogen and water, and CO_2 and water, as products. These are fed into the plant at a different point to be resynthesized (with excess grid electricity) into ammonia and methanol. Ammonia has advantages and disadvantages compared with carbon-based hydrogen-carriers for energy, but these are beyond the scope of this book. Furthermore, the RSOC would also perform co-electrolysis of water steam with CO_2, hence producing another crucial raw material of the chemical industry: syngas (CO and H_2). Crucially, the 2020 paper[26] illustrated how a practical facility could be built.

There are striking parallels with biology

Though not explicitly stated, the parallels with biology should now be clear: already in the title "reversibility" is mentioned, a key concept in biological metabolic pathways. The paper goes on to explore the hard realities of an energy storage and production facility (power-to-x-to-power) that not only stores energy in chemical intermediates, but also (1) does so in a reversible manner, using essentially the same device to make the storage product as to catabolize it (break it down) to produce electrical energy; and (2) integrates a variety of crucial feedstocks for chemical synthesis reactions as energy carriers. This would allow

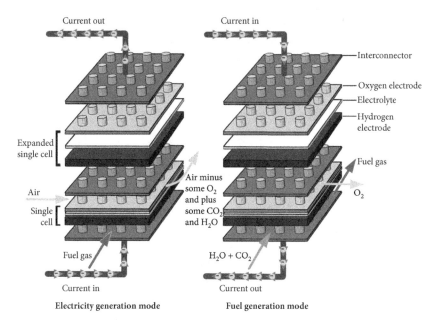

Figure 7.3 Schematic of the reversible solid oxide cell (RSOC). When working in the electricity generation mode, it is essentially a fuel cell and can "metabolize" hydrogen or any suitable hydrogen carrier (e.g., a range of organic molecules from alcohols to alkanes): at the hydrogen electrode—the anode in this mode—a hydrogen atom is taken from the fuel and catalytically "persuaded" to give up an electron. The electron is prevented from crossing to the cathode via the membranes that contain the electrolyte, which only permit H^+ ions to cross. Hence the electrons flow from the anode into whatever load is connected to the circuit (i.e., electric current flows). The H^+ ions react with oxygen on the other side of the electrolyte layer, at the oxygen electrode (the cathode in this mode): because the reaction requires electrons in order to be electrostatically neutral ($O_2 + 4H^+ + 4e^- \rightarrow 2H_2O$), the electrons ($e^-$) from the circuit incorporating the load device are taken back at the cathode, and the circuit is completed electronically and chemically via the formation of water (H_2O). Hydrogen can be burned pure, or it can be provided in the form of a suitable carbon-based H-carrier, which then produces CO_2 that can be captured and used in the fuel generation mode: in this mode, the directionality of the cell is reversed, and, instead of generating current, the cell consumes electricity, taking in water vapor and CO_2, and producing oxygen and C-H fuel in gaseous form.

Figure redrawn with minor modifications from Wang et al.[26]

for real-time adjustments between the proportions of each that are produced and either drawn off to be used for their material value or reacted with oxygen to produce electrical energy (catabolism)—indeed, exactly as the chloroplast/mitochondrion combination does. In the midst of all of this are storage facilities for the energy carriers: methane, methanol, ammonia, hydrogen, syngas—analogous to the many forms in which biological cells and organisms store energy carriers and materials.

Looking at Point 1 in more detail, it is clear that this is an analogy to an enzyme that sits in the middle of a chemical equilibrium and speeds up the transformation of A to B, generically speaking. This would be written $A \rightleftharpoons B$. It could also be $A \rightleftharpoons B + C$ or $A + B \rightleftharpoons C$, etc. The point is that the reaction is reversible, and, if the concentration of substances on the left and right of the equilibrium symbol is altered, then the reaction can be "driven" more to the right or more to the left. This is typically true of organic chemistry because the carbon compounds are easily interconvertible—"fungible"—thanks to the nature of the bonds that carbon makes with itself and many other elements. Burning carbon compounds takes us out of the organic realm, producing the inorganic CO_2, and that, by contrast, is very stable and quite hard to get back into the organic realm (i.e., by reducing it [with hydrogen] and turning it into, say, methane). That said, similar reactions are exactly what plants do via photosynthesis, as we have seen. The first enzymatic CO_2 fixation step is *ir*reversible—which is also what we want in our industrial synthesis—but subsequent steps of carbohydrate metabolism are reversible.

More explicitly, with the most prominent biochemical pathway on Earth

Crudely put, plants take up CO_2 and use sunlight and enzymes to reduce it with hydrogen (from water) to make carbohydrates (Figure 7.2 and Chapter 1, Box 1.4). Via these and simple sugar intermediates, they synthesize all the other materials that make up the plant that we see. The crucial "integrator" of this impressive economy of energy and material is the enzyme ribulose-1,5-bisphosphate-carboxylase/-oxygenase (RuBisCO). It was discovered in 1947, by Sam Wildman, a professor of plant physiology at the University of California at Los Angeles, as a major constituent of ground up plant leaves. But it took decades to unravel its role, substrates, products, and function in plant metabolism. In 1969, a key finding by Cooper and Filmer was published[27]: RuBisCO's principal substrate is CO_2! (It has another, secondary one, oxygen, but only CO_2 will concern us here.)

RuBisCO is known as the CO_2-fixing protein, and it exists in all photosynthetic organisms, right back to ancient photosynthesizing cyanobacteria (blue-green "algae" [sic]) that emerged 2.4 billion years ago.[28] It enables them to take up CO_2 from the atmosphere and is probably the most abundant protein on Earth: almost three-quarters of a gigatonne (0.7 in terrestrial plants and 0.03 in marine organisms).[28] Each year, this enormous global enzymatic capacity fixes around 120 gigatonnes of atmospheric CO_2—that's one-sixth of the total atmospheric CO_2 and between 10 and 20 times the anthropogenic CO_2 per year. What an enzyme! Despite this impressive feat, RuBisCO is a very slow worker, turning substrate into product at a rate of between 3 and 10 molecules per second (compare that with carbonic anhydrase—the fastest enzyme—at 600,000 molecules per second!).

As you might guess, this is the step that limits the rate at which plants can grow. The reasons for RuBisCO's sloth are still debated. Equally interesting, some researchers are trying to engineer mutants that are faster and can bind more CO_2 per unit time. This raises the tantalizing possibility of harnessing plants to remove excess CO_2 from our atmosphere more quickly, possibly also for fuel production. The strange thing is that whenever researchers create a mutant that increases speed, the enzyme's specificity for substrate declines. There is probably a very good reason why RuBisCO is so slow and has not "improved" during its billions of years of existence. And how RuBisCO gets the CO_2 that it "needs" is very interesting as well: preliminary findings suggest that plants use biogenic amines to trap CO_2 and pass it on to RuBisCO[29]—similarly to the most widely applied extraction methodologies that humans build for extracting CO_2 from plain air: organic amines.[30]

The facility mimics the biological interface between energy and material economies

As in industry, the living plant needs energy to fix CO_2 because it must chemically reduce it from its stable oxidized state. Initially, this energy comes from hydrogen produce in photosynthesis by the splitting of water (the equivalent of electrolysis, but where the electrons for the electrical energy emerge from sunlight-activated pigments in the plant's chloroplasts). The reducing energy of the hydrogen is (very simplistically put) then converted into the more manageable universal cellular energy currency, adenosine triphosphate (ATP). This acts as the "energy donor" for RuBisCO's carbon fixation reaction (an early step in carbohydrate synthesis): the first point at which energy and material economies interface with each other. Starting with this step of CO_2 fixation, all plant sugars and complex carbohydrates on Earth (starches and even cellulose itself) are made.

Now for an important distinction: many steps of carbohydrate formation are reversible and are, indeed, used by organisms (plant or animal) to generate energy from complex (stored) carbohydrates. The reversibility between glycolysis (splitting of glucose to generate the energy-carrier ATP) and gluconeogenesis (making of glucose) is an example: most of the enzymes in these pathways of central metabolism simply catalyze reactions in whichever direction is "physically indicated" (i.e., if the concentration of products on the right-hand side of the equation goes above a certain value, the enzyme produces more of the products on the left side of the equation, and vice versa). At the start of central metabolism, where the smallest molecules lie, reversibility essentially distinguishes between directly usable energy source and storage fuel (e.g., pyruvate and fructose: pyruvate is derived from fructose as a direct energy source for mitochondria, and it can be converted back to fructose for distribution or longer-term storage).

At various points in the pathway, intermediates are drawn off to make other crucial molecules of life: for protein synthesis the amino acid histidine is made from glucose-6-phosphate in four steps, for example; and the amino acid phenylalanine is derived from a sugar intermediate later in the glycolysis (glucose-splitting) pathway, namely 3-phosphoglycerate. Eighteen of the 20 fundamental amino acids found in nature are made from sugar intermediates of central metabolism. The building blocks of DNA and RNA (nucleosides: chemical bases, plus deoxyribose/ribose unit) are made from sugar derivatives (activated ribose—a pentose sugar) and derivatives of amino acids. During cachexia (extreme starvation), animals start converting protein into energy in a desperate attempt to survive until food supplies return. The biochemistry of life is riddled with reversibility: it saves on infrastructure.

But what are the hard figures surrounding efficiency?

According to the calculations in the paper,[26] the efficiencies of the "round-trip" processes (i.e., from using excess grid electricity to make storable fuel and then "burning" that fuel to regenerate electricity to feed back into the grid) are, depending on the storage fuel envisaged:

Methane: 47.5%
Syngas: 43.3%
Hydrogen: 42.6%
Methanol: 40.7%
Ammonia: 38.6%

(Note that these are substantially higher than those for a megabattery system in which the energy costs of battery manufacture are included.)

The attraction of this system is that it can either run as an almost perfectly closed energy/material recycler, storing and generating electrical energy equivalents, or, by topping it up with captured CO_2 or nitrogen, it can also (in parallel, with no modification) produce the most important basic feedstocks of many industries (methanol, syngas, methane, ammonia).

We must not shy away from complexity: It is probably the key to sustainability

This RSOC "contraption," you might say, seems rather complicated (though quite simple compared with a mitochondrion!). Yes, it is, but so is biological metabolism. There's no way around it because, in an economy of recycling, there should ideally never be an unconnected point (i.e., a part of the process that ends with something simply getting lost or disposed of permanently). This is also the principle of cradle-to-cradle design and management of goods and materials.[3] We need highly integrated systems, full of points (materials) that are interconnected with each other in networks and that react to each other in terms of feedback. That is the best way to ensure maximum recyclability: a cell that lacks glucose immediately reacts via gluconeogenesis (producing new glucose) because the enzymes are sensitive to the concentration of materials on the "glucose side of metabolism," and they reverse their chemical reactions to produce more of it. In biology, there is no emergency mining of cheap fossil fuel if supplies of energy run low; rather, metabolism reacts and makes interconversions and savings to keep the organism (and, in fact, whole ecosystems) "running."

Taking our lead from biological examples can give us a massive head-start in designing systems with feedback, robustness, and flexibility that result from (1) numerous links between the different nodes of the system, making them sensitive to changes in neighbors and able to pass on this "information" to other neighbors; and (2) redundancy (more than one way to get from A to B), so if route A is blocked, route B takes over. Organisms typically react to stimuli in ways that can quickly be reversed or that lead to a new stable way of life. If we want to continue living on Earth, as part of its biosphere, these are the kinds of systems that we must build. Even with the biological head-start, we need time, and, to grant us that time, we must reduce consumption immediately.

The exact "contraption" is less important than the principle of integration

Whether the RSOC becomes the key to energy and material storage and interconversion of the future is an open question: research and development still need to be done. However, it symbolizes a concept that I believe is of utmost importance: the *integrated*, and flexible management of energy and material resources without wastage of potential renewable energy. The major advance here is the realization of the interface between energy and material economies: precisely what biology does. The integration of these economies is not an option for our future on this planet: it's a must. The days of chemical plants at location X and power generation facilities at location Y are numbered because the material ↔ energy transformations implicit in both must be centralized to minimize losses—losses that, in one way or another, have a detrimental effect on the environment.

We must apply opportunism to prevent useful materials and energy becoming environmental problems

This is also a very biological concept, in the sense that I intend here. I mean taking advantage of circumstances, energy, and material that would otherwise be "wasted" or cause environmental degradation. Two greenhouse gases (GHGs) that escape opportunism far too often at present are CO_2 and methane. A classic example is the generation of methane (biogas) by decomposing organic material (mainly plant waste). It is far from clear that we can construct biogas facilities to supply a major part of portable energy needs (either via direct methane utilization or via conversion into e-fuels); however, it is indefensible simply to let the methane escape into the atmosphere, given that it is a much more powerful GHG than CO_2 (approximately 80 times over 10 years, and 30 times over 100 years) and ultimately converts to CO_2 in the atmosphere.

Microbially decomposable waste—much of which lingers in land-fill sites—is very unlikely to decrease on the background of an increasing world population. While this waste still contains substantial quantities of biological matter (i.e., it is still fermenting), it will decompose there, producing methane and CO_2. It would be better to capture that methane (and CO_2) for useful purposes and then feed as much nutritionally valuable waste (minerals and nitrogen compounds) back into the ground as fertilizer, thereby minimizing the need for chemical fertilizers from industry. Methane should also be captured from human and animal feces before their fermentation residue is applied as organic fertilizer. The overall environmental impact of such technology is greatly influenced by what we do with the fermentation residue[31]: if we plow it into fields, the whole technology (including the methane production) can have a net beneficial environmental impact because it reduces the use of chemical fertilizer (and hence energy use and CO_2 production). According to many, organic waste is a valid and justifiable source of methane, which can be divided between e-fuel generation, industrial application, and home heating, for example. A prime target for biogas should probably be the "hard-to-abate" cement industry, where use of biogas and "renewable waste" instead of fossil oil or natural gas (or unrecyclable plastic) could constitute a mere 6% by 2030.[32] A further large, and regrettable, proportion of anthropogenic methane comes from fossil oil extraction and coal mining: most of it is still "flared" off (burned), which, though better than allowing it to escape into the atmosphere, is surely not the best option.

You might also be wondering whether CO_2 collection from volcanic sources would be a good idea given that it emerges from the ground in a more concentrated form than in normal air. However, the highest estimate that I've read for global volcanic CO_2 emissions is 360 million tonnes per year: even if we could absorb all of it, that would only be one-hundredth of the annual anthropogenic emissions. Methane released from land, either naturally or as a result of global warming or anthropogenic degradation, is more of a problem, but it is too hard to collect and so we must try to reduce its cause instead.

An important geological source of CO_2 (albeit anthropogenic) that we *should* be capturing is that from the cement industry: fossil fuels and refuse are currently used to heat the kilns that convert calcium carbonate into calcium oxide, but even more CO_2 is given off by the chemical transformation itself—around 55% of total cement industry CO_2 emissions. We could envisage an economy where all that emitted CO_2 is captured. The 45% from fuel burning could be converted into e-methane and resupplied to the cement factory, thereby fairly replacing no more and no less of the energy carrier used; the 55% of CO_2 from calcium carbonate decomposition could be used to generate e-fuels for transport and feedstocks for the chemical industry. That would be better than simply letting it escape into the atmosphere. The alternative would be to pipe it to suitable places and sink it into rock, but that also requires large amounts of energy.

It's largely a question of what we wish to do with available regenerative energy sources. Around 8% of global CO_2 emissions are from cement making; 4.4% of them—that's 1.63 Gt per year—are from the calcium carbonate roasting process. To give you an impression of the scale, if all of that were converted into e-fuels, it would free more than half of the private cars worldwide from fossil fuels; not that I'm saying that that should necessarily be done. The major problem is finding the renewable energy to do the conversion. Even using the comparatively efficient RSOC plants (working at a minimum of 45% efficiency), it would require 1.7 times the world's current renewable electricity (Chapter 7, Calculation 5, online at https://doi.org/10.1093/oso/9780197664834.001.0001). But as discussed earlier, batteries for powering road transport are very likely to be a worse solution in terms of total environmental impact.

Finally, a very practical day-to-day example: more overlap between freight transport and passenger transport could be conceived, thereby minimizing the number of passenger trips with empty seats/accommodation, and the number of freight trips with empty cargo space. This infrastructural flexibility will surely come at a price for our personal flexibility: we might not be able to travel to and return from all destinations at exactly the times we wish, and we might not be able to travel long-distance as much as we currently do, but that might just be the price of sustainability.

Economy of scale and decentralized production are needed in appropriate measures

Economies of scale are a must in many areas because, per se, they result in greater efficiency of energy and material use than small-scale activities. The other side of the efficiency coin is decentralized production in cases where local resources make it more economical than centralized production followed by distribution. This is often financially and environmentally costly because of the building and maintenance of the distribution infrastructure. An interesting combination of the two would be the production of hydrogen and extraction of CO_2 from air in deserts using photovoltaic panels, or in windy places using wind turbines, or via hydroelectric power in regions with large columns of available water. Compressing hydrogen (and CO_2) for distribution in liquid form would entail building facilities and infrastructures that are very costly to create and maintain; in contrast, production of e-fuels on-site, followed by distribution using existing well-established infrastructures, would likely be more economical. Porsche and Siemens are already building a wind turbine–driven production facility for e-fuels in Chile, expecting to produce over half a billion liters of methanol by 2026[33,34]: enough to power 1.5 million average cars per year at average distance covered of 6,000 km. Wind power is greatly more efficient for hydrogen generation than photovoltaic, and local wind conditions in Chile's Magallanes Province are ideal: Punta Arenas has an average annual wind speed of 24 km/h[35] (the North Sea, where Europe is building large wind farms, has averages of around 22 km/h). Methanol is an industrially very important chemical, and it is currently generated largely from fossil fuels (mainly natural gas). In general, the development of e-fuels for shipping, aviation, and industrial processes such as steel and concrete production will automatically reduce the price for automotive use, hence reducing concerns about financial cost to car owners. but, as I will relate later, the cost will never reach the cheapness of fossil fuel.

Distribution networks can become political problems: there was a heated debate in Europe about the sense of the natural gas pipeline Nord Stream 2 (North Stream 2),[36] the largest ever built, stretching from Russia to Northern Germany. Had it not been made redundant by Russia's war in the Ukraine, it would have served energy needs in Northern Europe in a way that supposedly reduces climate impact. At the same time, millions of kilowatt hours of wind energy are being wasted that could be converted into methane. This squandered energy could represent a large part of the centralized, country-level gas supply for hard-to abate sectors of industry, such as cement, and for domestic heating. Biogas could be used locally, and the local fuel lines would be connected with a larger centralized synthetic methane distribution network. Thus local methane distribution facilities could contribute and draw off methane as demand required.

Incidentally, referring to CO_2 emissions from things like steel and cement production, aviation, and shipping as "hard-to-abate industry sectors" (domestic heating is also quite "hard") deserves further inspection. The very term is dangerously misleading, in my opinion, because it implies that the energy economy of road transportation is, in contrast, easy to abate. This clearly rests on a profoundly misleading piece of public "information"—I would go as far as calling it a "sham": it is that BEVs produce no emissions. Easy: we abated that CO_2 problem! Inconveniently, the truth is very different: what the customer sees is the tip of an iceberg, the lower reaches of which conceal enormous problems of environmental damage and challenges for recycling. In a sense, it is a great pity that it was so "easy" for the car industry to switch to BEV production because arguably a combination of different technologies and different user behavior would have done a more environmentally conscious abatement job. And that observation prepares us well for the next section.

Diversity of technology must be supported politically

The energy needs of human industry and society are so diverse in amount, power (energy delivery per unit time), and distribution in time that is almost impossible to envisage them all being addressed by a single technology in an efficient way. The history of human cultural evolution tells us that big mistakes can result from good intentions: for example, in the 1970s many countries entertained atomic energy as the unique solution to electricity demand—it also became viewed as a promising solution to powering ships, submarines, and even—amazingly—aircraft[37]; France, for example, built a nuclear energy capacity that today still produces more than 70% of its electricity. The problems that would result from large nuclear capacities were completely overlooked in the early years. We still have no satisfactory large-scale solution for nuclear waste. That is not to say that nuclear energy is, per se, bad, but that we clearly have not been able to develop it into the majority electricity-generating technology. On the other hand, at a time when we are under pressure to reduce CO_2 emissions quickly, we would be well-advised to reconsider using nuclear energy as *part* of the mix.

This is a classic example of a sensible tradeoff, or an advisable compromise, of which we will need many, many more. Promoting diversity of technologies requires a strong participation from non-industrial and non-government organizations: in short, independent researchers, who can analyze, do experiments, and give guidance in a way that is less subjective than industrial or political interests alone. Observing the developments in "green" energy and mobility, I conclude that this type of engagement is sorely lacking—due also to failings in political systems, which all too often lead to very one-sided decisions and one-dimensional technology development. On this note, I am convinced that we need hybrid solutions to our environmentally conscious energy economies (e.g., a mixture of electricity and organic chemistry). The hybrid system that biology has already made for us is remarkably similar, as I mentioned earlier. Carbon-based energy molecules store and transport energy at high density in relatively inert form within an individual cell and within a whole organism that can be a large as a blue whale. What a solution! Striving to go "all-XYZ" (as proponents of technology XYZ repeatedly call us to do) is not an intelligent approach in my opinion: it explicitly rules out hybrid solutions that can better react to challenges.

Experimentation, close monitoring, and adaptation are prerequisites for success

And so to the research that we continuously require to manage the sustainability of 8 billion humans on this planet. Experimentation is obviously needed to explore the multidimensional space of possible solutions, but it must not end there. I see an ever greater need to view technological applications as experiments as well, instead of seeing them as end-points of fundamental question-asking that merely undergo technological "evolution" and "incremental refinement." Of course, we need those, too, but if we don't continually analyze fundamental assumptions, we run the risk of failing to identify an early "unknown": the tendency for such an "unknown" or "underappreciated" effect to pop up is even greater in a situation where there are many uncertainties and emergent effects that are almost impossible to predict. The case in point is ecosystems and how they react to human-imposed stresses or simply to changes in human "strategies." This is all rather bad for politics, which seeks rather to communicate reliable, long-lasting solutions to the people. But politics must also change. I call explicitly for something novel: what I call "academic-political research alliances," where politicians simply must understand science and technology and closely follow their development instead of being mere receivers of "advice" tasked with making decisions as a reaction. This would require something else that I feel is important: a large increase in the number of scientists in politics.

In all political parties that I have investigated, scientists of any type are in a very small minority. Ironically, the representation of scientists in the Green parties is frequently the lowest. We need more social scientists and applied philosophers, as well as more natural scientists, in politics. The time has come when we will fail in our aims of sustainability if we rely on political systems consisting entirely of professional politicians (i.e., representatives of the people who, after a relatively brief training in a particular discipline, have devoted themselves to politics full-time, and consider *political acumen* to be their qualification). The environmental stakes are now so high that neither politicians nor scientists (who are also capable of overpromoting their particular corner of research) can entertain just one solution (either in policy or in technology).

The principle of adaptation is also very simple and yet hard to put into practice. It's obvious that we should quickly stop things that aren't working and quickly start new experiments or pilot projects that look more promising. However, there is a great deal of idealism in politics, intransigence in industry, and inertia in the general population that need to be overcome: culture is much stronger than strategy. We have been very good at researching the causes, extent, prognosis, and necessary countermeasures of global warming; we have been quite good at setting limits; we have been less good at making according laws and regulations; and finally, in my opinion, we have been least good at implementing strategies that demonstrate first-class intelligent thinking. Many ideas contributing to a new energy economy, some of them possibly brilliant, have been researched and developed in laboratories and production facilities worldwide. We have no lack of creativeness; instead we lack "steerage," anticipation, and flexibility to modify courses of action.

We must be prepared to pay more for almost everything

This is not about artificially raising prices in order to control consumer behavior; rather it refers to costs of materials that better reflect the true cost of more environmentally sustainable production: for example, are they produced in a highly recycling-dependent fashion, or are they made with very low environmental damage because better—more expensive—techniques are used? An obvious example is the price of e-fuels compared with fossil equivalents. In Table 7.2, next to European, I present US prices, because US prices contain much lower taxation than in Europe, and they are therefore closer to true economic costs.

This table of cost comparisons squares very well with latest analyses in the published literature of Fischer-Tropsch (FT) fuels and methanol, which are estimated, for example, at between 1.8 and 2.8 times the current prices of petroleum fuel and feedstock chemicals such as methanol.[44]

The comparison is interesting for the following reasons, among others:

- The smallest differential between natural and synthetic is for methane—because methane, being just one carbon with four hydrogens, is energetically less costly to produce.
- The largest differential between natural and synthetic is for ship's diesel, closely followed by kerosene. Neither of these fuels is taxed significantly, and so a switch to synthetic would incur large additional costs for transport.

Table 7.2 Cost comparisons between various hydrocarbon fuels from fossil and from synthetic sources.

Fuel	Fossil: Price	Per unit	Non-fossil: Price	Per unit	Ratio fossil to synthetic
Methane—global average	1.00 USD / 0.82 Euro	kg	3.3 USD / 4.0 Euro SYNTHETIC	kg	1:3.3
Methane—global average	1.00 USD / 0.82 Euro	kg	2.00 USD / 1.66 Euro BIO-METHANE	kg	1:2
Gasoline—USA	0.89 USD / 0.73 Euro	liter	2.8 USD / 2.3 Euro	liter	1:3.1
Gasoline—Europe	1.32 USD / 1.50 Euro	liter	2.8 USD / 2.3 Euro	liter	1:2.1
Automotive diesel—USA	0.84 USD / 0.69 Euro	liter	2.8 USD / 2.3 Euro	liter	1:3.3
Automotive diesel—Europe	1.65 USD / 1.35 Euro	liter	2.8 USD / 2.3 Euro	liter	1:1.7
Ship's diesel—global average	0.45 USD / 0.37 Euro	liter	2.8 USD / 2.3 Euro	liter	1:6.2
Aviation kerosene	0.47 USD / 0.39 Euro—US Gulf Coast type	liter	2.8 USD / 2.3 Euro	liter	1:6.0

Reference sources: Methane[38]; automotive gasoline prices[9]; automotive diesel prices[40]; European auto diesel prices: prices at pumps in May 2021[41]; ship's diesel prices[42] aviation fuel (kerosene).[43]

Environmentally conscious CO_2-neutral consumerism will be moderately to greatly more expensive for the individual consumer, and for society in general. The reason is that fossil reserves are basically large free energy reservoirs that are conveniently tucked away from potential environmental impact—that is, until we tap into them. Then come the small financial costs of extraction—these really are small because, in most European countries, 60–75% of the price of a liter of automotive diesel is tax: at €1.90 per liter diesel (August 2022), that leaves a between 76 and 48 cents for extraction, fractionation, purification, distribution, and oil company profit. Aviation fuel costs US$2.5 per gallon or €0.66 per liter (August 2022) but is hardly taxed. Fossil fuel is cheap, and synthetic fuel is expensive because, to make e-fuel, we need energy, whereas nature has made the fossil fuels "for free"; where we save over fossil fuel is in the separation and purification of the various fractions for different uses out of the thick black crude oil. Still, that is clearly a small cost. If governments (particularly in Europe) can subsidize what they *believe* to be an outright environmentally friendly solution to personal mobility—the BEV—with so many billions of Euros of taxpayer money, surely they can also subsidize e-fuels.

From my literature digging, I definitely believe that the prices of BEVs to consumers are greatly discounted at the expense of the environment. For some, they are likely only affordable with government subsidies (until the end of 2022, in Germany, as much as €9,000 per car, though that is now changing); if/when government subsidies fall away, and we additionally have to pay the real ecological price, who will pay the larger bill for this technology? Any talk of BEVs becoming cheaper than ICEVs is, I conclude, nothing more than a dangerous marketing stunt with the environment as hostage. Regarding aviation fuel, the near zero-taxation in many parts of the world cannot continue because it is in no relationship to the true cost to the environment of fuel production and use. Synthetic aviation fuel will cost up to seven times the usual price of aviation fuel at the pump. For heating and industrial gas, the relationship of production prices of around 7:1 synthetic-to-fossil is also likely. Taxation makes the gap between synthetic and fossil smaller. Heating is a necessity: much of the travel in which humans indulge is a luxury at the cost of the environment. It is good that vehicle gasoline and diesel are not cheaper because they approach the mark at which equivalent e-fuels are competitive (€2 per liter). As production increases, to replace fossil fuels, prices will drop, but never to the basic extraction-and-distribution price of diesel today.

The cost of energy is closely related to the cost of manufacture of most consumables, from food to phones. But there is more to it than that: complicated objects with hard-to-extract constituents require more energy to produce in an environmentally sympathetic way—I'll call that the "correct cost." I would not be surprised if the correct costs of most things that humans use and consume increased to up to three times their current price in an economy of full recycling and environmental conscientiousness. Researchers have already suggested that the price of meat in Europe should be 2.5 times its current average level in order to counter the climate impact of its production.[45]

It's time for a new kind of materialism: Valuing and caring more for the things we have

We need to derive more satisfaction from preserving what we have

It might (cue heretical laughter) make us happier, as the proverb goes, "to want what we have, rather than continually seeking to have what we want." We would be well-advised to redirect our materialism away from incessant overconsumption and a desire for more, better, bigger, faster, etc. and toward preserving material and hence using less of it. This is where things get really difficult in the stakes of culture versus strategy: only a small proportion of people in wealthy countries seem to live a culture based on being technically interested in, caring for, maintaining, and repairing objects to maximize their useful life-span. But they can be helped in many ways if legislators are willing. An important one is stipulating modular construction of products wherever possible. A great—sorry, I meant "appalling"—example in modern technology is the mobile phone, which has, despite enormous potential for modular construction, been made almost universally as a single block: something that is bought and then disposed of as soon as a significant part is either outdated or damaged. And cables and chargers have rarely been interchangeable between devices (legislators are now acting against that). What a waste! A few smartphone makers sell devices where modules (camera, battery, microphone, loudspeaker, screen) can easily be replaced or updated, but it's not a fast-growing sector.

Cars could, in principle, also be manufactured like that. It should be made as easy as possible for people to maintain and mend objects, or find others who can do so, because numerous studies have shown that large changes in behavior cannot be expected voluntarily, and one can't make legislation forcing people to repair things themselves. One can, however, support centers of expertise where people who like mending different kinds of things can be systematically networked. Yes, maker-/repair labs do already exist, but they are at an embryonic stage, and they are not usually networked with each other or with society. I won't go into more detail on the ways in which society could be changed to be more materialistically conservative: others have written whole books on the subject.[46]

We must develop sustainable personal behavior

Returning to the topic of this book, it is easy to believe that by getting away from carbon-based compounds as energy carriers we will get ourselves out of the present sustainability problems. To me, this seems like a desperate attempt to preserve an aspect of human behavior that wasn't working anyway: rampant overconsumption. Instead, we could take the danger signs given to us by CO_2 levels as the cue to reduce consumption and try to make a highly flexible energy conversion and recycling system based on carbon work for us and the environment after all. We have the scientific and technological capacity to do it, and billions of years of biology on Earth as a model to help us. On a geological, and even biological evolutionary, timescale, global warming and environmental degradation have become apparent to us very late: a mere second before the physical effects start to affect our own survival and reproduction chances. To avoid the worst of these physical effects, we are now trying to adapt. We seem to be attempting this very fast, but without enough thought for the longer-term future of human existence on Earth.

Does this mean a slow-down for life and industry?

This is a very sobering concept, stimulated by the question "If we preserve old things, surely we stifle technological innovation, right?" By extension, it would also mean "stifling" our industrial processes and means of energy generation, storage, and distribution because we would want to keep the cycling of energy-intensive materials to a minimum there as well. But there is an additional challenge looming ever greater, and that is the energy impact of electronics that communicate almost unbroken with networks and servers, shuttling data (increasingly, behavioral data, unknown to the user) to and fro. The "Internet-of-Things" (IoT) is estimated to have a CO_2 burden of 0.5–1 Gt $CO_{2eq.}$ per year by 2027[47] (i.e., 1.4–2.7% of total emissions). Some of this is very useful, but do we need *all* of it? Many readers won't like my thinking, but I am increasingly persuaded that it's very justifiable: we will have to get our technological evolution and industrial culture of so-called improvements much more under control and try to prevent "abuses" of technology that we do not really "need" and that are not environmentally sustainable. Sound a bit Luddite? Maybe, but in another sense, it's a very scientific idea, and, once again, I derive it from parallels with biology.

Much of what we see in biological evolution is very slow when compared with the rate of human social, cultural, and technological evolution (exceptions are viruses and some pathogenic microbes). This is because ecosystems are in a state of quite finely tuned equilibrium where interdependencies such as predator–prey and pathogen–host interactions keep things in check. Nevertheless, there is evolution going on as a dynamic equilibrium maintained by what we could regard as feedback loops and feedback inhibition. Here are a few generic examples:

- So-called *arms-races* between predator and prey, and between host and pathogen, where one side is selected for adaptations that increase predatory/pathogenic success, and the other for adaptations that increase defense. This is sometimes called the "Red Queen effect," after the Red Queen in Lewis Caroll's Alice in Wonderland, who tells Alice that, in the game that she is forced to play, she must run as fast as she can, just to stand still.
- Availability of food, where fluctuations in the abundance of different foods creates selection pressure favoring generalists—even omnivores (consuming plant and animal matter)—in many ecological niches. Generalists

–i.e., organisms that have not specialized in one [or a small group of] nutritional source(s)—are usually the animals that survive large and relatively fast ecological fluctuations; crises, you might say. Does this ring a bell with reference to supporting diversity of technology?
- Abiotic factors (related to the previous point) that change the niche, both on relatively short and relatively long timescales (e.g., temperature, rainfall, sunlight, geological phenomena).

The network of interconnectedness basically means that if any part of the system changes, it feeds on, or back, to a different part and affects that, too. Naturally, if the timescales of any of the above are too short, the affected species cannot adapt genetically or even epigenetically in a trans-generational sense: they end up reproducing increasingly less and finally go extinct. In many ecosystems on Earth, human activity has, directly or indirectly, accelerated the rate of niche change so much that many species are, indeed, going extinct—faster than at any time in Earth's history. What has not yet happened is a major negative effect on human reproduction—but it will, if we continue like this. The fastest known changes in biological diversity on Earth before the Anthropocene are the Cambrian Explosion,[48] which lasted between 13 and 15 million years, and the demise of (most of) the dinosaurs (and 75% of species in total) around 66 million years ago. That demise—despite the ferocity of the meteorite impact—is speculated by some to have taken up to 1 million years.[49] Some researchers believe that the dinosaurs were in steady and substantial demise even before the meteorite stuck. The cause: atmospheric changes caused by 30,000 years of massive volcanic activity in the Deccan Traps, covering around half of what is now India.[50] These are all enormous time spans compared with the period over which human activity is changing the environment (in many more ways that just GHG concentrations).

On the one hand yes, on the other no!

Cradle-to-cradle, in contrast to cradle-to-grave, must become the norm
So let's get practical: What would a more "biological" form of human technological evolution look like? First, we must realize that much of human industry on Earth is very unsustainable in terms of evolution and dynamic equilibria (the features of biological systems): we have factories, machines, and myriad devices and substances that work, or are created, in ways that have nothing to do with cyclical economies of energy or material and even less to do with feedback inhibition. That must change before we can envisage more sustainable human industry: we need to switch to cradle-to-cradle concepts (recycling that

is pre-engineered into materials and processes) and away from cradle-to-grave ones. Some things will be inhibited (i.e., slowed down) but only detrimental ones. We have no time left in the race against CO_2 and environmental degradation to burn fossil fuels and dig up massive amounts of the Earth's crust to feed this revolution. This leads to my next biological ingredient in the recipe, and one that will please people who are worried that everything will slow down to a snail's pace: modular design and interchangeability. But how are these biological?

Modular design, construction, and "component-exchangeability" must replace monoblock culture

Though selection of small, random mutations is a slow process, evolution also does something much faster, and that involves the moving around of individual genes and whole modules, often so-called *gene-regulatory networks* (GRNs) as reviewed in McLennan.[51] This happens in bacteria—classic examples being the exchange of antibiotic resistance genes and gene cassettes between species, and even phyla—through to vertebrates, where the moving around happens almost exclusively *within* a species' genome.

At its simplest, a gene is duplicated and inserts into the genome at a point where it is under the control of a different regulatory unit from its original one: a paradigm for this are the alpha-crystallins, proteins that indisputably evolved from ancient proteins that helped cells survive heat shock. As highly stress-resistant proteins, they were wonderfully suited to make the lens of the vertebrate eye, which is where they are presently also found—as alpha-crystallins. In the eye, they withstand ultraviolet radiation and chemical exposure for up to six decades before a cataract operation. Because the eye is developmentally and structurally a modular entity, it was possible for evolution to improve it in one fell swoop by plugging in a new protein—crudely put.

An even more amazing example is the existence of a type of placenta that nourishes developing fertilized eggs in the male seahorse (yes, you read correctly: in seahorses, the dad is the mum!): here, a whole set of genes (a GRN) was plugged into a new developmental location. One of the proteins encoded is highly likely to have arisen via duplication of a protein involved in the liver/kidney of all bony fishes. Other examples include the evolution of the tetrapod forelimb and courtship behavior in birds. One particular type of mobile genetic element, the transposon, is even believed to have created the rapid diversification of body-plans and emergence of species witnessed in the Cambrian explosion around 540 million years ago.

Biology is built in modular fashion and is highly opportunistic. It can do plug-and-play games with many components, adapt, and take advantage of changing situations quickly on evolutionary time-scales without major redesigning. We should strive for this principle wherever possible in human industry: from the interchangeability of components of smartphones and cars through to industrial facilities. It is at once conservative and innovative. Yes, your smartphone will continue to look the same for several years more than usual—just as species of plants and animals look the same, generation after generation. However, you will be able to increase its speed, or improve its camera (eye!), or dramatically increase its battery life by rapidly and simply plugging in the latest model: an ideal compromise between conservatism and progress! Industry need not be crippled by this strategy because a whole new domain would emerge, creating a great deal of positive activity and opportunity for entrepreneurial spirit: the maintenance and evolution of devices via component development and renewal.

And what can be done with smartphones can also be done with cars. Tesla, on the other hand, seems to be moving in a different direction, seeking radically to reduce the costs of its cars by installing the battery as a part of the frame of the car—welded in with glue—instead of as a separate component (departure from the "skateboard" design), as presented in 2021.[52] Might this be an example of the RyanAir strategy: beat the competition with rock-bottom prices? One commentator of this article suspects that these batteries will not be replaceable. He's not quite right, but it's definitely a major "ordeal" compared with previous designs, as noted in another article,[53] where the author states "it takes a total of 143 steps just to remove the structural battery pack . . . it involves basically removing a large part of the interior of the vehicle before actually starting to work on the pack." Further media reporting since casts doubt on the general recyclability of such cars, and in particular the prospects for second use of their batteries.

Negative feedback must be incorporated into economic systems to prevent damaging trends becoming self-reinforcing
If, against some of the tide, we succeed in reducing consumption and build new, cradle-to-cradle industrial metabolisms and modes of personal behavior based on modular design, they must be very sensitive to feedback inhibition. They must react as if they were parts of a large, global system of checks and balances that "corrects" itself—rather than waiting for some disaster to correct it. That is, we need to live in the same dynamic equilibrium as the natural world, using all the scientific and technological facilities that we can invent for monitoring our impact and adjusting accordingly. As I have mentioned before, the carbon economy lends itself to this endeavor for several reasons: (1) we have become experts at monitoring carbon fluxes on Earth; (2) we already have industrial metabolisms based on carbon compounds; (3) carbon compounds allow us a greater level of flexibility of conversion and transport of material and energy than any other substance that we have yet devised; and (4) the carbon compounds that are, and will be, typical of our industrial metabolism are closely related to—or identical

to—natural ones, hence enabling us to interdigitate into the larger global carbon metabolism.

Conclusion: Thinking more biologically, developing closed-loop economies, and getting back into Earth's "buffering" zone

A high-level conclusion that I reach from my literature synthesis is that there are so many unknowns and unsolved problems in alternative energy (including battery technology, e-fuel generation, and most other areas of energy management) that nothing is rapidly going to solve our CO_2-neutrality challenges; worse still, break-neck attempts to do so will rely on the (temporary) burning of even more fossil fuels, hence posing a very serious danger: let alone failing to stick to the +1.5°C maximum by 2050, we will generate an additional CO_2 bubble and increase average global temperature even more (possibly by 4–5°C by the end of the century). We desperately need to give the climate and environment a break—immediately—and greatly reduce consumption while we develop technology and human behavior in line with long-term solutions. As I write, TerraPower, the company founded by Bill Gates and Warren Buffet, is preparing to build a pilot nuclear energy plant ("Natrium").[54] Many have criticized this move as unsustainable, but it is a reaction to unsustainable consumption: either we cut consumption in half, or we might just be forced to go nuclear to provide urgently needed low-CO_2 energy—that's the bottom line.

We are currently trying to satisfy present demand and provide for ever-increasing demand (for everything, from energy to materials) using technologies that are not yet ripe; many of them are still in diapers. This was to be expected because the looming crisis didn't stimulate enough people to research, or put forward money to research, alternatives to fossil fuels. Now the pressure is on, but I am convinced that it is too late to expect technology to solve even half of the problem: most of the grand climate and environmental challenge will need to be solved by reduction in consumption—mainly in the wealthy countries of this world. True recyclability in either physically or conceptually closed loops should be our urgent aim. We must also build many more interfaces between energy economies and material economies, integrating between these two sectors wherever possible so as not to waste by-product energy or by-product material (one of the latter is CO_2 itself, escaping from industry or power plant chimneys).

We need to keep our eyes open for opportunities to benefit from what would otherwise be waste (either industrial or natural). We must create economies of scale where possible (e.g., energy storage from naturally fluctuating regenerative sources) and integrate local solutions into them where it makes sense (e.g., biogas for local heating from local waste). Technology diversity must be supported at political level to avoid one-dimensional developments that radically reduce our flexibility to react to unforeseen eventualities. Research and development need to be integrated with politics such that experimentation, monitoring, and adaptation are intimately linked with policy making. An important facet of this is to realize that current efforts to get GHG emissions under control are fundamentally a very large experiment, and we might need to change direction quite radically. We must be prepared to pay more for everything because that will also help support the development of more sustainable products and production methods.

Finally, we should value material possessions much more than at present, do our best to make them last considerably longer, and only change them if they no longer function well enough or they have more environmental impact than their replacement. We must try hard to take all environmental impacts into account, and not just CO_2 emissions. It is very unlikely that we can feed our energy demands, even with "green" technologies, without major environmental degradation. By extension, we must reasonably assume that the production and disposal of most material things that we have probably constitutes the largest environmental impact in the whole chain of their existence. That chain should not end with a heavy anchor of waste or pollution that drags the environment down, but be cyclical.

Any measure of success requires user-level changes in the cultures of mobility, nutrition, construction, and domestic use of energy and material. There are serious environmental consequences at stake because we have pushed the environment (in numerous ways, not just CO_2 concentration) way past its ability to buffer our impact. We will, therefore, have to push ourselves back onto the "right" side of the line, as it were. The "pushes" will inevitably be things like limitations on use and consumption. For example, I view it as very advisable to place a limit on the volume of fuel bought per private car each year and on flying. This amounts to rationing, of course. If countries are able to ration in times of war, they can certainly do it at a time of global crisis. There will need to be flexibility because not everyone has "equal" challenges in their lives: some degree of trading between areas of consumption and individuals—similarly to CO_2 emissions certificate trading that is commonplace nowadays. That would be a much more intelligent and interesting experiment than most pushes.

There are some "pulls" (incentives) at present, but I am not remotely impressed by many of them. For example, many governments give very generous subsidies to consumers for buying newer, more economical cars—above all, EVs. As an interesting side note, the so-called *Wirtschaftsweisen* (wise industrial thinkers), a group of German experts from industry and academia, firmly rated Germany's last attempt in this direction, *die Abwrackprämie* (the car-scrapping bonus in return for buying a new one), as a resounding failure in both environmental *and* economic terms. Seen from another angle, many people obviously prefer to buy a nice new electric car rather than have their house energetically upgraded (for a similar price). Subsidies for domestic energetic upgrading are a "pull" that I find very good. As I described in Chapter 4, an average domestic oil-fired central heating system can easily burn more than three times the fossil fuel (2,500 liters) as an average diesel car in a year. Insulating one's house and changing to triple-glazing can halve the energy needed to heat it (i.e., a saving 1,250 L of fuel per year). The final cost, minus government subsidies that many countries provide for such energetic renovations, would be comparable to the cost of a medium-sized BEV. And, even more to the point, domestic insulation has *real* environmental benefits, and those are much longer-term than any that might be provided by new cars. So, looking around oneself, one could ask whether consumer behavior is largely a value judgment.

Focusing on a single aspect of human behavior is not nearly as effective as taking in the larger picture: Where do people live? How do they live? Where do they work? How do they work? We have seen some interesting and potentially very beneficial (for environment, employees, and employers) changes in work practice as a reaction to COVID-19. We must pursue these with proactive strategies now. For example, companies that require employees to come in to work regularly must help make the use of cars more efficient by supporting journey-sharing; individuals must also think more about this, and digital services that make it easier to journey-share in private cars need to be expanded and improved greatly. Yes, some degree of flexibility will be lost: not everyone will be able to arrive at work and leave work at their preferred times, but they would benefit by having less driving stress. There will, naturally, be consequences for those with children of kindergarten or school age. I am optimistic that we can find socially acceptable solutions to such challenges, and I believe that we must.

Where and how we live are areas crying out for ideas, too. Instead of providing generous subsidies for consumption (e.g., buying new cars), governments could reroute money to helping people relocate to lessen home ↔ work distance or otherwise tailor their lives to reduce regular travel. Perhaps subsidies for downsizing and living in energetically more economical multifamily houses would work in some cases. The concept of the *Mehrgenerationenhaus* (German: house accommodating several generations of inhabitants) could work very well in conjunction with a smaller number of cars and volunteer drivers (people living in the same house). Volunteers would take elderly inhabitants with them on selected journeys. Yes, this might be more organizational work, but electronic applications for organizing all manner of human behavior have been created: surely we can crack this nut. People might have to forsake something, but they will gain something more valuable (e.g., time, well-being, simple enjoyment, or health). This is an ambitious endeavor, and we will need the best creative brains of applied sociologists and philosophers to help us. Patience, endurance, and compromise will also have to be practiced by more people than at present. Living our lives at an increasingly fast pace will serve only the interests of industry and the few extremely wealthy people on this planet who have managed to maneuver themselves out of the rat race—and out of environmental harm's way, as it were.

A cradle-to-cradle material system should be conceived as a global industrial metabolism that does not go all-out for more, faster, better, but rather produces what we *really* need; it would adapt and create new things if and when their creation and recycling could likely be done at zero net negative ecological impact. I have outlined the profound parallels between biological economies and human economies of energy and material, and I strongly believe that we must embrace carbon rather than trying desperately, but only seemingly, to decarbonize. I hope that I have illustrated the fascinating parallels and opposites between the biological harvesting of the Sun's energy (photosynthesis) and the burning of the resulting fuel (carbohydrates) in cellular respiration. This system, which also embodies the biochemical hubs that produce, recycle and alter carbon compounds for energy and material purposes, is intrinsically sustainable in the ways in which biological organisms express it in nature.

We must understand industry in terms of integration of material and energy economies similarly to those that have sustained species, ecosystems, life, and our living Earth for 3.5 billion years. This evolved with cyclical systems using carbon as the crucial component of energy carriers and materials. Applying similar principles in human civilization would support a vibrant industry of new thinking, improvements, updates, and replacement of components in larger objects (from ships down to telephones). It would not be the industry that the richest entrepreneurs on the planet are used to, but the Earth won't allow us to pursue that kind of industry much longer anyway. The developments that include massive quantities of mineral mining and battery manufacture—often referred to as the "clean energy transitions" (e.g., by the IEA)—are anything but clean. We must do much, much better than that.

List of figures, tables and information boxes

1.1	Carbon bucky ball and bucky tube	9
1.2	Impression of a soot particle from an asymptotic giant branch (AGB) star	10
1.3	A large anthracene-based molecule with a variety of chemical groups	11
1.4	Examples of smaller organic compounds formed in the universe before the origin of life and also found in certain meteorites	12
1.5	s and p orbitals represented as 95% probability volumes in which the respective electrons are to be found	14
1.6	The sp hybrid orbitals in methane (CH_4)	16
1.7	2D projection of a typical alkane (pentane)	16
1.8	Sigma and pi bonds between carbon atoms	17
1.9	Diamond and graphite	20
1.10	A potential origin of life in alkaline thermal vents on the sea bed	28
1.11	Chemical structure of glucose	33
2.1	The principal constituents of living organisms	46
2.2	Ketones circulating in humans undergoing ketosis because of hypoglycemia	49
2.3	Glucose and palmitic acid (as a linear Fischer projection)	54
2.4	Linear Fischer projection of one form of hexadecatrienoic acid (HTA), an omega-3 unsaturated fatty acid (FA)	56
2.5	How the chemistry of trans-unsaturated fatty acids (FAs) makes them potentially unhealthy	56
2.6	Average milk product consumption in the United States produces $CO_{2eq.}$ emissions in the same ballpark as conventional car-driving	66
2.7	Average beef-eating in Europe produces $CO_{2eq.}$ emissions comparable with average car-driving	69
2.8	The paleolithic barbecue: principal components of wood smoke and charred meat, showing only generic compounds to represent larger classes	72
3.1	Stretching and bending modes of CO_2	94
3.2	Short-wavelength radiation has more energy than long-wave radiation	95

3.3	Superposition of infrared (IR) absorption spectra of H_2O and CO_2	96
3.4	Atmospheric concentrations of CO_2 in ppm since pre-industrial times	98
3.5	CO_2 absorbs infrared (IR) radiation, but most of the energy is converted to heat before it can be re-emitted as IR radiation again	104
3.6	Distribution of biomass of carbon on Earth in gigatonnes	108
3.7	Projected change in atmospheric CO_2 concentrations and ocean pH	112
3.8	Schematic of coal formation	115
3.9	The intriguing chemistry of coal	119
3.10	The compound that betrayed oil's organic origin	121
3.11	Lipids that are abundant in cell membranes	122
3.12	Alkanes that are gaseous or liquid at ambient temperature and pressure	123
4.1	Alcohols, from C_1 to C_5	134
4.2	Dimethylether (DME) compared with hexadecane to emphasize the different chemistry	143
4.3	Inverse relationship between the ratio of carbon atoms to hydrogen atoms and energy density in fully saturated hydrocarbons	148
4.4	Polymerization of sugars into complex carbohydrates	154
4.5	How far do particles go when we breath them in?	170
5.1	Anglicized version of the 1970s anti–nuclear power sticker in Germany	180
5.2	The repeating unit of polyethylene terephthalate (PET)	183
5.3	Polyethylene (PE), polyvinyl chloride (PVC), and polypropylene (PP)	188
5.4	Essential principle of a blast furnace	193
5.5	Principle of a rotary cement kiln	195
5.6	Basic principles of direct air capture (DAC) and CarbFix technology developed by Climeworks and applied on Iceland	209
6.1	Plot of cumulative difference in CO_2 emissions from a single car, either battery electric vehicle (BEV) or internal combustion engine vehicle (ICEV)	222
6.2	Plot of the cumulative difference in CO_2 emissions that result from the scenario of replacing a well-functioning and relatively economical internal combustion engine vehicle (ICEV) with a comparable new battery electric vehicle (BEV)	224
6.3	Plot of total cumulative energy consumption car economies of battery electric vehicles (BEVs) and internal combustion engine vehicles (ICEVs) using e-fuels; all energies renewable	233

6.4	Plot of total cumulative environmental impact of car economies of battery electric vehicles (BEVs) and internal combustion engine vehicles (ICEVs) using e-fuels; all energies renewable	236
6.5	Comparison of impact of vehicle fuel technologies measured as human toxicity potential over 100 years in terms of kg of 1,4-DB-eq	237
6.6	Plot of total cumulative human health impact as disability-adjusted life years (DALY) of car economies of battery electric vehicles (BEVs) and internal combustion engine vehicles (ICEVs) using e-fuels; all energies renewable	238
6.7	How geopolitical instability can rapidly favor e-fuels in countries where oil and minerals for battery/electronics manufacture are scarce or hard to mine	256
7.1	Schematic of energy and material transitions in a plant cell	279
7.2	The mitochondrion as an interchangeable hub of energy and material economies in the cell	281
7.3	Schematic of the reversible solid oxide cell (RSOC)	283

List of Boxes

1.1	More on isotopes: from hydrogen to heavy elements	2
1.2	Carbon isotopes: one stable, one unstable, both useful	7
1.3	More on the Miller-Urey experiment	24
1.4	How green plants make the substrates of everything they need via photosynthesis, including figures: schematic of adenosine triphosphate (ATP); electron transport chain and the Calvin cycle in plant cells	35
1.5	How plant and animal cells biologically burn carbohydrates: from glycolysis to oxphos, including figure: metabolism of pyruvate in the tricarboxylic acid (TCA) cycle and oxidative phosphorylation	38
2.1	Triglycerides produce large ratios of adenosine triphosphate (ATP) per unit mass by containing fatty acids (FA) where all available carbons are "tanked up" with hydrogen	53
2.2	The carnivore's quandary: growing a whole animal over several years	64
3.1	Short-wavelength radiation	86
3.2	Long-wavelength (black body) radiation	88

3.3   Long-wavelength radiation in the context of "fingerprint" radiation, including figure: bond stretching and bending modes of molecules; bond stretching modes in H_2O compared with CO_2, and the relationship with the wavelength of radiation absorbed/emitted   91
4.1   The Fischer-Tropsch process   142
4.2   Non-alcoholic fuels from a variety of sources   147

List of Tables

3.1   Permanent gases   92
3.2   Variable gases   93
4.1   Average (between higher heating value [HHV] and lower heating value [LHV]) energy densities of wood, methane, liquid natural gas (LNG), and heating oil   152
5.1   Petroleum products by type and volume consumed per day in the United States (2021)   191
7.1   Advantages and disadvantages of Li-ion battery and e-fuel technology   276
7.2   Cost comparisons between various hydrocarbon fuels from fossil and from synthetic sources   294

References

Chapter 1

1. Stellar Nucleosynthesis. Wikipedia. https://en.wikipedia.org/wiki/Stellar_nucleosynthesis
2. Vangioni-Flam E, Cassé M. Cosmic lithium-beryllium-boron story. In: Spite M, ed. *Galaxy Evolution: Connecting the Distant Universe with the Local Fossil Record*. Springer; 2012:77–86.
3. Trimble V. The origin and evolution of the chemical elements. In: Malkan MA, Zuckertoman B, eds. *Origin and Evolution of the Universe: From Big Bang to ExoBiology*. 2nd ed. World Scientific, Jones and Bartlett Publishers; 1996: 63–94.
4. Abundance of the Chemical Elements. Wikipedia. https://en.wikipedia.org/wiki/Abundance_of_the_chemical_elements
5. Furukawa Y, Chikaraishi Y, Ohkouchi N, Nakamura T. Extraterrestrial ribose and other sugars in primitive meteorites. *Proc Natl Acad Sci USA*. 2019;116:24440–24445. doi:10.1073/pnas.1907169116
6. Are diamonds really forever? Stackexchange. https://chemistry.stackexchange.com/questions/34193/are-diamonds-really-forever
7. Lab-grown diamond global production by region 2020. *Statista*. 2020. https://www.statista.com/statistics/1204042/global-lab-grown-diamond-production-by-region/
8. Schreck M, Gsell S, Brescia R, Fischer M. Ion bombardment induced buried lateral growth: the key mechanism for the synthesis of single crystal diamond wafers. *Sci Rep*. 2017;7:article number 44462. doi:10.1038/srep44462
9. Miller SL. A production of amino acids under possible primitive Earth conditions. *Science*. 1953;117:528–529. doi:10.1126/science.117.3046.528
10. Miller SL, Urey HC. Organic compound synthesis on primitive Earth. *Science*. 1959;130:245–251. doi:10.1126/science.130.3370.245
11. Sossi PA, Burnham AD, Badro J, Newville M, O'Neil HStC. Redox state of Earth's magma ocean and its Venus-like early atmosphere. *Sci Adv*. 2020;6. doi:10.1126/sciadv.abd1387
12. Martin W, Baross J, Kelley D, Russel MJ. Hydrothermal vents and the origin of life. *Nature Rev Microbiol*. 1991;6:805–814. doi:10.1038/nrmicro1991
13. McCollom TM, Ritter G, Simoneit BRT. Lipid synthesis under hydrothermal conditions by Fischer-Tropsch reactions. *Origins of Life and Evolution of the Biosphere*. 1999;29:153–166.
14. Jordan SF, Rammu H, Zheludev IN, Hartley AM, Maréchal A, Lane N. Promotion of protocell self-assembly from mixed amphiphiles at the origin of life. *Nature Ecol Evol*. 2019;3:1705–1714. doi:10.1038/s41559-019-1015-y
15. Martin W. Early evolution without a tree of life. *BMC Biol Direct*. 2011;6:article number 36. doi:10.1186/1745-6150-6-36

16. Halevy I, Bachan A. The geologic history of seawater pH. *Science*. 2017;355:1069–1071. doi:10.1126/science.aal4151
17. Russell MJ. Green rust: the simple organizing "seed" of all life? *MDPI Life*. 2018;8. doi:10.3390/life8030035
18. Barge LM, Cardoso SSS, Cartwright JHE, et al. From chemical gardens to chemobrionics. *Chem Rev*. 2015;115:8652–8703. doi:10.1021/acs.chemrev.5b00014
19. Herschy B, Whicher A, Camprubi E, et al. An origin-of-life reactor to simulate alkaline hydrothermal vents. *J Mole Evol*. 2014;79:213–227. doi:10.1007/s00239-014-9658-4
20. Cavalazzi Barbara, Lemelle Laurence, Simionovici Alexandre, et al. Cellular remains in a ~3.42-billion-year-old subseafloor hydrothermal environment. *Sci Adv*. 2021;29:eabf3963. doi:10.1126/sciadv.abf3963
21. Sánchez-Baracaldo P, Bianchini G, Wilson JD, Knoll AH. Cyanobacteria and biogeochemical cycles through Earth history. *Trends Microbiol*. 2022;30(2):143–157. doi:10.1016/j.tim.2021.05.008
22. Sánchez-Baracaldo P, Raven JA, Pisani D, Knoll AH. Early photosynthetic eukaryotes inhabited low-salinity habitats. *Proc Natl Acad Sci*. 2017;114(37):E7737–E7745. doi:10.1073/pnas.1620089114
23. Schreiber M, Rensing SA, Gould SB. The greening ashore. *Trends Plant Sci*. Jun 20, 2022. doi:10.1016/j.tplants.2022.05.005
24. Canfield DE, Rosing MT, Bjerrum C. Early anaerobic metabolisms. *Phil Trans R Soc B: Biol Sci*. 2006;361(1474):1819–1836. doi:10.1098/rstb.2006.1906
25. Moore A. Life defined. *BioEssays*. 2012;34:253–254. doi:10.1002/bies.201290011

Chapter 2

2. Linquist B, Van Groenigen KJ, Adviento-Borbe MA, Pittelkow C, Van Kessel C. An agronomic assessment of greenhouse gas emissions from major cereal crops. *Glob Chang Biol*. 2012;18(1):194–209. doi:10.1111/j.1365-2486.2011.02502.x
3. Yue Q, Xu X, Hillier J, Cheng K, Pan G. Mitigating greenhouse gas emissions in agriculture: from farm production to food consumption. *J Clean Prod*. 2017;149:1011–1019. doi:10.1016/j.jclepro.2017.02.172
4. Morgavi DP, Forano E, Martin C, Newbold CJ. Microbial ecosystem and methanogenesis in ruminants. *Animal*. 2010;4(7):1024–1036. doi:10.1017/S1751731110000546
5. Ketogenesis. Wikipedia. https://en.wikipedia.org/wiki/Ketogenesis
6. Guelpa G, Marie A. Lutte contre l'épilepsie par la désintoxication. *Bull gen de therap*. 1910:616–624.
7. Bailey EE, Pfeifer HH, Thiele EA. The use of diet in the treatment of epilepsy. *Epilepsy Behav*. 2005;6(1):4–8. doi:10.1016/j.yebeh.2004.10.006
8. Kim JM. Ketogenic diet: old treatment, new beginning. *Clin Neurophysiol Pract*. 2017;2:161–162. doi:10.1016/j.cnp.2017.07.001
9. Beta Oxidation. Wikipedia. https://en.wikipedia.org/wiki/Beta_oxidation
10. Dąbek A, Wojtala M, Pirola L, Balcerczyk A. Modulation of cellular biochemistry, epigenetics and metabolomics by ketone bodies. Implications of the ketogenic diet in the physiology of the organism and pathological states. *Nutrients*. 2020;12(3). doi:10.3390/nu12030788

11. Murugan M, Boison D. Ketogenic diet, neuroprotection, and antiepileptogenesis. *Epilepsy Res.* 2020;167:106444. doi:10.1016/j.eplepsyres.2020.106444
12. Richter M, Baerlocher K, Bauer JM, et al. Revised reference values for the intake of protein. *Ann Nutr Metab.* 2019;74(3):242–250. doi:10.1159/000499374
13. List of Foods by Protein Content. Wikipedia. https://en.wikipedia.org/wiki/List_of_foods_by_protein_content
14. Nutrition and Healthy Eating. Mayo Clinic. 2022. https://www.mayoclinic.org/healthy-lifestyle/nutrition-and-healthy-eating/in-depth/carbohydrates/art-20045705
15. Cooper GM. *The Cell: A Molecular Approach.* 2nd ed. Sinauer Associates; 2000. https://www.ncbi.nlm.nih.gov/books/NBK9903/
16. Bhattacharya A. *Effect of High Temperature on Crop Productivity and Metabolism of Macro Molecules.* Academic Press, Elsevier Inc.; 2019. https://www.sciencedirect.com/book/9780128175620/effect-of-high-temperature-on-crop-productivity-and-metabolism-of-macro-molecules
17. Gallagher ND, Playoust MR. Absorption of saturated and unsaturated fatty acids by rat jejunum and ileum. *Gastroenterology.* 1969;57(1):9–18. doi:10.1016/S0016-5085(19)33954-X
18. Oteng AB, Kersten S. Mechanisms of action of trans fatty acids. *Adv Nutr.* 2020;11(3):697–708. doi:10.1093/advances/nmz125
19. Lehninger AL, Greville DG. The enzymatic oxidation of D- and L-β-hydroxybutyrate. *Biochem Biophys Acta.* 1953;12:188–202. doi:10.1021/ja01102a538
20. Kashiwaya Y, Sato K, Tsuchiya N, et al. Control of glucose utilization in working perfused rat heart. *J Biol Chem.* 1994;269(41):25502–25514. doi:0.1016/S0021-9258(18)47278-X
21. Murray AJ, Knight NS, Cole MA, et al. Novel ketone diet enhances physical and cognitive performance. *FASEB J.* 2016;30(12):4021–4032. doi:10.1096/fj.201600773R
22. Dedkova EN, Blatter LA. Role of β-hydroxybutyrate, its polymer poly-β-hydroxybutyrate and inorganic polyphosphate in mammalian health and disease. *Front Physiol.* 2014;5. doi:10.3389/fphys.2014.00260
23. Tieu K, Perier C, Caspersen C, et al. D-β-Hydroxybutyrate rescues mitochondrial respiration and mitigates features of Parkinson disease. *J Clin Invest.* 2003;112(6):892–901. doi:10.1172/JCI18797
24. Yang H, Shan W, Zhu F, Wu J, Wang Q. Ketone bodies in neurological diseases: focus on neuroprotection and underlying mechanisms. *Front Neurol.* 2019;10. doi:10.3389/fneur.2019.00585
25. Norwitz NG, Hu MT, Clarke K. The mechanisms by which the ketone body D-β-hydroxybutyrate may improve the multiple cellular pathologies of Parkinson's disease. *Front Nutr.* 2019;6. doi:10.3389/fnut.2019.00063
26. Cotter DG, Ercal B, Huang X, et al. Ketogenesis prevents diet-induced fatty liver injury and hyperglycemia. *J Clin Invest.* 2014;124(12):5175–5190. doi:10.1172/JCI76388
27. Veech RL, Bradshaw PC, Clarke K, Curtis W, Pawlosky R, King MT. Ketone bodies mimic the life span extending properties of caloric restriction. *IUBMB Life.* 2017;69(5):305–314. doi:10.1002/iub.1627
28. Glucogenic Amino Acid. Wikipedia. https://en.wikipedia.org/wiki/Glucogenic_amino_acid
29. Godfray HCJ, Aveyard P, Garnett T, et al. Meat consumption, health, and the environment. *Science.* 2018;361(6399). doi:10.1126/science.aam5324

30. White RR, Hall MB. Nutritional and greenhouse gas impacts of removing animals from US agriculture. *Proc Natl Acad Sci.* 2017;114(48):E10301–E10308. doi:10.1073/pnas.1707322114
31. Food Outlook. FAO: Food and Agriculture Organization of the United Nations. 2018. https://www.fao.org/3/CA2320EN/ca2320en.pdf
32. Global Food: Waste Not, Want Not. Institution of Mechanical Engineers. 2013. https://www.imeche.org/docs/default-source/reports/Global_Food_Report.pdf
33. Horrillo A, Gaspar P, Escribano M. Organic farming as a strategy to reduce carbon footprint in Dehesa agroecosystems: a case study comparing different livestock products. *Animals.* 2020;10(1). doi:10.3390/ani10010162
34. American Food Production Requires More Energy Than You'd Think. Save on Energy. 2022. https://www.saveonenergy.com/resources/food-production-requires-energy/
35. Rotz CA, Asem-Hiablie S, Place S, Thoma G. Environmental footprints of beef cattle production in the United States. *Agric Syst.* 2019;169:1–13. doi:10.1016/j.agsy.2018.11.005
36. How Much Meat Can You Expect from a Fed Steer? South Dakota State University Extension. 2020. https://extension.sdstate.edu/how-much-meat-can-you-expect-fed-steer
37. An Overview of Meat Consumption in the United States. farmdocDAILY. 2021. https://farmdocdaily.illinois.edu/2021/05/an-overview-of-meat-consumption-in-the-united-states.html
38. The History of Beef Consumption in the U.S. and How It Is Changing. Sentient Media. 2021. https://sentientmedia.org/beef-consumption-in-the-us/
39. PerCapitaConsumptionofBonelessRedMeatintheUnitedStatesin2021,byType.*Statista.* 2021. https://www.statista.com/statistics/184378/per-capita-consumption-of-red-meat-in-the-us-2009-by-type
40. Greenhouse Gas Emissions from the Dairy Sector: A Life Cycle Assessment. FAO. 2010. https://www.fao.org/3/k7930e/k7930e00.pdf
41. List of Countries by Milk Consumption per Capita. Wikipedia. https://en.wikipedia.org/wiki/List_of_countries_by_milk_consumption_per_capita
42. Countries Who Drink the Most Milk. World Atlas. 2018. https://www.worldatlas.com/articles/countries-who-drink-the-most-milk.html
43. Per Capita Milk Consumption, 2017. Our World in Data. 2018. https://ourworldindata.org/grapher/per-capita-milk-consumption
44. Brade W. CO2-Fußabdrücke für Milch und Milchprodukte (CO2-footprints of milk and milk products). *Zeitschrift für Agrarpolitik und Landwirtschaft (Journal of farming politics and farming).* 2014;92:1–16. doi:10.12767/buel.v92i1.43
45. Flysjö A. *Greenhouse Gas Emissions in Milk and Dairy Product Chains: Improving the Carbon Footprint of Dairy Products.* Aarhus University 2012. https://pure.au.dk/ws/files/45485022/Anna_20Flusj_.pdf
46. Average Miles Driven per Year by Americans. Smartfinancial. 2022. https://smartfinancial.com/average-miles-driven-per-year
47. Well-to-Tank. Zemo Partnership. 2021. https://www.zemo.org.uk/work-with-us/buses-coaches/low-emission-buses/well-to-tank.htm
48. Well-to-Wheels Analyses. EU Science Hub of the Joint Research Centre of the European Union. 2016. https://joint-research-centre.ec.europa.eu/welcome-jec-website/jec-activities/well-wheels-analyses_en

49. Rated Real-World Well-to-Wheel Greenhouse Gas Emissions of New Light-Duty Vehicle Sales Worldwide by Size Segment, 2019. International Energy Agency (IEA). 2021. https://www.iea.org/data-and-statistics/charts/rated-real-world-well-to-wheel-greenhouse-gas-emissions-of-new-light-duty-vehicle-sales-worldwide-by-size-segment-2019
50. Foley JA. Can we feed the world and sustain the planet? *Sci Am*. 2011;305(5):60–65.
51. Farming for Failure. Greenpeace. 2020. https://storage.googleapis.com/planet4-eu-unit-stateless/2020/09/20200922-Greenpeace-report-Farming-for-Failure.pdf
52. Livestock and Climate Change. Worldwatch. 2009. https://awellfedworld.org/wp-content/uploads/Livestock-Climate-Change-Anhang-Goodland.pdf
53. Xu X, Sharma P, Shu S, et al. Global greenhouse gas emissions from animal-based foods are twice those of plant-based foods. *Nature Food*. 2021;2(9):724–732. doi:10.1038/s43016-021-00358-x
54. Agri-Environmental Indicator: Greenhouse Gas Emissions. Eurostat (Statistics department of the European Commission). 2018. https://ec.europa.eu/eurostat/statistics-explained/index.php?title=Archive:Agri-environmental_indicator_-_greenhouse_gas_emissions&oldid=374989
55. Harchaoui S, Chatzimpiros P. Can agriculture balance its energy consumption and continue to produce food? A framework for assessing energy neutrality applied to French agriculture. *Sustainability*. 2018;10(12). doi:10.3390/su10124624
56. Grossi G, Goglio P, Vitali A, Williams AG. Livestock and climate change: impact of livestock on climate and mitigation strategies. *Anim Front*. 2019;9(1):69–76. doi:10.1093/af/vfy034
57. Ripple WJ, Wolf C, Newsome TM, Barnard P, Moomaw WR. World scientists' warning of a climate emergency. *BioScience*. 2020;70(1):8–12. doi:10.1093/biosci/biz088
58. Energy Efficiency of Meat and Dairy Production. Our World in Data. 2016. https://ourworldindata.org/grapher/energy-efficiency-of-meat-and-dairy-production
59. Tilman D, Clark M. Global diets link environmental sustainability and human health. *Nature*. 2014;515(7528):518–522. doi:10.1038/nature13959
60. Environmental Impacts of Food Production. Our World in Data. 2020. https://ourworldindata.org/environmental-impacts-of-food?
61. EU Agricultural Outlook 2019–30: African Swine Fever Continues to Affect Global Meat Market. European Commission. 2019. https://ec.europa.eu/info/news/eu-agricultural-outlook-2019-2030-african-swine-fever-continues-impact-global-meat-market-2019-dec-10_en
62. Kustar A, Patino-Echeverri D. A review of environmental life cycle assessments of diets: plant-based solutions are truly sustainable, even in the form of fast foods. *Sustainability*. 2021;13(17). doi:10.3390/su13179926
63. How Much of Global Greenhouse Gas Emissions Come from Food? Our World in Data. 2021. https://ourworldindata.org/greenhouse-gas-emissions-food
64. Greenhouse Gas Emissions from Agriculture in Europe. European Environment Agency. 2021. https://www.eea.europa.eu/ims/greenhouse-gas-emissions-from-agriculture
65. Reisinger A, Clark H. How much do direct livestock emissions actually contribute to global warming? *Glob Chang Biol*. 2018;24(4):1749–1761. doi:10.1111/gcb.13975

66. Lynch J, Cain M, Frame D, Pierrehumbert R. Agriculture's contribution to climate change and role in mitigation is distinct from predominantly fossil CO2-emitting sectors. *Front Sustain Food Syst*. 2021;4. doi:10.3389/fsufs.2020.518039
67. Winkler K, Fuchs R, Rounsevell M, Herold M. Global land use changes are four times greater than previously estimated. *Nat Commun*. 2021;12(1):2501. doi:10.1038/s41467-021-22702-2
68. Haber Process. Wikipedia. https://en.wikipedia.org/wiki/Haber_process
69. Production Volume of Ammonium Nitrate Worldwide from 2009 to 2020. Statista. 2022. https://www.statista.com/statistics/1287049/global-ammonium-nitrate-production/
70. Understanding Global Warming Potentials. US EPA. https://www.epa.gov/ghgemissions/understanding-global-warming-potentials
71. Ripple WJ, Wolf C, Newsome TM, et al. World scientists' warning of a climate emergency 2021. *BioScience*. 2021;71(9):894–898. doi:10.1093/biosci/biab079
72. Ripple WJ, Smith P, Haberl H, Montzka SA, McAlpine C, Boucher DH. Ruminants, climate change and climate policy. *Nature Climate Change*. 2014;4(1):2–5. doi:10.1038/nclimate2081
73. Clark Michael A, Springmann Marco, Hill Jason, Tilman David. Multiple health and environmental impacts of foods. *Proc Natl Acad Sci*. 2019;116(46):23357–23362. doi:10.1073/pnas.1906908116
74. Seiwert N, Heylmann D, Hasselwander S, Fahrer J. Mechanism of colorectal carcinogenesis triggered by heme iron from red meat. *Biochim Biophys Acta Rev Cancer*. 2020;1873(1):188334. doi:10.1016/j.bbcan.2019.188334
75. Cena H, Calder PC. Defining a healthy diet: evidence for the role of contemporary dietary patterns in health and disease. *Nutrients*. 2020;12(2). doi:10.3390/nu12020334
76. Food Waste: The Problem in the EU in Numbers. European Parliament. 2017. https://www.europarl.europa.eu/news/en/headlines/society/20170505STO73528/food-waste-the-problem-in-the-eu-in-numbers-infographic
77. Notarnicola B, Tassielli G, Renzulli PA, Castellani V, Sala S. Environmental impacts of food consumption in Europe. *J Clean Prod*. 2017;140:753–765. doi:10.1016/j.jclepro.2016.06.080
78. Food Waste in America: Facts and Statistics. Rubicon. 2020. https://www.rubicon.com/blog/food-waste-facts/
79. Worldwide Food Waste. Think – Eat – Save. Website of the United Nations Environment Programme. https://www.unep.org/thinkeatsave/get-informed/worldwide-food-waste
80. Amicarelli V, Lagioia G, Bux C. Global warming potential of food waste through the life cycle assessment: an analytical review. *Environ Impact Assess Rev*. 2021;91:106677. doi:10.1016/j.eiar.2021.106677
81. Kleeman MJ, Schauer JJ, Cass GR. Size and composition distribution of fine particulate matter emitted from wood burning, meat charbroiling, and cigarettes. *Environ Sci Technol*. 1999;33(20):3516–3523. doi:10.1021/es981277q
82. Moorthy B, Chu C, Carlin DJ. Polycyclic aromatic hydrocarbons: from metabolism to lung cancer. *Toxicol Sci*. 2015;145(1):5–15. doi:10.1093/toxsci/kfv040
83. Ames BN, Gold LS. Chemical carcinogenesis: too many rodent carcinogens. *Proc Natl Acad Sci*. 1990;87(19):7772–7776. doi:10.1073/pnas.87.19.7772

84. Jacob J. The significance of polycyclic aromatic hydrocarbons as environmental carcinogens. 35 years research on PAH: a retrospective. *Polycycl Aromat Compd.* 2008;28(4–5):242–272. doi:10.1080/10406630802373772
85. Rhomberg LR, Goodman JE, Prueitt RL. The weight of evidence does not support the listing of styrene as "reasonably anticipated to be a human carcinogen" in NTP's twelfth report on carcinogens. *Hum Ecol Risk Assess.* 2013;19(1):4–27. doi:10.1080/10807039.2012.650577
86. Arif JM, Dresler C, Clapper ML, et al. Lung DNA adducts detected in human smokers are unrelated to typical polyaromatic carcinogens. *Chem Res Toxicol.* 2006;19(2):295–299. doi:10.1021/tx0502443
87. Weng Mao-wen, Lee Hyun-Wook, Park Sung-Hyun, et al. Aldehydes are the predominant forces inducing DNA damage and inhibiting DNA repair in tobacco smoke carcinogenesis. *Proc Natl Acad Sci.* 2018;115(27):E6152–E6161. doi:10.1073/pnas.1804869115
88. Tang MS, Lee HW, Weng MW, et al. DNA damage, DNA repair and carcinogenicity: tobacco smoke versus electronic cigarette aerosol. *Mutat Res Rev Mutat Res.* 2022;789:108409. doi:10.1016/j.mrrev.2021.108409
89. Cancer Stat Facts: Lung and Bronchus Cancer. National Cancer Institute of the US National Institute of Health. 2021. https://seer.cancer.gov/statfacts/html/lungb.html
90. Johansson M, Relton C, Ueland PM, et al. Serum B vitamin levels and risk of lung cancer. *JAMA.* 2010;303(23):2377–2385. doi:10.1001/jama.2010.808
91. Levine ME, Crimmins EM. A genetic network associated with stress resistance, longevity, and cancer in humans. *J Gerontol: Series A.* 2016;71(6):703–712. doi:10.1093/gerona/glv141
92. Lea-Langton AR, Spracklen DV, Arnold SR, et al. PAH emissions from an African cookstove. *Journal of the Energy Institute.* 2019;92(3):587–593. doi:10.1016/j.joei.2018.03.014
93. Yadav VK, Prasad S, Patel DK, Khan AH, Tripathi M, Shukla Y. Identification of polycyclic aromatic hydrocarbons in unleaded petrol and diesel exhaust emission. *Environ Monit Assess.* 2010;168(1):173–178. doi:10.1007/s10661-009-1101-8
94. Dawkins R. *The Blind Watchmaker.* WW Norton and Company; 1986.

Chapter 3

1. Global CO2 Emissions in 2019. International Energy Agency. 2020. https://www.iea.org/articles/global-co2-emissions-in-2019
2. Fyfe JC, Kharin VV, Swart N, Flato GM, Sigmond M, Gillett NP. Quantifying the influence of short-term emission reductions on climate. *Sci Adv.* 2012;7. doi:10.1126/sciadv.abf7133
3. Global CO2 Emissions Rebounded to Their Highest Level in History in 2021. International Energy Agency. 2022. https://www.iea.org/news/global-co2-emissions-rebounded-to-their-highest-level-in-history-in-2021
4. Remaining Carbon Budget. MCC. https://www.mcc-berlin.net/en/research/co2-budget.html
5. CO2-earth. https://www.co2.earth/global-co2-emissions

6. Climate Change 2021: The Physical Science Basis. IPCC. 2022. https://www.ipcc.ch/report/ar6/wg1/
7. Krishnamurthy V. Predictability of weather and climate. *Earth Space Sci.* 2019;6(7):1043–1056.
8. Three Body Problem. Wikipedia. https://en.wikipedia.org/wiki/Three-body_problem
9. Black-Body Radiation. Wikipedia. https://en.wikipedia.org/wiki/Black-body_radiation
10. Force Constants for Bond Stretching. chemnetbase.com. 2022. https://hbcp.chemnetbase.com/faces/documents/09_06/09_06_0001.xhtml
11. *Composition of the Atmosphere: Course Material.* University of Arizona. 2016. http://www.atmo.arizona.edu/students/courselinks/fall16/atmo336/lectures/sec1/composition.html
12. Anthropogenic and Natural Radiative Forcing. IPCC Working Group I. 2018. https://www.ipcc.ch/site/assets/uploads/2018/02/WG1AR5_Chapter08_FINAL.pdf
13. Clausius–Clapeyron Relation. Wikipedia. https://en.wikipedia.org/wiki/Clausius–Clapeyron_relation
14. Dlugokencky E, Houweling S, Dirksen R, et al. *Observing Water Vapour.* World Meteorological Association. 2016. https://public.wmo.int/en/resources/bulletin/observing-water-vapour
15. Plyler EK. The infrared emission spectra of CO2 and H_2O molecules. *Science.* 1948;107:48. doi:10.1126/science.107.2767
16. Zhong W, Haigh JD. The greenhouse effect and carbon dioxide. *Weather.* 2013;68:100–105. doi:10.1002/wea.2072
17. Absorption Spectra of Atmospheric Gases in the IR, Visible and UK Regions. Georgia Institute of Technology. http://irina.eas.gatech.edu/EAS8803_SPRING2012/Lec7.pdf
18. Tran H, Turbet M, Chelin P, Landsheere X. Measurements and modeling of absorption by CO2+H_2O mixtures in the spectral region beyond the CO2 v3-band head. *arXiv.* 2018. doi:10.48550/arXiv.1802.01352
19. MacFarling Meure C, Etheridge D, Trudinger C, et al. Law Dome CO2, CH_4 and N_2O ice core records extended to 2000 years BP. *Geophys Res Lett.* 2006;33(14).
20. Keeling DC, Piper SC, Bacastow M, et al. *Exchanges of Atmospheric CO2 and 13CO2 with the Terrestrial Biosphere and Oceans from 1978 to 2000.* Scripps Institution of Oceanography. 2001:88. doi:10.1029/2006GL026152
21. Infrared Absorption by CH4, H2O and CO2. Air Force Geophysics Laboratory. 1967. https://apps.dtic.mil/dtic/tr/fulltext/u2/a039380.pdf
22. ExxonMobil Climate Change Controversy. Wikipedia. https://en.wikipedia.org/wiki/ExxonMobil_climate_change_controversy
23. Arrhenius S. On the influence of carbonic acid in the air upon the temperature of the ground. *Philosophical Magazine and Journal of Science.* 1896;5:237–276.
24. Solomon S, Daniel JS, Sanford TJ, et al. Persistence of climate changes due to a range of greenhouse gases. *Proc Natl Acad Sci.* 2010;107(43):18354–18359.
25. Buckley S. Detecting methane emissions: how spectroscopy is contributing to sustainability efforts. *Spectroscopy.* 2022;37(4):22–26.
26. AR5 Synthesis Report: Climate Change 2014. Intergovernmental Panel on Climate Change (IPCC). 2014. https://www.ipcc.ch/report/ar5/syr/

27. Etminan M, Myhre G, Highwood E, Shine K. Radiative forcing of carbon dioxide, methane, and nitrous oxide: a significant revision of the methane radiative forcing. *Geophys Res Lett*. 2016;43(24):12–614.
28. Archer D, Brovkin V. The millennial atmospheric lifetime of anthropogenic CO2. *Clim Change*. 2008;90(3):283–297.
29. Bar-On YM, Phillips R, Milo R. The biomass distribution on Earth. *Proc Natl Acad Sci*. 2018;115(25):6506–6511.
30. Watson AJ, Schuster U, Shutler JD, et al. Revised estimates of ocean-atmosphere CO2 flux are consistent with ocean carbon inventory. *Nat Commun*. 2020;11(1):1–6.
31. Le Quéré C, Andrew RM, Friedlingstein P, et al. Global Carbon Budget 2018. *Earth Syst Sci Data*. 2018;10:2141–2194. doi:10.5194/essd-10-2141-2018
32. Khatiwala S, Primeau F, Hall T. Reconstruction of the history of anthropogenic CO2 concentrations in the ocean. *Nature*. 2009;462(7271):346–349.
33. Oxygen: Solubility in Fresh and Sea Water vs. Temperature. Engineering Toolbox. https://www.engineeringtoolbox.com/oxygen-solubility-water-d_841.html
34. Krogh A. *On the Tension of Carbonic Acid in Natural Waters and Especially in the Sea*. CA Reitzel; 1904.
35. Jiang LQ, Carter BR, Feely RA, Lauvset SK, Olsen A. Surface ocean pH and buffer capacity: past, present and future. *Sci Rep*. 2019;9(1):1–11.
36. Takahashi T, Sutherland SC, Chipman DW, et al. Climatological distributions of pH, pCO2, total CO2, alkalinity, and CaCO3 saturation in the global surface ocean, and temporal changes at selected locations. *Mar Chem*. 2014;164:95–125.
37. Robbins LL, Wynn JG, Lisle JT, et al. Baseline monitoring of the Western Arctic Ocean estimates 20% of Canadian Basin surface waters are undersaturated with respect to aragonite. *PloS One*. 2013;8(9):e73796.
38. Watson RT, Zinyowera MC, Moss RH. *Climate Change 1995. Impacts, Adaptations and Mitigation of Climate Change: Scientific-Technical Analyses*. IPCC;1996.
39. Wiebe R, Gaddy V. The solubility of carbon dioxide in water at various temperatures from 12 to 40° and at pressures to 500 atmospheres. Critical phenomena. *J Am Chem Soc*. 1940;62(4):815–817.
40. Berner RA. The long-term carbon cycle, fossil fuels and atmospheric composition. *Nature*. 2003;426(6964):323–326.
41. Bolin B, Degens ET, Kempe S, Ketner P. *The Global Carbon Cycle*. Wiley;1979.
42. Westneat MW, Betz O, Blob RW, Fezzaa K, Cooper WJ, Lee WK. Tracheal respiration in insects visualized with synchrotron X-ray imaging. *Science*. 2003;299(5606):558–560.
43. Anthracite: An Overview. ScienceDirect Topics. 2003. https://www.sciencedirect.com/topics/chemistry/anthracite
44. Casagrande DJ, Gronli K, Sutton N. The distribution of sulfur and organic matter in various fractions of peat: origins of sulfur in coal. *Geochim Cosmochim Acta*. 1980;44(1):25–32.
45. Braunkohle: Förderung bis 2021 (Lignite: Mining until 2021). Statista. 2022. https://de.statista.com/statistik/daten/studie/156258/umfrage/braunkohlefoerderung-in-deutschland-seit-1990/
46. Soft Brown Coal Production Top Countries. Statista. 2021. https://www.statista.com/statistics/264779/countries-with-the-largest-soft-brown-coal-production/
47. Straka P, Sýkorová I. Coalification and coal alteration under mild thermal conditions. *Int J Coal Sci Technol*. 2018;5(3):358–373.

48. Wang S, Shao LY, Yan ZM, Shi MJ, Zhang YH. Characteristics of early cretaceous wildfires in peat-forming environment, NE China. *J Palaeogeography*. 2019;8(1):17. doi:10.1186/s42501-019-0035-5
49. Coal. Chemistry Explained. http://www.chemistryexplained.com/Ce-Co/Coal.html
50. Coal. Wikipedia. https://en.wikipedia.org/wiki/Coal
51. Petroleum. National Geographic Society. https://www.nationalgeographic.org/encyclopedia/petroleum/
52. Eigenbrode JL, Summons RE, Steele A, et al. Organic matter preserved in 3-billion-year-old mudstones at Gale crater, Mars. *Science*. 2018;360(6393):1096–1101.
53. Kissin Y. Hydrocarbon components in carbonaceous meteorites. *Geochim Cosmochim Acta*. 2003;67(9):1723–1735.
54. Pehr K, Bisquera R, Bishop AN, et al. Preservation and distributions of covalently bound polyaromatic hydrocarbons in ancient biogenic kerogens and insoluble organic macromolecules. *Astrobiology*. 2021;21(9):1049–1075.
55. 7 Billion-Year-Old Stardust Is Oldest Material Found on Earth: Some of These Ancient Grains Are Billions of Years Older Than Our Sun. Live Science. 2020. https://www.livescience.com/oldest-material-on-earth.html
56. Statistical Review of World Energy 2021. British Petroleum p.l.c. 2021. https://www.bp.com/content/dam/bp/business-sites/en/global/corporate/pdfs/energy-economics/statistical-review/bp-stats-review-2021-full-report.pdf
57. Statistical Review of World Energy June 2011. Stanford University. 2011. http://large.stanford.edu/courses/2011/ph240/goldenstein1/docs/bp2011.pdf
58. Supply and Demand of Natural Graphite. Deutsche Rohstoffagentur (German Agency for Natural Resources). 2020. https://www.deutsche-rohstoffagentur.de/DERA/DE/Downloads/Studie%20Graphite%20eng%202020.pdf

Chapter 4

1. Oil and Petroleum Products Explained: Use of Oil. U.S. Energy Information Administration (EIA). 2022. https://www.eia.gov/energyexplained/oil-and-petroleum-products/use-of-oil.php
2. Butera G, Gadsbøll RØ, Ravenni G, Ahrenfeldt J, Henriksen UB, Clausen LR. Thermodynamic analysis of methanol synthesis combining straw gasification and electrolysis via the low temperature circulating fluid bed gasifier and a char bed gas cleaning unit. *Energy*. 2020;199:117405. doi:10.1016/j.energy.2020.117405
3. Bos MJ, Kersten SRA, Brilman DWF. Wind power to methanol: renewable methanol production using electricity, electrolysis of water and CO2 air capture. *Applied Energy*. 2020;264:114672. doi:10.1016/j.apenergy.2020.114672
4. Methanol Production Capacity Globally 2030. Statista. 2022. https://www.statista.com/statistics/1065891/global-methanol-production-capacity/
5. Energy Content of Fuels. Appropedia. https://www.appropedia.org/Energy_content_of_fuels
6. Few Transport Fuels Surpass the Energy Density of Gasoline and Diesel. U.S. Energy Information Administration (EIA). 2013. https://www.eia.gov/todayinenergy/detail.php?id=9991

7. Verhelst S, Turner JW, Sileghem L, Vancoillie J. Methanol as a fuel for internal combustion engines. *Prog Energy Combust Sci*. 2019;70:43–88. doi:10.1016/j.pecs.2018.10.001
8. When California Had 15,000 Methanol Cars. Fuel Freedom Foundation. 2013. https://www.fuelfreedom.org/when-california-had-15000-methanol-cars/
9. H.R. 2493 (113th): Open Fuel Standard Act of 2013. 2013. https://www.govtrack.us/congress/bills/113/hr2493/summary
10. Could Methanol Replace Diesel? China Dialogue. 2017. https://chinadialogue.net/en/pollution/10129-could-methanol-replace-diesel/
11. A Brief Review of China's Methanol Vehicle Pilot and Policy. Methanol Institute. 2019. https://www.methanol.org/wp-content/uploads/2019/03/A-Brief-Review-of-Chinas-Methanol-Vehicle-Pilot-and-Policy-20-March-2019.pdf
12. dos Santos RG, Alencar AC. Biomass-derived syngas production via gasification process and its catalytic conversion into fuels by Fischer Tropsch synthesis: a review. *Int J Hydrogen Energy*. 2020;45(36):18114–18132. doi:10.1016/j.ijhydene.2019.07.133
13. Beijing to accelerate deployment of methanol vehicles under carbon-neutral drive. South China Morning Post. 2022. https://www.scmp.com/business/article/3193161/beijing-accelerate-deployment-methanol-vehicles-under-carbon-neutral-drive.
14. Methanol, a Future-Proof Fuel–A primer for the Methanol Institute. futurefuelstrategies.com. 2020. https://www.methanol.org/wp-content/uploads/2020/03/Future-Fuel-Strategies-Methanol-Automotive-Fuel-Primer.pdf
15. Innovation Outlook: Renewable Methanol. International Renewable Energy Agency in partnership with the Methanol Institute. 2021. https://www.irena.org/-/media/Files/IRENA/Agency/Publication/2021/Jan/IRENA_Innovation_Renewable_Methanol_2021.pdf
16. A.P. Moller–Maersk Accelerates Fleet Decarbonisation with 8 Large Ocean-Going Vessels to Operate on Carbon Neutral Methanol. Maersk. 2021. https://www.maersk.com/news/articles/2021/08/24/maersk-accelerates-fleet-decarbonisation
17. Ethanol Blend Mandates for Gasoline in Selected Countries in South America as of May 2022. Statista. 2022. https://www.statista.com/statistics/1113015/ethanol-bleding-gasoline-south-america-country/
18. 10 Years of EU Fuels Policy Increased EU's Reliance on Unsustainable Biofuels. Transport & Environment. 2021. https://www.transportenvironment.org/wp-content/uploads/2021/08/Biofuels-briefing-072021.pdf
19. Feedstock Used in Fuel Ethanol Production in the European Union from 2012 to 2020, with a Forecast for 2021, by Type. Statista. 2020. https://www.statista.com/statistics/1295918/eu-ethanol-fuel-feedstock-consumption/
20. Distribution of Ethanol Production in the United States in 2021, by Feedstock Type. Statista. 2022. https://www.statista.com/statistics/1106316/us-share-ethanol-production-by-feedstock-type/
21. Searle S, Malins C. A reassessment of global bioenergy potential in 2050. *GCB Bioenergy*. 2015;7(2):328–336. doi:10.1111/gcbb.12141
22. Sustainable Production of Second-Generation Biofuels: Potential and Perspectives in Major Economies and Developing Countries. International Energy Agency (IEA). 2010. https://iea.blob.core.windows.net/assets/42da53f9-d2ce-499d-9d65-31d0effa13d0/second_generation_biofuels.pdf

23. Mariano AP, Keshtkar MJ, Atala DIP, et al. Energy requirements for butanol recovery using the flash fermentation technology. *Energy Fuels.* 2011;25(5):2347–2355. doi:10.1021/ef200279v
24. Pătrașcu I, Bîldea CS, Kiss AA. Eco-efficient downstream processing of biobutanol by enhanced process intensification and integration. *ACS Sustainable Chem Eng.* 2018;6(4):5452–5461. doi:10.1021/acssuschemeng.8b00320
25. Kumakiri I, Yokota M, Tanaka R, et al. Process intensification in bio-ethanol production–recent developments in membrane separation. *Processes.* 2021;9(6). doi:10.3390/pr9061028
26. Liu S, Li H, Kruber B, Skiborowski M, Gao X. Process intensification by integration of distillation and vapor permeation or pervaporation: an academic and industrial perspective. *Results Eng.* 2022;15:100527. doi:10.1016/j.rineng.2022.100527
27. Song C, Qiu Y, Liu Q, et al. Process intensification of cellulosic ethanol production by waste heat integration. *Chem Eng Res Des.* 2018;132:115–122. doi:10.1016/j.cherd.2018.01.016
28. Susmozas A, Martín-Sampedro R, Ibarra D, et al. Process strategies for the transition of 1G to advanced bioethanol production. *Processes.* 2020;8(10). doi:10.3390/pr8101310
29. Veza I, Said MFM, Latiff ZA. Progress of acetone-butanol-ethanol (ABE) as biofuel in gasoline and diesel engine: a review. *Fuel Processing Technology.* 2019;196:106179. doi:10.1016/j.fuproc.2019.106179
30. Grisales Diaz VH, Olivar Tost G. Energy efficiency of acetone, butanol, and ethanol (ABE) recovery by heat-integrated distillation. *Bioprocess Biosyst Eng.* 2018;41(3):395–405. doi:10.1007/s00449-017-1874-z
31. Hossain N, Zaini J, Indra Mahlia TM. Life cycle assessment, energy balance and sensitivity analysis of bioethanol production from microalgae in a tropical country. *Renew Sustain Energy Rev.* 2019;115:109371. doi:10.1016/j.rser.2019.109371
32. Overview of Biofuels Policies and Markets Across the EU-27 and the UK. ePURE. 2021. https://www.epure.org/wp-content/uploads/2021/01/201104-DEF-REP-Overview-of-biofuels-policies-and-markets-across-the-EU-Nov.-2020.pdf
33. How Is Ethanol Produced? ePURE. https://www.epure.org/about-ethanol/how-is-ethanol-produced/
34. Mishra A, Kumar A, Ghosh S. Energy assessment of second generation (2G) ethanol production from wheat straw in Indian scenario. *3 Biotech.* 2018;8(3):142. doi:10.1007/s13205-018-1135-0
35. Energy Balance Guide. European Commission. 2019. https://ec.europa.eu/eurostat/documents/38154/4956218/ENERGY-BALANCE-GUIDE-DRAFT-31JANUARY2019.pdf/cf121393-919f-4b84-9059-cdf0f69ec045
36. Mayer FD, Brondani M, Vasquez Carrillo MC, Hoffmann R, Silva Lora EE. Revisiting energy efficiency, renewability, and sustainability indicators in biofuels life cycle: analysis and standardization proposal. *J Clean Prod.* 2020;252:119850. doi:10.1016/j.jclepro.2019.119850
37. Feedgrains Sector at a Glance. United States Department of Agriculture (USDA) Economic Research Service. 2022. https://www.ers.usda.gov/topics/crops/corn-and-other-feedgrains/feedgrains-sector-at-a-glance/
38. Lark TJ, Hendricks NP, Smith A, et al. Environmental outcomes of the US renewable fuel standard. *Proc Natl Acad Sci.* 2022;119(9):e2101084119. doi:10.1073/pnas.2101084119

39. Cadillo-Benalcazar JJ, Bukkens SGF, Ripa M, Giampietro M. Why does the European Union produce biofuels? Examining consistency and plausibility in prevailing narratives with quantitative storytelling. *Energy Res Soc Sci.* 2021;71:101810. doi:10.1016/j.erss.2020.101810
40. Frequently Asked Questions [about DME]. Aboutdme. https://www.aboutdme.org/FAQ
41. Mota N, Millán Ordoñez E, Pawelec B, Fierro JLG, Navarro RM. Direct synthesis of dimethyl ether from CO2: recent advances in bifunctional/hybrid catalytic systems. *Catalysts.* 2021;11(4). doi:10.3390/catal11040411
42. Wang C, Yang WCD, Raciti D, et al. Endothermic reaction at room temperature enabled by deep-ultraviolet plasmons. *Nat Mater.* 2021;20(3):346–352. doi:10.1038/s41563-020-00851-x
43. Michailos S, McCord S, Sick V, Stokes G, Styring P. Dimethyl ether synthesis via captured CO2 hydrogenation within the power to liquids concept: a techno-economic assessment. *Energy Convers Manag.* 2019;184:262–276. doi:10.1016/j.enconman.2019.01.046
44. Bîldea CS, Győrgy R, Brunchi CC, Kiss AA. Optimal design of intensified processes for DME synthesis. *Comput Chem Eng.* 2017;105:142–151. doi:10.1016/j.compchemeng.2017.01.004
45. Wodołażski A, Smoliński A. Modelling and process integration study of dimethyl ether synthesis from syngas derived from biomass gasification: flowsheet simulation. *Alexandria Engineering Journal.* 2020;59(6):4441–4448. doi:10.1016/j.aej.2020.07.050
46. Aviation Fuel. Wikipedia. https://en.wikipedia.org/wiki/Aviation_fuel
47. Boehm RC, Yang Z, Bell DC, Feldhausen J, Heyne JS. Lower heating value of jet fuel from hydrocarbon class concentration data and thermo-chemical reference data: an uncertainty quantification. *Fuel.* 2022;311:122542. doi:10.1016/j.fuel.2021.122542
48. Blakey S, Rye L, Wilson CW. Aviation gas turbine alternative fuels: A review. *Proc Combust Inst.* 2011;33(2):2863–2885. doi:10.1016/j.proci.2010.09.011
49. Electric Synthetic Fuels on Their Way to Industrialisation. AVL. 2018. https://www.avl.com/documents/1982862/5383001/Session+7+Electric+synthetic+fuels+on+their+way+to+industrialisation+Olshausen.pdf
50. Sustainable Aviation Fuel: Review of Technical Pathways. US Department of Energy. 2020. https://www.energy.gov/sites/prod/files/2020/09/f78/beto-sust-aviation-fuel-sep-2020.pdf
51. *Energy and Power of Flying (Course Material).* Stanford University. 2013. http://large.stanford.edu/courses/2013/ph240/eller1/
52. Alternative Fuels and Their Potential Impact on Aviation. National Aeronautics and Space Administration (NASA). 2006. https://ntrs.nasa.gov/api/citations/20060051881/downloads/20060051881.pdf
53. Scheelhaase J, Maertens S, Grimme W. Synthetic fuels in aviation: current barriers and potential political measures. *Transportation Research Procedia.* 2019;43:21–30. doi:10.1016/j.trpro.2019.12.015
54. More Than Half of All Passenger Flights in Germany in 2020 Were Short-Haul Flights: Press Release N037, 4 June 2021. DE STATIS (German government office of statistics). 2021. https://www.destatis.de/EN/Press/2021/06/PE21_N037_462.html
55. FuelsEurope Statistical Report 2021. FuelsEurope. 2021. https://www.fuelseurope.eu/publication/fuelseurope-statistical-report-2021/

56. Vela-García N, Bolonio D, Mosquera AM, Ortega MF, García-Martínez MJ, Canoira L. Techno-economic and life cycle assessment of triisobutane production and its suitability as biojet fuel. *Appl Energy*. 2020;268:114897. doi:10.1016/j.apenergy.2020.114897
57. Kim J, Sovacool BK, Bazilian M, et al. Decarbonizing the iron and steel industry: a systematic review of sociotechnical systems, technological innovations, and policy options. *Energy Res Soc Sci*. 2022;89:102565. doi:10.1016/j.erss.2022.102565
58. Nhuchhen DR, Sit SP, Layzell DB. Decarbonization of cement production in a hydrogen economy. *Appl Energy*. 2022;317:119180. doi:10.1016/j.apenergy.2022.119180
59. Kumar Verma Y, Mazumdar B, Ghosh P. Thermal energy consumption and its conservation for a cement production unit. *Environ Engineer Res*. 2020;26(3):200111–0. doi:10.4491/eer.2020.111
60. Heating Oil Usage Calculator: How Much Heating Oil Am I Actually Using? Rix Petroleum Products. 2020. https://www.rix.co.uk/blog/heating-oil-usage-calculator-how-much-heating-oil-am-i-actually-using/
61. Number of Households in the United Kingdom in 2021, by Region. Statista. 2022. https://www.statista.com/statistics/295297/households-in-uk-by-region/
62. Geiser F, Stawski C, Wacker CB, Nowack J. Phoenix from the ashes: fire, torpor, and the evolution of mammalian endothermy. *Front Physiol*. 2017;8:842. doi:10.3389/fphys.2017.00842
63. Bartholomew GA. Energy metabolism. In: Gordon MS, ed. *Animal Physiology: Principles and Adaptations*. 4th ed. MacMillan; 1982:46–93.
64. Arciero PJ, Goran MI, Poehlman ET. Resting metabolic rate is lower in women than in men. *J Appl Physiol*. 1993;75(6):2514–2520. doi:10.1152/jappl.1993.75.6.2514
65. Bassett DR, Vachon JA, Kirkland AO, Howley ET, Duncan GE, Johnson KR. Energy cost of stair climbing and descending on the college alumnus questionnaire. *Med Sci Sports Exerc*. 1997;29(9). https://journals.lww.com/acsm-msse/Fulltext/1997/09000/Energy_cost_of_stair_climbing_and_descending_on.19.aspx
66. Chapman JB, Gibbs CL. An energetic model of muscle contraction. *Biophys J*. 1972;12(3):227–236. doi:10.1016/S0006-3495(72)86082-X
67. Carey FG, Teal JM. Heat conservation in tuna fish muscle. *Proc Natl Acad Sci*. 1966;56(5):1464–1469. doi:10.1073/pnas.56.5.1464
68. Exhaust Heat Recovery System. Wikipedia. https://en.wikipedia.org/wiki/Exhaust_heat_recovery_system#_blank
69. Fossil Fuel Power Station. Wikipedia. https://en.wikipedia.org/wiki/Fossil_fuel_power_station
70. Ringler J, Seifert M, Guyotot V, Hübner W. Rankine cycle for waste heat recovery of IC engines. *SAE Int J Engines*. 2009;2(1):67–76. doi:10.4271/2009-01-0174
71. Mansour C, Bou Nader W, Dumand C, Nemer M. Waste heat recovery from engine coolant on mild hybrid vehicle using organic Rankine cycle. *Proc Inst Mech Eng D J Auto Eng*. 2019;233(10):2502–2517. doi:10.1177/0954407018797819
72. Broekaert S, Grigoratos T, Savvidis D, Fontaras G. Assessment of waste heat recovery for heavy-duty vehicles during on-road operation. *Appl Therm Eng*. 2021;191:116891. doi:10.1016/j.applthermaleng.2021.116891
73. Aliahmadi M, Moosavi A, Sadrhosseini H. Multi-objective optimization of regenerative ORC system integrated with thermoelectric generators for low-temperature waste heat recovery. *Energy Rep*. 2021;7:300–313. doi:10.1016/j.egyr.2020.12.035

74. Valencia G, Fontalvo A, Cárdenas Y, Duarte J, Isaza C. Energy and exergy analysis of different exhaust waste heat recovery systems for natural gas engine based on ORC. *Energies.* 2019;12(12). doi:10.3390/en12122378
75. Zhang W, Yang J, Zhang W, Ma F. Research on regenerative braking of pure electric mining dump truck. *World Electric Vehicle J.* 2019;10(2). doi:10.3390/wevj10020039
76. Björnsson LH, Karlsson S. The potential for brake energy regeneration under Swedish conditions. *Appl Energy.* 2016;168:75–84. doi:10.1016/j.apenergy.2016.01.051
77. Regenerative Brake. Wikipedia. https://en.wikipedia.org/wiki/Regenerative_brake#_blank
78. Interview Dr. Pfeifer, Mahle: "Drei bis fünf Prozent Potential" (Three to Five Percent Potential). Eurotransport. https://www.eurotransport.de/artikel/interview-dr-andreas-pfeifer-mahle-drei-bis-fuenf-prozent-verbrauchspotenzial-10406604.html
79. Project Report HD-TEG (on Current Status and Prospects for Thermoelectric Energy Recovery). Deutsches Zentrum für Luft und Raumfahrt e.V. Institute of Vehicle Concepts. 2020. https://elib.dlr.de/142027/2/2020-12-31_Projektbericht_HD-TEG_V1_%5Bengl%5D.pdf
80. Malmgren E, Brynolf S, Fridell E, Grahn M, Andersson K. The environmental performance of a fossil-free ship propulsion system with onboard carbon capture: a life cycle assessment of the HyMethShip concept. *Sustain Energy Fuels.* 2021;5(10):2753–2770. doi:10.1039/D1SE00105A
81. Sharma S, Maréchal F. Carbon dioxide capture from internal combustion engine exhaust using temperature swing adsorption. *Front Energy Res.* 2019;7:143. doi:10.3389/fenrg.2019.00143
82. Vance D, Nimbalkar S, Thekdi A, et al. Estimation of and barriers to waste heat recovery from harsh environments in industrial processes. *J Clean Prod.* 2019;222:539–549. doi:10.1016/j.jclepro.2019.03.011
83. Comeback of the Combustion Engine: Clean Exhaust Gases Thanks to New Adsorption Technology. LISA – Life Science Austria. 2022. https://www.lifescienceaustria.at/resources/news/detail/comeback-of-the-combustion-engine-clean-exhaust-gases-thanks-to-new-adsorption-technology
84. Leach F, Kalghatgi G, Stone R, Miles P. The scope for improving the efficiency and environmental impact of internal combustion engines. *Transport Engineer.* 2020;1:100005. doi:10.1016/j.treng.2020.100005
85. Kalghatgi G. Is it really the end of internal combustion engines and petroleum in transport? *Appl Energy.* 2018;225:965–974. doi:10.1016/j.apenergy.2018.05.076
86. Internal Combustion Engines: Performance and Fuel Economy Trends. University Lafayette. https://sites.lafayette.edu/egrs352-sp14-cars/performance-and-fuel-economy-trends/
87. Improved Efficiency of Aircraft Engines. Fraunhofer. 2018. https://www.fraunhofer.de/en/press/research-news/2018/May/improved-efficiency-of-aircraft-engines.html#_blank
88. Fuel Efficiency Trends for New Commercial Jet Aircraft: 1960 to 2014. International Council on Clean Transportation (ICCT). 2015. https://theicct.org/publication/fuel-efficiency-trends-for-new-commercial-jet-aircraft-1960-to-2014/
89. Catalytic Converter. Wikipedia. https://en.wikipedia.org/wiki/Catalytic_converter
90. Hammadi MQ, Yassen RS, Abid KN. Recovery of platinum and palladium from scrap automotive catalytic converters. *Al-Khwarizmi Engineer J.* 2017;13(3):131–141. doi:10.22153/kej.2017.04.002

91. Orellano P, Reynoso J, Quaranta N, Bardach A, Ciapponi A. Short-term exposure to particulate matter (PM10 and PM2.5), nitrogen dioxide (NO2), and ozone (O3) and all-cause and cause-specific mortality: systematic review and meta-analysis. *Environ Int.* 2020;142:105876. doi:10.1016/j.envint.2020.105876
92. Chen J, Hoek G. Long-term exposure to PM and all-cause and cause-specific mortality: A systematic review and meta-analysis. *Environ Int.* 2020;143:105974. doi:10.1016/j.envint.2020.105974
93. Kleinhenz M, Fiedler A, Lauer P, Döring A. SCR coated DPF for marine engine applications. *Top Catal.* 2019;62(1):282–287. doi:10.1007/s11244-018-1115-y
94. Rs 140 to Solve Air Pollution: TN Man's Invention Cuts Emissions by 40%! The Better India. 2019. https://www.thebetterindia.com/204212/tamil-nadu-man-sivakasi-low-cost-vehicle-filter-innovation-air-pollution-india/
95. European Emission Standards. Wikipedia. https://en.wikipedia.org/wiki/European_emission_standards
96. Matějka V, Leonardi M, Praus P, Straffelini G, Gialanella S. The role of graphitic carbon nitride in the formulation of copper-free friction composites designed for automotive brake pads. *Metals.* 2022;12(1). doi:10.3390/met12010123
97. Beddows DCS, Harrison RM. PM10 and PM2.5 emission factors for non-exhaust particles from road vehicles: dependence upon vehicle mass and implications for battery electric vehicles. *Atmos Environ.* 2021;244:117886. doi:10.1016/j.atmosenv.2020.117886
98. Hicks W, Beevers S, Tremper AH, et al. Quantification of non-exhaust particulate matter traffic emissions and the impact of COVID-19 lockdown at London Marylebone Road. *Atmosphere.* 2021;12(2). doi:10.3390/atmos12020190
99. Grigoratos T, Gustafsson M, Eriksson O, Martini G. Experimental investigation of tread wear and particle emission from tyres with different treadwear marking. *Atmos Environ.* 2018;182:200–212. doi:10.1016/j.atmosenv.2018.03.049
100. 1.A.3.b.vi-Vii Road Tyre and Brake Wear 2019. European Environment Agency. 2019. https://www.eea.europa.eu/publications/emep-eea-guidebook-2019/part-b-sectoral-guidance-chapters/1-energy/1-a-combustion/1-a-3-b-vi/view
101. Breuer JL, Samsun RC, Peters R, Stolten D. The impact of diesel vehicles on NOx and PM10 emissions from road transport in urban morphological zones: a case study in North Rhine-Westphalia, Germany. *Sci Total Environ.* 2020;727:138583. doi:10.1016/j.scitotenv.2020.138583
102. Pollution Warning Over Car Tyre and Brake Dust. BBC. 2019. https://www.bbc.com/news/business-48944561
103. Selley L, Schuster L, Marbach H, et al. Brake dust exposure exacerbates inflammation and transiently compromises phagocytosis in macrophages. *Metallomics.* 2020;12(3):371–386. doi:10.1039/C9MT00253G
104. Posselt KP, Neuberger M, Köhler D. Fine and ultrafine particle exposure during commuting by subway in Vienna. *Wien Klin Wochenschr.* 2019;131(15):374–380. doi:10.1007/s00508-019-1516-3
105. Jonsdottir HR, Delaval M, Leni Z, et al. Non-volatile particle emissions from aircraft turbine engines at ground-idle induce oxidative stress in bronchial cells. *Commun Biol.* 2019;2(1):90. doi:10.1038/s42003-019-0332-7

106. Yu Z, Liscinsky DS, Fortner EC, et al. Evaluation of PM emissions from two in-service gas turbine general aviation aircraft engines. *Atmos Environ*. 2017;160:9–18. doi:10.1016/j.atmosenv.2017.04.007
107. Boyle KA. Evaluating particulate emissions from jet engines: analysis of chemical and physical characteristics and potential impacts on coastal environments and human health. *Transp Res Rec*. 1996;1517(1):1–9. doi:10.1177/0361198196151700101
108. Pope CA. Epidemiology of fine particulate air pollution and human health: biologic mechanisms and who's at risk? *Environ Health Perspect*. 2000;108(suppl 4):713–723. doi:10.1289/ehp.108-1637679
109. Donaldson K, Stone V, Seaton A, MacNee W. Ambient particle inhalation and the cardiovascular system: potential mechanisms. *Environ Health Perspect*. 2001;109(suppl 4):523–527. doi:10.1289/ehp.01109s4523
110. Hudda N, Gould T, Hartin K, Larson TV, Fruin SA. Emissions from an international airport increase particle number concentrations 4-fold at 10 km downwind. *Environ Sci Technol*. 2014;48(12):6628–6635. doi:10.1021/es5001566
111. Kwon HS, Ryu MH, Carlsten C. Ultrafine particles: unique physicochemical properties relevant to health and disease. *Exp Mol Med*. 2020;52(3):318–328. doi:10.1038/s12276-020-0405-1
112. Leikauf GD, Kim SH, Jang AS. Mechanisms of ultrafine particle-induced respiratory health effects. *Exp Mol Med*. 2020;52(3):329–337. doi:10.1038/s12276-020-0394-0
113. Dennekamp M, Howarth S, Dick CAJ, Cherrie JW, Donaldson K, Seaton A. Ultrafine particles and nitrogen oxides generated by gas and electric cooking. *Occup Environ Med*. 2001;58(8):511. doi:10.1136/oem.58.8.511
114. Li Y, Wu A, Wu Y, et al. Morphological characterization and chemical composition of PM2.5 and PM10 collected from four typical Chinese restaurants. *Aerosol Sci Technol*. 2019;53(10):1186–1196. doi:10.1080/02786826.2019.1645292
115. Junkermann W, Hacker JM. Ultrafine particles in the lower troposphere: major sources, invisible plumes, and meteorological transport processes. *Bull Am Meteorol Soc*. 2018;99(12):2587–2602. doi:0.1175/BAMS-D-18-0075.1
116. McDuffie EE, Martin RV, Spadaro JV, et al. Source sector and fuel contributions to ambient PM2.5 and attributable mortality across multiple spatial scales. *Nat Commun*. 2021;12(1):3594. doi:10.1038/s41467-021-23853-y
117. Karagulian F, Belis CA, Dora CFC, et al. Contributions to cities' ambient particulate matter (PM): systematic review of local source contributions at global level. *Atmos Environ*. 2015;120:475–483. doi:10.1016/j.atmosenv.2015.08.087
118. Supeni A, Permadi DA, Gunawan D, Dayantolis W, Suwarman R. Variability of PM10 in a global atmosphere watch station near the equator. *IOP Conf Ser Earth Environ Sci*. 2021;724(1):012051. doi:10.1088/1755-1315/724/1/012051
119. Daellenbach KR, Uzu G, Jiang J, et al. Sources of particulate-matter air pollution and its oxidative potential in Europe. *Nature*. 2020;587(7834):414–419. doi:10.1038/s41586-020-2902-8
120. Emissions of the Main Air Pollutants in Europe. European Environment Agency. 2021. https://www.eea.europa.eu/ims/emissions-of-the-main-air
121. Sources and Emissions of Air Pollutants in Europe. European Environment Agency. 2021. https://www.eea.europa.eu/publications/air-quality-in-europe-2021/sources-and-emissions-of-air
122. Weagle CL, Snider G, Li C, et al. Global sources of fine particulate matter: interpretation of PM2.5 chemical composition observed by SPARTAN using a global chemical

transport model. *Environ Sci Technol.* 2018;52(20):11670–11681. doi:10.1021/acs.est.8b01658
123. Samek L, Stegowski Z, Styszko K, et al. Seasonal variations of chemical composition of PM2.5 fraction in the urban area of Krakow, Poland: PMF source attribution. *Air Qual Atmos Health.* 2020;13(1):89–96. doi:10.1007/s11869-019-00773-x
124. Zhao N, Li B, Chen D, et al. The effect of coal size on PM2.5 and PM-bound polycyclic aromatic hydrocarbon (PAH) emissions from a domestic natural cross-draft stove. *J Energy Inst.* 2020;93(2):542–551. doi:10.1016/j.joei.2019.06.010
125. Chafe Zoë A., Brauer Michael, Klimont Zbigniew, et al. Household cooking with solid fuels contributes to ambient PM2.5 air pollution and the burden of disease. *Environ Health Perspect.* 2014;122(12):1314–1320. doi:10.1289/ehp.1206340
126. Abdal Dayem A, Hossain MK, Lee SB, et al. The role of reactive oxygen species (ROS) in the biological activities of metallic nanoparticles. *Int J Mol Sci.* 2017;18(1). doi:10.3390/ijms18010120
127. Abbas I, Badran G, Verdin A, et al. Polycyclic aromatic hydrocarbon derivatives in airborne particulate matter: sources, analysis and toxicity. *Environ Chem Lett.* 2018;16(2):439–475. doi:10.1007/s10311-017-0697-0
128. Stec AA, Dickens KE, Salden M, et al. Occupational exposure to polycyclic aromatic hydrocarbons and elevated cancer incidence in firefighters. *Sci Rep.* 2018;8(1):2476. doi:10.1038/s41598-018-20616-6
129. Wolhuter K, Arora M, Kovacic JC. Air pollution and cardiovascular disease: can the Australian bushfires and global COVID-19 pandemic of 2020 convince us to change our ways? *BioEssays.* 2021;43(9):2100046. doi:10.1002/bies.202100046
130. Namork E, Johansen BV, Løvik M. Detection of allergens adsorbed to ambient air particles collected in four European cities. *Toxicol Lett.* 2006;165(1):71–78. doi:10.1016/j.toxlet.2006.01.016
131. De Grove KC, Provoost S, Brusselle GG, Joos GF, Maes T. Insights in particulate matter-induced allergic airway inflammation: Focus on the epithelium. *Clin Exp Allergy.* 2018;48(7):773–786. doi:10.1111/cea.13178
132. Nor NSM, Yip CW, Ibrahim N, et al. Particulate matter (PM2.5) as a potential SARS-CoV-2 carrier. *Sci Rep.* 2021;11(1):2508. doi:10.1038/s41598-021-81935-9
133. Linak WP, Yoo JI, Wasson SJ, et al. Ultrafine ash aerosols from coal combustion: characterization and health effects. *Proc Combust Inst.* 2007;31(2):1929–1937. doi:10.1016/j.proci.2006.08.086
134. Klosterköther A, Kurtenbach R, Wiesen P, Kleffmann J. Determination of the emission indices for NO, NO2, HONO, HCHO, CO, and particles emitted from candles. *Indoor Air.* 2021;31(1):116–127. doi:10.1111/ina.12714

Chapter 5

1. VW ID.4: So schlägt sich der Elektro-SUV im ADAC Test (Test performance of the VW ID.4 SUV by the German Automobile Club – ADAC). ADAC (German Automobile Club). 2022. https://www.adac.de/rund-ums-fahrzeug/autokatalog/marken-modelle/vw/vw-id-4/
2. Plastics Use in Vehicles to Grow 75% by 2020, Says Industry Watcher. Plastics Today. 2015. https://www.plasticstoday.com/automotive-and-mobility/plastics-use-vehicles-grow-75-2020-says-industry-watcher

3. Plastics in the Automotive Industry: Which Materials Will Be the Winners and Losers? NexantECA. 2018. https://www.nexanteca.com/reports/plastics-automotive-industry-which-materials-will-be-winners-and-losers
4. The Plastics Used in Automotives. AZO Materials. 2018. https://www.azom.com/article.aspx?ArticleID=17014
5. Where Your Car Goes to Die. Popular Mechanics. 2009. https://www.popularmechanics.com/cars/a1481/4213384/
6. Abullah Z, Id A. Assessment of end-of-life vehicle recycling: remanufacturing waste sheet steel into mesh sheet. *PLoS ONE*. 2021;16:1–17. doi:10.1371/journal.pone.0261079
7. Recycling (of Automotive Steel). WorldAutoSteel. 2021. https://www.worldautosteel.org/life-cycle-thinking/recycling/
8. Simic V. Fuzzy risk explicit interval linear programming model for end-of-life vehicle recycling planning in the EU. *Waste Management*. 2015;35:265–282. doi:10.1016/j.wasman.2014.09.013
9. Six Times More Plastic Waste Is Burned in US Than Is Recycled. Plastic Pollution Coalition. 2019. https://www.plasticpollutioncoalition.org/blog/2019/4/29/six-times-more-plastic-waste-is-burned-in-us-than-is-recycled
10. Plastic Waste in the United States. Statista. 2021. https://www.statista.com/topics/5127/plastic-waste-in-the-united-states/
11. Exposing the Myth of Plastic Recycling: Why a Majority Is Burned or Thrown in a Landfill. wbur Here & Now. 2019. https://www.wbur.org/hereandnow/2019/09/20/how-to-recycle-plastic
12. How Circular Is PET? A Report on the Circularity of PET Bottles, Using Europe as a Case Study. Zero Waste Europe. 2022. https://zerowasteeurope.eu/wp-content/uploads/2022/02/HCIP_V13-1.pdf
13. Global Plastic Waste Management Projections 2016–2040. Statista. 2021. https://www.statista.com/statistics/1270110/plastic-waste-management-projections-worldwide/
14. The New Plastics Economy: Rethinking the Future of Plastics. World Economic Forum. 2016. https://www3.weforum.org/docs/WEF_The_New_Plastics_Economy.pdf
15. The Impact of Waste-to-Energy Incineration on Climate. Zero Waste Europe. 2019. https://zerowasteeurope.eu/wp-content/uploads/edd/2019/09/ZWE_Policy-briefing_The-impact-of-Waste-to-Energy-incineration-on-Climate.pdf
16. Muller R, Muller E. Fugitive methane and the role of atmospheric half-life. *Geoinfor Geostat An Overview*. 2017;5(1). doi:10.4172/2327-4581.1000162
17. Hmiel B, Petrenko VV, Dyonisius MN, et al. Preindustrial 14CH4 indicates greater anthropogenic fossil CH4 emissions. *Nature*. 2020;578(7795):409–412. doi:10.1038/s41586-020-1991-8
18. Dermatas D, Georganti-Ntaliape A. Plastic waste trafficking: an ever-growing environmental crime that needs to be tackled. *Waste Manag Res*. 2020;38(11):1187–1188. doi:10.1177/0734242X20966250
19. Synthetic Carpet Fibers. Polymer Database. https://polymerdatabase.com/Fibers/Carpets.html
20. Oil Companies Are Going All-In on Petrochemicals: and Green Chemistry Needs Help to Compete. The Conversation. 2021. https://theconversation.com/oil-compan

ies-are-going-all-in-on-petrochemicals-and-green-chemistry-needs-help-to-compete-153598
21. Oil and Petroleum Products Explained: Use of Oil. U.S. Energy Information Administration (EIA). 2022. https://www.eia.gov/energyexplained/oil-and-petroleum-products/use-of-oil.php
22. Deutlich weniger erneuerbarer strom im jahr 2021 (Significantly Less Regenerative Electricity in 2021). Umwelt Bundesamt (German Ministry of the Environment). 2021. https://www.umweltbundesamt.de/presse/pressemitteilungen/deutlich-weniger-erneuerbarer-strom-im-jahr-2021
23. Nettostromerzeugung in Deutschland 2021: erneuerbare Energien witterungsbedingt schwächer (Electricity Generation in Germany 2021: Regenerative Energy Suffers from Weather Conditions). Fraunhofer Institut. 2022. https://www.ise.fraunhofer.de/de/presse-und-medien/news/2022/nettostromerzeugung-in-deutschland-2021-erneuerbare-energien-witterungsbedingt-schwaecher.html
24. Primärenergiebedarf 2021: Mehr Kohle, weniger Erneuerbare (Primary Energy Demand 2021: More Coal, Less Regenerative). ECOreporter. 2021. https://www.ecoreporter.de/artikel/prim%C3%A4renergiebedarf-2021-mehr-kohle-weniger-erneuerbare/
25. Patisson F, Mirgaux O. Hydrogen ironmaking: how it works. *Metals*. 2020;10(7). doi:10.3390/met10070922
26. Alternative Sustainable Carbon Sources as Substitutes for Metallurgical Coal. International Energy Agency (IEA). 2019. https://www.ieabioenergy.com/wp-content/uploads/2020/01/IEA-Bioenergy-Task-Lignin-as-a-met-coal-substitute-December-2019-Final-191218-1.pdf
27. Cloete S, Ruhnau O, Cloete JH, Hirth L. Blue hydrogen and industrial base products: the future of fossil fuel exporters in a net-zero world. *J Cleaner Production*. 2022:132347.
28. Energieverbrauch in der Industrie 2020 um 1,9 % gegenüber dem Vorjahr gesunken (1.9% Decrease in Energy Demand from Industry 2020 Compared with Previous Year). De Statis (German Government Office of Statistics). 2021. https://www.destatis.de/DE/Presse/Pressemitteilungen/2021/12/PD21_551_435.html
29. Bhagath Singh GVP, Subramaniam KVL. Production and characterization of low-energy Portland composite cement from post-industrial waste. *J Cleaner Production*. 2019;239:118024. doi:10.1016/j.jclepro.2019.118024
30. Biogasanlagen: anzahl in Deutschland bis 2021(Number of Biogas Production Facilities in Germany up till 2021). Statista. 2022. https://de.statista.com/statistik/daten/studie/167671/umfrage/anzahl-der-biogasanlagen-in-deutschland-seit-1992/
31. Dokumentarfilm im Ersten: die recyclinglüge (Documentary from Channel 1: The Recycling Lie). Das erste deutsche Fernsehen (Channel 1 Germany TV). 2022. https://www.ardmediathek.de/video/dokus-im-ersten/dokumentarfilm-im-ersten-die-recyclingluege/das-erste/Y3JpZDovL2Rhc2Vyc3RlLmRlL3JlcG9ydGFnZSBfIGRva3VtZW50YXRpb24gaW0gZXJjdGVuL2YwMTFjNmY0LTc1MGUtNDc5Mi1iZDgyLWRkZDM4YTNhMWU4Yw
32. Larsson J, Elofsson A, Sterner T, Åkerman J. International and national climate policies for aviation: a review. *Climate Policy*. 2019;19(6):787–799. doi:10.1080/14693062.2018.1562871

33. Ripple WJ, Wolf C, Newsome TM, Barnard P, Moomaw WR. World scientists' warning of a climate emergency. *BioScience*. 2020;70(1):8–12. doi:10.1093/biosci/biz088
34. Sausen R, Isaksen I, Grewe V, et al. Aviation radiative forcing in 2000: an update on IPCC (1999). *Meteorologische Zeitschrift*. 2005;14(4):555–561. doi:10.1127/0941-2948/2005/0049
35. Environmental Impact of Aviation. Wikipedia. https://en.wikipedia.org/wiki/Environmental_impact_of_aviation
36. Timko MT, Herndon SC, de la Rosa Blanco E, et al. Combustion products of petroleum jet fuel, a Fischer–Tropsch synthetic fuel, and a biomass fatty acid methyl ester fuel for a gas turbine engine. *Combustion Science and Technology*. 2011;183(10):1039–1068. doi:10.1080/00102202.2011.581717
37. Saffaripour M, Veshkini A, Kholghy M, Thomson MJ. Experimental investigation and detailed modeling of soot aggregate formation and size distribution in laminar coflow diffusion flames of Jet A-1, a synthetic kerosene, and n-decane. *Combust Flame*. 2014;161(3):848–863. doi:10.1016/j.combustflame.2013.10.016
38. Bester N, Yates A. Assessment of the operational performance of Fischer-Tropsch synthetic-paraffinic kerosene in a T63 gas turbine compared to conventional jet A-1 fuel. *Am Soc Mech Engineer*. 2010:1063–1077. doi:10.1115/GT2009-60333
39. Jet Fuel Price Monitor. IATA (International Air Transport Association). https://www.iata.org/en/publications/economics/fuel-monitor/
40. Notice to Members: Petition No 0560/2019 by Timothée Galvaire (French) on Behalf of Fairosene on the Introduction of a Tax on Aviation Fuel (Kerosene) for Flights within the Union. European Parliament. 2021. https://www.europarl.europa.eu/doceo/document/PETI-CM-641353_EN.pdf
41. Matsuda S, Ono M, Yamaguchi S, Uosaki K. Criteria for evaluating lithium–air batteries in academia to correctly predict their practical performance in industry. *Mater Horiz*. 2022;9(3):856–863. doi:10.1039/D1MH01546J
42. Turbofan. Wikipedia. https://en.wikipedia.org/wiki/Turbofan
43. Ali ARA, Janajreh I. Numerical simulation of turbine blade cooling via jet impingement. *Energy Procedia*. 2015;75:3220–3229. doi:10.1016/j.egypro.2015.07.683
44. From Bean to Cup, What Goes into the Cost of Your Coffee? *Financial Times*. 2019. https://www.ft.com/content/44bd6a8e-83a5-11e9-9935-ad75bb96c849.
45. Hilakari M. *Carbon Footprint Calculation of Shipbuilding*. Turku University of Applied Science; 2019. https://www.theseus.fi/bitstream/handle/10024/179517/hilakari_marianna.pdf
46. How Many Gallons of Fuel Does a Container Ship Carry? 2020. https://www.freightwaves.com/news/how-many-gallons-of-fuel-does-a-container-ship-carry
47. I Didn't Want to Fly: So I Took a Cargo Ship from Germany to Canada. *The Guardian*. 2020. https://www.theguardian.com/travel/2020/jan/07/cargo-ship-train-rail-to-vancouver-canada-low-carbon-travel-europe.
48. Volvo Promises Full Electric HGV Range for 2021. TU Automotive. 2020. https://www.tu-auto.com/volvo-promises-full-electric-hgv-range-for-2021/
49. Billionaires' Single Space Flight Produces a Lifetime's Worth of Carbon Footprint: Report. Earth.Org. 2021. https://earth.org/billionaires-single-space-flight-produces-a-lifetimes-worth-of-carbon-footprint-report/

50. Barros B, Wilk R. The outsized carbon footprints of the super-rich. *Sustainability: Science, Practice and Policy.* 2021;17(1):316–322. doi:10.1080/15487733.2021.1949847
51. Global Crude Steel Output Decreases by 0.9% in 2020. World Steel. 2021. https://worldsteel.org/media-centre/press-releases/2021/global-crude-steel-output-decreases-by-0-9-in-2020/
52. China Cement Production capacity 2021. Statista. 2022. https://www.statista.com/statistics/1291494/production-capacity-of-cement-in-china/
53. China Production Volume of Cement 1970–2020. Statista. 2021. https://www.statista.com/statistics/307647/china-production-volume-of-cement/
54. Abouhamad M, Abu-Hamd M. Life cycle assessment framework for embodied environmental impacts of building construction systems. *Sustainability.* 2021;13(2). doi:10.3390/su13020461
55. Revealed: The Climate Cost of "Disposable Smartphones." European Environmental Bureau. 2019. https://eeb.org/revealed-the-climate-cost-of-disposable-smartphones/
56. Marland G, Oda T, Boden TA. Per capita carbon emissions must fall to 1955 levels. *Nature.* 2019;565(7737):567–568. doi:10.1038/d41586-019-00325-4
57. Electricity Information. International Energy Agency (IEA). 2022. https://www.iea.org/data-and-statistics/data-product/electricity-information
58. Global Energy Review 2019: Renewables. International Energy Agency (IEA). 2019. https://www.iea.org/reports/global-energy-review-2019/renewables
59. Qatar EO2 Emissions. Worldometer. https://www.worldometers.info/co2-emissions/qatar-co2-emissions/
60. The Evidence Is Clear: The Time for Action Is Now. We Can Halve Emissions by 2030. Intergovernmental Panel on Climate Change (IPCC). 2022. https://www.ipcc.ch/2022/04/04/ipcc-ar6-wgiii-pressrelease/
61. Moore A. We must use less: a year of climate crisis, a glimpse of hope, but not in mere technology. *BioEssays.* 2019;41(12):1900214. doi:10.1002/bies.201900214
62. Domke GM, Oswalt SN, Walters BF, Morin RS. Tree planting has the potential to increase carbon sequestration capacity of forests in the United States. *Proc Natl Acad Sci.* 2020;117(40):24649–24651. doi:10.1073/pnas.2010840117
63. Nowak DJ, McBride JR, Beatty RA. Newly planted street tree growth and mortality. *J Arboriculture.* 1990;16(5):124–130. https://www.nrs.fs.fed.us/pubs/jrnl/1990/ne_1990_nowak_001.pdf
64. Crisis in the Brazilian Amazon. Human Rights Watch. 2022. https://www.hrw.org/news/2022/04/19/crisis-brazilian-amazon
65. Zhang Y, Song C, Band LE, Sun G. No proportional increase of terrestrial gross carbon sequestration from the greening earth. *J Geophys Res Biogeosci.* 2019;124(8):2540–2553. doi:10.1029/2018JG004917
66. Deng L, Yuan H, Xie J, Ge L, Chen Y. Herbaceous plants are better than woody plants for carbon sequestration. *Resour Conserv Recycl.* 2022;184:106431. doi:10.1016/j.resconrec.2022.106431
67. Harris NL, Gibbs DA, Baccini A, et al. Global maps of twenty-first century forest carbon fluxes. *Nat Clim Chang.* 2021;11(3):234–240. doi:10.1038/s41558-020-00976-6
68. Hlásny T, Barka I, Roessiger J, et al. Conversion of Norway spruce forests in the face of climate change: a case study in Central Europe. *Eur J For Res.* 2017;136(5):1013–1028. doi:10.1007/s10342-017-1028-5

69. Czybulka D, Köck W. ed. Forstwirtschaft und Biodiversitätsschutz im Wald: Beiträge zum 14. Deutschen Naturschutzrechtstag (Forestry Science and Biodiversity Conservation in Forests: Contributions to the 14th German Conference on Nature Conservation Rights). In: Vol 14. *Beiträge zum Landwirtschaftsrecht und zur Biodiversität*. Nomos Verlagsgesellschaft mbH & Co. KG; 2022. doi:10.5771/9783748921165
70. Where Can Peatlands Be Found? International Peatland Society. https://peatlands.org/peatlands/where-can-peatlands-be-found/
71. Ribeiro K, Pacheco FS, Ferreira JW, et al. Tropical peatlands and their contribution to the global carbon cycle and climate change. *Glob Chan Biol*. 2021;27(3):489–505. doi:10.1111/gcb.15408
72. Krause L, McCullough KJ, Kane ES, Kolka RK, Chimner RA, Lilleskov EA. Impacts of historical ditching on peat volume and carbon in northern Minnesota USA peatlands. *J Environ Manage*. 2021;296:113090. doi:10.1016/j.jenvman.2021.113090
73. Pogge von Strandmann PAE, Burton KW, Snæbjörnsdóttir SO, et al. Rapid CO2 mineralisation into calcite at the CarbFix storage site quantified using calcium isotopes. *Nat Commun*. 2019;10(1):1983. doi:10.1038/s41467-019-10003-8
74. Matter JM, Broecker WS, Gislason SR, et al. The CarbFix Pilot Project–Storing carbon dioxide in basalt. *Energy Procedia*. 2011;4:5579–5585. doi:10.1016/j.egypro.2011.02.546
75. Orca: The First Large-Scale Plant. Climeworks. https://climeworks.com/roadmap/orca
76. CO2SINK (European Union Project to Sequester CO2 Underground in Ketzin, Germany). https://www.co2sink.org/
77. Pilotstandort Ketzin Forschungsprojekt CO2MAN (Pilot Plant Ketzin Research Project CO2MAN). https://www.co2ketzin.de/en/pilot-site-ketzin/summary
78. CO2-Verpressung in Brandenburg unerwünscht (CO2 Sinking in Brandenburg Not Wanted). Deutschland Funk (German Radio). 2016. https://www.deutschlandfunkkultur.de/themenwoche-energiewende-co2-verpressung-in-brandenburg-100.html
79. Keine CO2-Endlager in Brandenburg! (No CO2 Sequestration in Brandenburg!). Greenpeace. 2014. https://www.greenpeace.de/klimaschutz/energiewende/kohleausstieg/co2-endlager-brandenburg
80. Moore als CO2-speicher – vom klimakiller zum hoffnungsträger (Moorland as CO2 Storage: From Climate-Killer to Hope for the Future). Tagesschau (News from German Channel 1 TV). 2021. https://www.tagesschau.de/inland/gesellschaft/moore-klimaschutz-101.html
81. Wang Y, Vuik C, Hajibeygi H. CO2 Storage in deep saline aquifers: impacts of fractures on hydrodynamic trapping. *Int J Greenhouse Gas Control*. 2022;113:103552. doi:10.1016/j.ijggc.2021.103552
82. Ocean Storage of CO2. The Maritime Executive. 2018. https://www.maritime-executive.com/features/ocean-storage-of-co2
83. Large-Scale CCS Projects Globally by Status 2021. Statista. 2022. https://www.statista.com/statistics/726624/large-scale-carbon-capture-and-storage-projects-worldwide-by-status/
84. Harari YN. *Sapiens: A Brief History of Humankind*. 1st US edition. Harper; 2015. https://search.library.wisc.edu/catalog/9910419687402121

Chapter 6

1. Carbon dioxide emissions from passenger cars worldwide from 2000 to 2020. *Statista*. https://www.statista.com/statistics/1107970/carbon-dioxide-emissions-passenger-transport/
2. Global carbon dioxide emissions from 1970 to 2021, by sector. *Statista*. https://www.statista.com/statistics/276480/world-carbon-dioxide-emissions-by-sector/
3. Emissions Due to Agriculture. FAO (Food and Agriculture Organization of the United Nations). https://www.fao.org/3/cb3808en/cb3808en.pdf
4. Wietschel M, Kühnbach M, Rüdiger D. Working Paper Sustainability and Innovation No. S 02/2019. Die aktuelle Treibhausgas emissionsbilanz von Elektrofahrzeugen in Deutschland (The Current Greenhouse Gas Emissions Balance of EVs in Germany). 2019. https://www.isi.fraunhofer.de/content/dam/isi/dokumente/sustainability-innovation/2019/WP02-2019_Treibhausgasemissionsbilanz_von_Fahrzeugen.pdf
5. VDI Wissensforum GmbH. Anwendungsfallabhängige CO2-Bilanzen elektrifizierter Fahrzeugantriebe (Application-dependent CO2 balances of drive trains) (Ludwig O. et al.) In: *Innovative Antriebe 2018: der Ausblick auf die Fahrzeugantriebe für die kommenden Dekaden (Innovative Drive Trains 2018: The Outlook for Vehicle Drives for the Coming Decades)*. Vol. 2334, 1st ed. VDI-Berichte/VDI Verlag; 2018. doi:10.51202/9783181023341
6. Is It Ethical to Purchase an EV Lithium Battery Powered Vehicle? Eurasiareview – News and Analysis. 2022. https://www.eurasiareview.com/07062022-is-it-ethical-to-purchase-an-ev-lithium-battery-powered-vehicle-oped/
7. US Largest Lithium Mining Project Faces Strong Resistance. English.news.cn, Roundup. 2022. https://english.news.cn/northamerica/20220708/0d9153bcb3df4792b1f56827be760e75/c.html
8. Environmentalists Have Turned on the Lithium Industry. OilPrice.com. 2022. https://oilprice.com/Energy/Energy-General/Environmentalists-Have-Turned-On-The-Lithium-Industry.html
9. Green Activists Stage Tent Protest to Halt Lithium Exploration in Serbia. Reuters. 2022. https://www.reuters.com/business/environment/green-activists-stage-tent-protest-halt-lithium-exploration-serbia-2022-02-11/
10. Zeng D, Dong Y, Cao H, et al. Are the electric vehicles more sustainable than the conventional ones? Influences of the assumptions and modeling approaches in the case of typical cars in China. *Resour, Conserv Recycl*. 2021;167:105210. doi:10.1016/j.resconrec.2020.105210
11. Wu D, Guo F, Field FR, et al. Regional heterogeneity in the emissions benefits of electrified and lightweighted light-duty vehicles. *Environ Sci Technol*. 2019;53(18):10560–10570. doi:10.1021/acs.est.9b00648
12. Mayyas A, Omar M, Hayajneh M, Mayyas AR. Vehicle's lightweight design vs. electrification from life cycle assessment perspective. *J Clean Prod*. 2017;167:687–701. doi:10.1016/j.jclepro.2017.08.145
13. The Potential of Electric Vehicles. ADEME. 2016. https://librairie.ademe.fr/mobilite-et-transport/1922-the-potential-of-electric-vehicles.html

14. He X, Hu Y. Understanding the role of emotions in consumer adoption of electric vehicles: the mediating effect of perceived value. *J Environ Plan Manage.* 2022;65(1):84–104. doi:10.1080/09640568.2021.1878018
15. Changing Mobility Patterns After Buying an Electric Vehicle: One-Click Survey. FlipTheFleet. https://flipthefleet.org/2018/1-click-survey-20/
16. China as Reliant as Ever on Fossil Fuels Despite Renewables Green Push. Bloomberg.com. 2022. https://www.bloomberg.com/news/articles/2022-01-19/two-charts-that-show-china-is-as-reliant-as-ever-on-fossil-fuels
17. Longest-Owned Car Models. Germain Cars. 2022. https://www.germaincars.com/average-length-of-car-ownership/
18. Britons Are Keeping Their Cars Longer Than Ever, Says New Report. This Is Money. 2020. https://www.thisismoney.co.uk/money/cars/article-8739771/Britons-keeping-cars-longer-says-new-report.html
19. DAT Report 2016. 2016. https://www.dat.de/fileadmin/media/download/DAT-Report/DAT-Report-2016.pdf
20. What Can 6,000 Electric Vehicles Tell Us About EV Battery Health? Geotab. 2020. https://www.geotab.com/blog/ev-battery-health/
21. Saxena S, Le Floch C, MacDonald J, Moura S. Quantifying EV battery end-of-life through analysis of travel needs with vehicle powertrain models. *J Power Sources.* 2015;282:265–276. doi:10.1016/j.jpowsour.2015.01.072
22. The Rise and Precarious Reign of China's Battery King. Wired. 2022. https://www.wired.com/story/catl-china-battery-production-evs/
23. Change in Distance Travelled by Car. Odyssee-Mure. 2020. https://www.odyssee-mure.eu/publications/efficiency-by-sector/transport/distance-travelled-by-car.html
24. Prenner S, Allesch A, Staudner M, et al. Static modelling of the material flows of micro-and nanoplastic particles caused by the use of vehicle tyres. *Environ Pollut.* 2021;290:118102. doi:10.1016/j.envpol.2021.118102
25. Tyrewearmapping: reifenabrieb in Deutschland (Tire Wear in Germany). Fraunhofer Institut. 2020. https://www.umsicht.fraunhofer.de/de/projekte/tyrewearmapping.html
26. Daten und fakten reifenabrieb – Fraunhofer UMSICHT (Data and Facts on Tire Wear – Fraunhofer UMSICHT Project). Fraunhofer Institut. 2021. https://www.umsicht.fraunhofer.de/de/presse-medien/glossare-faqs/glossar-reifenabrieb.html#faq_faqitem_1987518866-answer
27. EV Reliability as Varied as Gasoline Cars, Consumer Reports Finds. Forbes Wheels. 2021. https://www.forbes.com/wheels/news/ev-reliability-varied-as-gasoline-cars-consumer-reports/
28. Reliability Problems Plague Newer Electric Cars. Consumer Reports. 2020. https://www.consumerreports.org/hybrids-evs/reliability-problems-plague-newer-electric-cars-a2176942694/
29. Peters JF, Baumann M, Zimmermann B, Braun J, Weil M. The environmental impact of Li-Ion batteries and the role of key parameters: A review. *Renew Sustain Energy Rev.* 2017;67:491–506. doi:10.1016/j.rser.2016.08.039
30. Gohlke D, Zhou Y. Assessment of Light-Duty Plug-in Electric Vehicles in the United States, 2010–2020. Argonne National Laboratory. 2021. https://publications.anl.gov/anlpubs/2021/06/167626.pdf
31. Liu B, Li L, Ni X. *Research on Life Cycle of Typical Passenger Vehicles Based on Energy Structure: Technical Report at the SAE 2020 Vehicle Electrification and Autonomous*

Vehicle Technology Forum. Tongji University; 2020. https://www.sae.org/publications/technical-papers/content/2020-01-5187/
32. The Monster Footprint of Digital Technology. Low-Tech Magazine. 2009. https://www.lowtechmagazine.com/2009/06/embodied-energy-of-digital-technology.html
33. Irimia-Vladu M. "Green" electronics: biodegradable and biocompatible materials and devices for sustainable future. *Chem Soc Rev*. 2014;43(2):588–610. doi:10.1039/C3CS60235D
34. Williams ED. Revisiting energy used to manufacture a desktop computer: hybrid analysis combining process and economic input-output methods. *IEEE International Symposium on Electronics and the Environment, 2004. Conference Record. 2004*. 2004:80–85. doi:10.1109/ISEE.2004.1299692
35. Tracking Appliances and Equipment 2020. International Energy Agency (IEA). 2020. https://www.iea.org/reports/tracking-appliances-and-equipment-2020
36. 20 Staggering E-Waste Facts in 2021. Earth 911. 2021. https://earth911.com/eco-tech/20-e-waste-facts/
37. Nithya R, Sivasankari C, Thirunavukkarasu A. Electronic waste generation, regulation and metal recovery: a review. *Environ Chem Lett*. 2021;19(2):1347–1368. doi:10.1007/s10311-020-01111-9
38. Chordia M, Nordelöf A, Ellingsen LAW. Environmental life cycle implications of upscaling lithium-ion battery production. *Int J Life Cycle Assess*. 2021;26(10):2024–2039. doi:10.1007/s11367-021-01976-0
39. *Global EV Outlook 2020*. International Energy Agency (IEA). 2020. https://www.iea.org/reports/global-ev-outlook-2020
40. Global EV Sales Growth Leads Industry in 2020. TechNewsWorld. 2021. https://www.technewsworld.com/story/global-ev-sales-growth-leads-industry-in-2020-87013.html
41. ReCiPe 2016 v1.1. Pre-sustainability.com. 2017. https://pre-sustainability.com/legacy/download/Report_ReCiPe_2017.pdf
42. Ellingsen LA, Majeau-Bettez G, Singh B, Srivastava AK, Valøen LO, Strømman AH. Life cycle assessment of a lithium-ion battery vehicle pack. *J Ind Ecol*. 2014;18(1):113–124. doi:10.1111/jiec.12072
43. Karasu H. *Life Cycle Assessment of Conventional and Alternative Fuels for Vehicles*. University of Ontario Institute of Technology; 2018. https://ir.library.dc-uoit.ca/bitstream/10155/952/1/Karasu_Huseyin.pdf
44. Electric Vehicles: Recycled Batteries and the Search for a Circular Economy. *Financial Times*. 2021. https://www.ft.com/content/e88e00e3-0a0c-469a-986b-1ffda60b6aee
45. Jones B, Elliott RJR, Nguyen-Tien V. The EV revolution: The road ahead for critical raw materials demand. *Applied Energy*. 2020;280:115072. doi:10.1016/j.apenergy.2020.115072
46. Kelly JC, Wang M, Dai Q, Winjobi O. Energy, greenhouse gas, and water life cycle analysis of lithium carbonate and lithium hydroxide monohydrate from brine and ore resources and their use in lithium ion battery cathodes and lithium ion batteries. *Resour, Conserv Recycl*. 2021;174:105762.doi:10.1016/j.resconrec.2021.105762
47. Lithium Sustainability Information. Bundesanstalt für Geowissenschaften und Rohstoffe (German Federal Agency for Geosciences and Natural Resources). 2020. https://www.bgr.bund.de/EN/Gemeinsames/Produkte/Downloads/Informationen_Nachhaltigkeit/lithium_en.pdf

48. Di Maria A, Elghoul Z, Van Acker, K. Environmental assessment of an innovative lithium production process. *Procedia CIRP.* 2022;105:672–677. doi:10.1016/j.procir.2022.02.112
49. Brandenburg: Größte Lithium-Fabrik Europas liefert ab 2024 Rohstoff für 500.000 E-Autos jährlich (Brandenburg: Largest Lithium Factory in Europe Will Produce Material for 500,000 EVs per Year from 2024). t3n news. https://t3n.de/news/lithium-fabrik-brandenburg-1415635/
50. Hydrologen warnen: Deutschland trocknet aus (Warning from Hydrologists: Germany's Drying Up). National Geographic Society. https://www.nationalgeographic.de/umwelt/2022/03/hydrologen-warnen-deutschland-trocknet-aus
51. Average Age of the EU Vehicle Fleet, by Country. European Automobile Manufacturers' Association (ACEA). 2022. https://www.acea.auto/figure/average-age-of-eu-vehicle-fleet-by-country/
52. Study: EVs Cost More to Service than ICE Cars. Autonews. 2021. https://www.autonews.com/retail/study-evs-cost-more-service-ice-vehicles
53. Nissan LEAF Teardown: Lithium-Ion Battery Pack Structure. MarkLines. 2018. https://www.marklines.com/en/report_all/rep1786_201811
54. Tesla Model S. Wikipedia. https://en.wikipedia.org/wiki/Tesla_Model_S
55. So sieht Teslas 100-kWh-Akku von Innen aus (This Is What the Tesla 100 kWh Battery Looks Like Inside). Ecomento. https://ecomento.de/2017/01/31/so-sieht-teslas-100-kwh-akku-von-innen-aus/
56. Most Lithium Batteries End Up in a Landfill: A New Bill Aims to Change That. Grist. 2020. https://grist.org/politics/most-lithium-batteries-end-up-in-a-landfill-a-new-bill-aims-to-change-that/
57. Skeete JP, Wells P, Dong X, Heidrich O, Harper G. Beyond the EVent horizon: battery waste, recycling, and sustainability in the United Kingdom electric vehicle transition. *Energy Res Soc Sci.* 2020;69:101581. doi:10.1016/j.erss.2020.101581
58. Doose S, Mayer JK, Michalowski P, Kwade A. Challenges in ecofriendly battery recycling and closed material cycles: a perspective on future lithium battery generations. *Metals.* 2021;11(2):291. doi:10.3390/met11020291
59. An Analysis of Lithium-Ion Battery Fires in Waste Management and Recycling. US Environmental Protection Agency (EPA). 2021. https://www.epa.gov/system/files/documents/2021-08/lithium-ion-battery-report-update-7.01_508.pdf
60. Recycling Efficiency for Lithium-Ion Batteries: How to Meet the EU-Requirements. Reneos. https://www.reneos.eu/case/recycling-efficiency-for-lithium-ion-batteries-how-to-meet-the-eu-requirements
61. Proposal for a Regulation of the European Parliament and of the Council Concerning Batteries and Waste Batteries, Repealing Directive 2006/66/EC and Amending Regulation (EU) No. 2019/1020. Eur-Lex. https://eur-lex.europa.eu/legal-content/EN/TXT/?uri=CELEX%3A52020PC0798
62. Helmers E, Dietz J, Weiss M. Sensitivity analysis in the life-cycle assessment of electric vs. combustion engine cars under approximate real-world conditions. *Sustainability.* 2020;12(3):1241. doi:10.3390/su12031241
63. Held M, Rosat N, Georges G, Pengg H, Boulouchos K. Lifespans of passenger cars in Europe: empirical modelling of fleet turnover dynamics. *Eur Transport Res Rev.* 2021;13(1):9. doi:10.1186/s12544-020-00464-0

64. Neugebauer M, Żebrowski A, Esmer O. Cumulative emissions of CO2 for electric and combustion cars: a case study on specific models. *Energies*. 2022;15(7). doi:10.3390/en15072703
65. CO_2 Emission Performance Standards for Cars and Vans. European Commission. https://ec.europa.eu/clima/eu-action/transport-emissions/road-transport-reducing-co2-emissions-vehicles/co2-emission-performance-standards-cars-and-vans_en
66. Volkswagen Lupo. Wikipedia. https://en.wikipedia.org/wiki/Volkswagen_Lupo
67. Lupo 3l TDI Once Again Enters Guinness World Records' Book. VW Press UK. 2022. https://www.vwpress.co.uk/en-gb/releases/855
68. Volkswagen Emissions Scandal. Wikipedia. https://en.wikipedia.org/wiki/Volkswagen_emissions_scandal
69. Change in US Car Demand by Vehicle Type. Statista. 2022. https://www.statista.com/statistics/276506/change-in-us-car-demand-by-vehicle-type/
70. SUVs Make Up 40 Percent of Total Car Registrations as the European Market Records Its Best October Result Since 2009. JATO. 2019. https://www.jato.com/suvs-make-up-40-of-total-car-registrations-as-the-european-market-records-its-best-october-result-since-2009/
71. In 2020, Pickup Trucks Outsell Cars for the First Time . . . Because Cars Are Disappearing. Kelly Blue Book. 2021. https://www.kbb.com/car-news/in-2020-pickup-trucks-outsell-cars-for-the-first-timebecause-cars-are-disappearing/
72. Global Car Sales by Size and Powertrain 2020. Statista. 2021. https://www.statista.com/statistics/1256707/global-car-sales-by-powertrain-and-size/
73. Elektroautos: Reichweite & Stromverbrauch im Vergleich (Electric Cars: Range and Electricity Consumption Compared). German Automobile Club (ADAC). 2022. https://www.adac.de/rund-ums-fahrzeug/tests/elektromobilitaet/stromverbrauch-elektroautos-adac-test/
74. BMW iX (i20): Motoren & Technische Daten (BMW iX (i20): Motors & Technical Data). BMW. 2022. https://www.bmw.de/de/neufahrzeuge/bmw-i/bmw-ix/2021/bmw-ix-technische-daten.html#tab-0
75. Smart EQ forfour (2018–2019) Preise und technische Daten (Smart EQ Forfour (2018–2019) Prices and Technical Data. EV-Database. 2020. https://ev-database.de/pkw/1134/Smart-EQ-forfour
76. Porsche Delivers 68,426 Vehicles in the First Quarter. Porsche Newsroom. 2022. https://newsroom.porsche.com/en/2022/company/porsche-deliveries-2022-first-quarter-28019.html
77. Audi e-tron GE. Wikipedia. https://insideevs.com/news/498568/audi-etron-sales-us-q12021/
78. Audi e-tron Sales in US Surge to Record Level in Q1 2021. InsideEVs. 2021. https://insideevs.com/news/498568/audi-etron-sales-us-q12021/
79. The Adjustment Factor Tesla Uses to Get Its Big EPA Range Numbers. Car and Driver. 2020. https://www.caranddriver.com/features/a33824052/adjustment-factor-tesla-uses-for-big-epa-range-numbers/
80. How Much Does a Tesla Model 3 Battery Replacement Cost? Current Automotive. 2020. https://www.currentautomotive.com/how-much-does-a-tesla-model-3-battery-replacement-cost/
81. Sector by Sector: Where Do Global Greenhouse Gas Emissions Come From? Our World in Data. 2020. https://ourworldindata.org/ghg-emissions-by-sector

82. Cars, Planes, Trains: Where Do CO2 Emissions From Transport Come From? Our World in Data. 2020. https://ourworldindata.org/co2-emissions-from-transport
83. How Many Cars Are There in the World in 2022? Hedges & Company. 2022. https://hedgescompany.com/blog/2021/06/how-many-cars-are-there-in-the-world/
84. Car Industry Employee Numbers in Germany 2010–2020. Statista. 2021. https://www.statista.com/statistics/587576/number-employees-german-car-industry/
85. Germany: Employment 2013–2023. Statista. 2021. https://www.statista.com/statistics/795244/employment-in-germany/
86. Crashing the Climate. Greenpeace. 2019. https://www.greenpeace.de/publikationen/Crashing%20the%20Climate%20engl%20LF.pdf
87. Renewable Fuels. Wikipedia. https://en.wikipedia.org/wiki/Renewable_fuels
88. Bosch: Renewable Synthetic Fuels for Less CO_2. Bosch-Presse. 2019. https://www.bosch-presse.de/pressportal/de/en/bosch-renewable-synthetic-fuels-for-less-co%E2%82%82-200070.html
89. Methanol Fuel. Wikipedia. https://en.wikipedia.org/wiki/Methanol_fuel
90. It's Time to Get Serious About Recycling Lithium-Ion Batteries. Chemical and Engineering News. 2019. https://cen.acs.org/materials/energy-storage/time-serious-recycling-lithium/97/i28
91. Morse I. A dead battery dilemma. *Science.* 2021:780–783. doi:10.1126/science.abj5426
92. Electric and Hybrid Cars: New Rules on Noise Emitting to Protect Vulnerable Road Users. European Commission. 2019. https://ec.europa.eu/growth/news/electric-and-hybrid-cars-new-rules-noise-emitting-protect-vulnerable-road-users-2019-07-03_en
93. Noise Generators for EVs Mandatory From July 1, 2019 in Europe. AutoEvolution. 2019. https://www.autoevolution.com/news/noise-generators-for-evs-mandatory-from-july-1-2019-in-europe-135022.html
94. Führende Professoren im Bereich Kfz-Bau: Falsche Stoppsignale bei Verkehrswende - E-Autos erfüllen nicht alle Anforderungen an die Mobilität der Zukunft (Leading Professors of Vehicle Construction: Wrong Stop Signals in the Mobility Revolution: EVs Do Not Fulfil All the Requirements of Future Mobility). *Stuttgarter Zeitung.* 2021. https://www.stuttgarter-zeitung.de/inhalt.fuehrende-professoren-im-bereich-kfz-bau-falsche-stoppsignale-bei-verkehrswende.e5c8c92e-26c3-49ea-a6fb-f7d6597a47ef.html?

Chapter 7

1. What Is the Climate Impact of Eating Meat and Dairy? Interactive Carbon Brief. 2020. https://interactive.carbonbrief.org/what-is-the-climate-impact-of-eating-meat-and-dairy/
2. Wer sein Smartphone länger nutzt, spart viel CO2 (Using a Mobile Phone for Longer Saves a Lot of CO2). Energiezukunft. 2019. https://www.energiezukunft.eu/klimawandel/wer-sein-smartphone-laenger-nutzt-spart-viel-co2/
3. Cradle-to-Cradle Design. Wikipedia. https://en.wikipedia.org/wiki/Cradle-to-cradle_design

4. Huijbregts MAJ, Steinmann ZJN, Elshout PMF, et al. ReCiPe2016: a harmonised life cycle impact assessment method at midpoint and endpoint level. *Int J Life Cycle Assess*. 2017;22(2):138–147. doi:10.1007/s11367-016-1246-y
5. Additional 20GWh of Battery Storage Could Cut Wasted Wind Power by 50%. The Energist. 2021. https://theenergyst.com/additional-20gwh-of-battery-storage-could-cut-wasted-wind-power-by-50/
6. As Electric Vehicles Take Off, We'll Need to Recycle Their Batteries. National Geographic. 2021. https://www.nationalgeographic.com/environment/article/electric-vehicles-take-off-recycling-ev-batteries
7. Baum ZJ, Bird RE, Yu X, Ma J. Lithium-ion battery recycling: overview of techniques and trends. *ACS Energy Lett*. 2022;7(2):712–719. doi:10.1021/acsenergylett.1c02602
8. How Big Are Power Line Losses? Schneider Electric Blog. 2013. https://blog.se.com/energy-management-energy-efficiency/2013/03/25/how-big-are-power-line-losses/
9. How Much Power Loss in Transmission Lines. Chint Global. 2021. https://chintglobal.com/blog/how-much-power-loss-in-transmission-lines/
10. Lost in Transmission: How Much Electricity Disappears Between a Power Plant and Your Plug? Inside Energy. 2015. http://insideenergy.org/2015/11/06/lost-in-transmission-how-much-electricity-disappears-between-a-power-plant-and-your-plug/
11. DOE: 77%–82% of Energy Put into an Electric Car Is Used to Move the Car Down the Road. Green Car Congress. https://www.greencarcongress.com/2018/09/20180905-fotw.html
12. Energy Efficiency Evaluation of a Stationary Lithium-Ion Battery Container Storage System via Electro-Thermal Modeling and Detailed Component Analysis. Appl Energy. 2018. https://www.sciencedirect.com/science/article/abs/pii/S0306261917315696
13. Michaelides EE. Thermodynamics, energy dissipation, and figures of merit of energy storage systems: a critical review. *Energies*. 2021;14(19). doi:10.3390/en14196121
14. Biswas S, Kulkarni AP, Giddey S, Bhattacharya S. A review on synthesis of methane as a pathway for renewable energy storage with a focus on solid oxide electrolytic cell-based processes. *Front Energy Res*. 2020;8. https://www.frontiersin.org/articles/10.3389/fenrg.2020.570112
15. Reversible Solid-Oxide Cell Stack Based Power-to-x-to-Power Systems: Comparison of Thermodynamic Performance. Appl Energy. (2020). https://www.sciencedirect.com/science/article/pii/S0306261920308424
16. Mogensen MB, Chen M, Frandsen HL, et al. Reversible solid-oxide cells for clean and sustainable energy. *Clean Energy*. 2019;3(3):175–201. doi:10.1093/ce/zkz023
17. Strom in Deutschland im Frühjahr überwiegend aus Kohle statt Windkraft (Electricity in Germany in Spring Was Largely from Coal Rather Than Wind). Der Spiegel. https://www.spiegel.de/wirtschaft/service/strom-in-deutschland-im-fruehjahr-ueberwiegend-aus-kohle-statt-windkraft-a-56174011-cfae-4219-a089-24b5499ff535. 2021.
18. Gas Cylinder. Wikipedia. https://en.wikipedia.org/wiki/Gas_cylinder
19. Global Electric Car Sales Have Continued Their Strong Growth in 2022 After Breaking Records Last Year. International Energy Agency (IEA). 2022. https://www.iea.org/news/global-electric-car-sales-have-continued-their-strong-growth-in-2022-after-breaking-records-last-year

20. Harper G, Sommerville R, Kendrick E, et al. Recycling lithium-ion batteries from electric vehicles. *Nature*. 2019;575(7781):75–86. doi:10.1038/s41586-019-1682-5
21. Piątek J, Afyon S, Budnyak TM, Budnyk S, Sipponen MH, Slabon A. Sustainable li-ion batteries: chemistry and recycling. *Adv Energy Mater*. 2021;11(43):2003456. doi:10.1002/aenm.202003456
22. Yang Y, Okonkwo EG, Huang G, Xu S, Sun W, He Y. On the sustainability of lithium ion battery industry: a review and perspective. *Energy Storage Mater*. 2021;36:186–212. doi:10.1016/j.ensm.2020.12.019
23. Titirici MM. Sustainable batteries: quo vadis? *Adv Energy Mater*. 2021;11(10):2003700. doi:10.1002/aenm.202003700
24. Innovation Outlook: Renewable Methanol. International Renewable Energy Agency (IRENA) in partnership with The Methanol Institute. 2021. https://www.irena.org/-/media/Files/IRENA/Agency/Publication/2021/Jan/IRENA_Innovation_Renewable_Methanol_2021.pdf
25. Adnan MA, Kibria MG. Comparative techno-economic and life-cycle assessment of power-to-methanol synthesis pathways. *Appl Energy*. 2020;278:115614. doi:10.1016/j.apenergy.2020.115614
26. Wang L, Zhang Y, Pérez-Fortes M, et al. Reversible solid-oxide cell stack based power-to-x-to-power systems: comparison of thermodynamic performance. *Appl Energy*. 2020;275:115330. doi:10.1016/j.apenergy.2020.115330
27. Cooper T, Filmer D, Wishnick M, Lane MD. The active species of "CO2" utilized by ribulose diphosphate carboxylase. *J Biol Chem*. 1969;244(4):1081–1083. doi:10.1016/S0021-9258(18)91899-5
28. Erb TJ, Zarzycki J. A short history of RubisCO: the rise and fall (?) of Nature's predominant CO2 fixing enzyme. *Curr Opin Biotechnol*. 2018;49:100–107. doi:10.1016/j.copbio.2017.07.017
29. Yasumoto K, Sakata T, Yasumoto J, et al. Atmospheric CO2 captured by biogenic polyamines is transferred as a possible substrate to Rubisco for the carboxylation reaction. *Sci Rep*. 2018;8(1):17724. doi:10.1038/s41598-018-35641-8
30. Yamada H. Amine-based capture of CO2 for utilization and storage. *Polym J*. 2021;53(1):93–102. doi:10.1038/s41428-020-00400-y
31. Florio C, Fiorentino G, Corcelli F, et al. A life cycle assessment of biomethane production from waste feedstock through different upgrading technologies. *Energies*. 2019;12(4). doi:10.3390/en12040718
32. Cement: Tracking Report. International Energy Agency (IEA). 2021. https://www.iea.org/reports/cement
33. Will Electricity or e-Fuel Power the Cars of the Future? OilPrice.com. 2021. https://oilprice.com/Alternative-Energy/Renewable-Energy/Will-Electricity-Or-E-Fuel-Power-The-Cars-Of-The-Future.html
34. The Silyzer 300 Electrolyzer Arrives in Punta Arenas. H2 Energy News. 2022. https://energynews.biz/the-silyzer-300-electrolyzer-arrives-in-punta-arenas/
35. The Windiest Places on Planet Earth. Surfer Today. https://www.surfertoday.com/windsurfing/the-windiest-places-on-planet-earth
36. Nord Stream AG. Wikipedia. https://de.wikipedia.org/wiki/Nord_Stream_AG
37. Nuclear-Powered Aircraft. Wikipedia. https://en.wikipedia.org/wiki/Nuclear-powered_aircraft
38. Availability and Costs of Liquefied Bio- and Synthetic Methane: The Maritime Shipping Perspective. CE Delft. 2020. https://cedelft.eu/wp-content/uploads/sites/2/

2021/03/CE_Delft_190236_Availability_and_costs_of_liquefied_bio-_and_synthetic_methane_Def.pdf
39. Vereinigte Staaten benzinpreise, 31-Mai-2021 (US Gasoline Prices, 31 May 2021). GlobalPetrolPrices.com. 2021. https://de.globalpetrolprices.com/USA/gasoline_prices/
40. Vereinigte Staaten dieselpreise, 31-Mai-2021 (US Diesel Prices, 31 May 2021). GlobalPetrolPrices.com. 2021. https://de.globalpetrolprices.com/USA/diesel_prices/
41. Fuel Prices in Europe (2021–2023). https://www.tolls.eu/fuel-prices
42. World Bunker Prices. Ship & Bunker. 2021. https://shipandbunker.com/prices
43. US Gulf Coast Kerosene-Type Jet Fuel Spot Price. Ycharts. 2021. https://ycharts.com/indicators/gulf_coast_jet_fuel_spot_price
44. Zang G, Sun P, Yoo E, et al. Synthetic methanol/Fischer–Tropsch fuel production capacity, cost, and carbon intensity utilizing CO2 from industrial and power plants in the United States. *Environ Sci Technol.* 2021;55(11):7595–7604. doi:10.1021/acs.est.0c08674
45. Pieper M, Michalke A, Gaugler T. Calculation of external climate costs for food highlights inadequate pricing of animal products. *Nat Commun.* 2020;11(1):6117. doi:10.1038/s41467-020-19474-6
46. Folkers M, Paech N. *All You Need Is Less.* Oekom Verlag; 2020. https://www.oekom.de/buch/all-you-need-is-less-9783962380588
47. Pirson T, Bol D. Assessing the embodied carbon footprint of IoT edge devices with a bottom-up life-cycle approach. *J Clean Prod.* 2021;322:128966. doi:10.1016/j.jclepro.2021.128966
48. Cambrian Explosion. Wikipedia. https://en.wikipedia.org/wiki/Cambrian_explosion
49. Follows M. End of an era: how long did it take the dinosaurs to die out? *New Scientist.* 2019;(3251). https://www.newscientist.com/lastword/mg24432511-200-end-of-an-era-how-long-did-it-take-the-dinosaurs-to-die-out/
50. Deccan Traps. Wikipedia. https://en.wikipedia.org/wiki/Deccan_Traps
51. McLennan DA. The concept of co-option: why evolution often looks miraculous. *Evo Edu Outreach.* 2008;1(3):247–258. doi:10.1007/s12052-008-0053-8
52. Future Teslas Will Have Batteries That Double as Structure, Making Them Extra Stiff While Improving Efficiency, Safety and Cost. TechCrunch. 2020. https://techcrunch.com/2020/09/22/future-teslas-will-have-batteries-that-double-as-structure-making-them-extra-stiff-while-improving-efficiency-safety-and-cost/
53. Tesla Structural Battery Pack Is Removable, But It's Quite an Ordeal. Electrek. 2022. https://electrek.co/2022/05/20/tesla-structural-battery-pack-removable/
54. Bill Gates' TerraPower Aims to Build Its First Advanced Nuclear Reactor in a Coal Town in Wyoming. CNBC News. 2021. https://www.cnbc.com/2021/11/17/bill-gates-terrapower-builds-its-first-nuclear-reactor-in-a-coal-town.html

Index of topics

For the benefit of digital users, indexed terms that span two pages (e.g., 52–53) may, on occasion, appear on only one of those pages

Tables, figures, and boxes are indicated by t, f, and b following the page number

ABE (fermentate)
  production of, 137
ABS (plastic)
  properties/uses of, 182
acetogens (bacteria)
  microbial fermentation of syngas, 139
ADEME
  drive-train studies by, 203, 219–20, 228, 239, 246–47
adenine
  in ATP, 31, 35
  chemical structure of, 12f
  as a component of genetic material, 30–31
  in compounds of energy and material metabolism, 281f
  in NADPH, 32–34
adenosine triphosphate (ATP)
  in context of photosynthesis and cell respiration, 35, 36
  in driving polymerization of sugars, 154
  in mitochondrial energy/material metabolism, 281f
  in primitive metabolism, 30, 31
  primordial evolution of energy-carrier, 28f, 29–30
  production in glycolysis, 38–39
  production via oxidative phosphorylation (oxphos), 38–39, 40
  production from fat/fatty acid metabolism, 52–54, 53b, 54f, 55
  as proxy for energy content of sugars/fats, 52
aerobic
  ~respiration, 40, 52
agriculture
  animals in, 44–45
  bioethanol from, 134–35
  carbohydrates from crops, 47
  GHG emissions from (general), 66–67
    methane and nitrous oxide emissions, 69
  GHG emissions from livestock, 67, 68, 70
  meat from, 63
  overall environmental impact (livestock), 67–68, 81
alcohols
  chemical structures, C_1 to C_5 (methanol – pentanol), 134f
  E-methanol, 139–40
  ethanol from fermentation, energy balances, 135–36
  from Fischer-Tropsch process, 142b
  glycerol, 51–52
  methanol as an ICE fuel, 129–31
  research landscape, 136–37
  from RSOC technology, 283f
  synthesis from biogas, 139
  various, from (bio)synthetic sources, 134
aldehydes
  formaldehyde in atmospheric methane degradation, 107–8
  formation in Miller-Urey experiment, 31–32
  as potential carcinogen (in tobacco smoke), 77, 173–74
algae
  biofermenters and biofuel production, 137, 147b
  evolution of, 32, 35
  in petroleum formation, 120
  phytoplankton, 110–11
  sinking of CO_2, 208
alkanes
  basic chemical structure, 16f
  chemical structures, from methane to butane, 123f
  comparison with dimethyl ether (DME), 143f
  in context of bond orbitals, 15
  from Fischer-Tropsch process, 142b
  plot of energy densities, 148f
  from RSOC cell, 283f
  similarity to fatty acids, 54f

Ames test
  modification of, 74–75
  principle of, 74
amino acids
  in certain meteorites, 123–24
  chemical conversion during grilling, 72f, 73
  chemical structure, 12f, 46f
  chirality in, 33f, 58–59
  conversion to ketone bodies, 50
  formation at alkaline ocean vents, 29–31
  formation in the Miller-Urey experiment, 24b, 25
  functional groups in, 46f
  generation in central metabolism, 36, 61, 281f, 286
  metabolic interconversion with glucose, 61
  in peptides/proteins, 46f
  presence on primordial Earth, hypothesized, 23
  presence in space, 11, 12f
ammonia (NH_3)
  in amino acid synthesis, 61
  in cyclical energy economies, 282
  in fertilizer production, 70
  as jet engine additive, 168–69
  in Miller-Urey experiment, 23
  in energy storage, 272
amylase
  function and discovery, 47–48
anabolism
  in central metabolism of plants, 32–34, 279f
  in context of body temperature, 155–56
  in context of mitochondrial processes, 281f
anaerobic
  ~decomposition (conditions of), 120
  ~energy metabolism, 38–39
  ~microbes, 48, 116
  ~organisms, 37–40
anhydrous ethanol
  definition and production, 136
  unsuitability as aviation fuel, 146
anthracenes/anthrenes/anthenes
  occurence in molecules formed in space, 11f
  occurrence in chemical species in coal, 119f
  production in meat searing and wood smoke, 72f
anthracite
  coal formation, 117
  typical chemical species found in, 119f
antioxidants (and ~enzymes)
  antioxidant defense in context of aging, 75
  antioxidant enzymes, 60
  neutralization of radicals, 16–17
  origin of sulfur in coal, 119f
  protection against ROS from PMs, 169
aragonite
  production in marine organisms, 111–12
  relative solubility, 111–12
archaea
  in breakdown of cellulose, 48
  in evolution of eukaryotic cell, 37–41
  fossil record of, 32
  production of hydrogenase enzymes, 81
  as proportion of global biomass, 108–9
aromatic (compound)
  in aviation fuel, 196–97
  in carbon-based molecules in space, 11f
  in context of coal constituents, 119f
  from meat searing, 73
  in polyaromatic compounds, 73, 172
asymptotic giant branch star/AGB star
  Buckminster fullerenes/soot particles from, 8, 9–10, 10f
  definition, 8
atomic energy/power. *See* nuclear energy/power
ATP. *See* adenosine triphosphate
aviation (fuel, economy, emissions). *See also* kerosene
  alternatives to fossil kerosene, 146, 149
  CO_2 emissions compared with car-driving, 198
  in context of hard-to-abate sectors, 196
  ~fuel, energy-density considerations, 144, 145, 146
  ~fuel, cost considerations, 148–49, 294t, 294–95
  ~fuel taxation, 295
  health implications of, 168–69
  improvements in efficiency, 162

bacteria
  exchange of genes in, 299
  in the evolution of photosynthesis, 35, 37–40
  in the evolution of the eukaryotic cell, 37–41
  in fermentative production of ethanol, 135–36
  fossil record of, 32
  in hydrogen metabolism, 81
  as proportion of global biomass, 108–9
  sinking of CO_2, 208
basal metabolic rate
  definition, 156
  endotherms compared with ectotherms, 155–56

INDEX OF TOPICS   343

beta oxidation
  in fatty acid metabolism in general, 52, 53*b*, 56*f*, 57, 281*f*
  in ketogenic metabolism, 49–50
Big Bang
  background radiation from, 4–5
  formation of atomic nuclei, 4
  formation of subatomic particles, 3–4
biodiesel. *See* diesel
bioethanol. *See* ethanol
biofuel. *See also* bioethanol/biodiesel/ABE
  in aviation fuel, 144
  in context of agriculture, 175–76
  non-alcoholic, 147*b*
  policy development, 137–38, 139
biogas. *See also* methane
  in local energy infrastructures, 290
  methanol from, 133
  potential for fueling cement industry, 195–96
  potential for powering variety of sectors, 288
black body radiation
  nature of, 87
  physics of, 88*b*
blast furnace
  chemical/physical principles of, 193*f*
  iron/steel production, 192
brines (also lithium-containing ~)
  lithium extraction from, 240–41, 257
  for sequestration of CO_2, 211
Buckminster fullerenes
  designation and physical properties, 9
butterfly effect
  in context of emergent effects, 84–85

calcite
  in living organisms, 111–12
CarbFix technology
  energy requirement, 212
  principle, 209–10, 209*f*
carbohydrates
  chemical nature of energy metabolism, 34–37
  complex carbohydrates, 48
  in context of ketosis, 60
  in contrast to fats/oils, 52, 55, 57
  depletion during starvation, 49
  details of energy metabolism, 35
  energy content from a chemical perspective, 51
  in health and disease, 82
  importance in agricultural/energy terms, 47
  in material metabolism, 61, 155, 281*f*
  production via photosynthesis, 32, 279*f*, 285

proportion of nutrition, 44–45, 57
reaction with amino acids during meat searing, 73
recommended intake, 50
reducing potential of, 25
role in nutrition, 45
storage in animals, 48, 79–80
carbon capture. *See also* CarbFix; carbon sequestration; direct air capture(DAC)
  principle and examples, 208–13
carbon fiber
  as bucky tubes, 9–10
  in car manufacture, 182
Carboniferous (period)
  coal formation during, 114–15, 117
  oil formation during, 114–15
carbon sequestration. *See also* carbon capture
  principle and examples, 208–13
  rate of progress, 215–16
catabolism
  in central metabolism, 281*f*
catalytic converter
  principle and applications, 163–64
cellulose
  alteration during coal formation, 118
  in biofuel production, 137, 175–76
  in carbohydrate metabolism, 32–34, 36
  in context of energy/CO_2 balance of wood, 151–52
  microbial digestion of, 48
  in thermo-swing filters for CO_2, 209*f*
cement (and industry)
  challenges of de-carbonizing, 194–96
  CO_2 emissions from $CaCO_3$, 289
  energy sources/challenges, 150, 191, 194–96
  manufacturing method/chemistry of, 195*f*
  PM emissions from industry, 172–73
  pre-combustion energy, 159
  quantities produced, 202
chirality
  in metabolism, 58
chondrite meteorites
  content of organic compounds, 123–24
cis-unsaturated fatty acids
  biochemistry of, 55
  chemical structure, 56*f*
Climeworks
  company, history, technology, 209–11
  impact of, 213
clinker (cement)
  in cement production, 194–95, 195*f*
CO_2 compensation
  schemes, 205–7, 213

CO$_2$ fixation/CO$_2$ sequestration
  biological, 36, 177, 207
  technological, 209–10, 209f, 211–13
CO$_2$ sinking. *See* CO$_2$ sequestration
coke (coal)
  in blast furnace, 193f
  from coal-roasting, 117
  from petroleum, 191t
  source of graphite, 125–26
cradle-to-cradle
  design of products, 287, 298–99
  economy, 264, 300–1, 304
cradle-to-grave
  economy in general, 298–99
  study of car economy, 246–47
creatine
  taste of fried meat, 72f, 73
Cretaceous (period)
  coal from, 117

D-beta-hydroxybutyric acid / D-beta-hydroxybutyrate (DBHB)
  chemical structure/characteristics, 49f
  evolutionary significance, 57
  in ketosis, 50
  in life-span extension, 58
  metabolism and chirality, 58–59
diamond
  formation, physical properties, artificial synthesis, uses, 19–21
  kinetic versus thermodynamic stability, 21, 118
  origin of physical properties, 9
diesel
  biodiesel in automobile fuel blends, 131
  biodiesel feedstocks, 137–38
  biodiesel, cost to consumer, 250–51, 255, 294t, 294–95
  chemical composition of, 123f, 143f
  consumption, 191t
  economic considerations, 250–51, 255, 294t, 294–95
  energy density of, 130
  ~engine, efficiency of, 158, 248–49
  ~engine, improvements, 161
  from Fischer-Tropsch process, 142b, 197
  synthetic replacements for, 176
  ~trucks, economic considerations of, 201
Dieselgate
  scandal, 248–49
diesel particulate filter (DPF)
  emissions regulations related to, 165
  principles of action, 164–65
  for ships' engines, 165

dimethyl ether (DME)
  chemical and combustion characteristics, 141–42, 143–44
  chemical structure, 143f
  chemical synthesis, 143
  as a truck fuel, 176
  uses, 142–43
direct air capture (DAC) of CO$_2$
  in CarbFix technology, 209f
  from flue gas, 208, 215–16
DNA (Deoxyribonucleic acid)
  biological order/information, 22
  chemical constituents of, 12f
  chemical structure of nitrogenous bases in, 46f
  components formed in the Miller-Urey experiment, 23, 24b
  damage/repair, 75, 76, 77, 78–79, 173
  evolution of, 30–31
DNA/RNA bases
  chemical structures, 46f
  found in meteorites, 123–24
  produced in the Miller-Urey experiment, 24b, 25

E5, E10, E85 gasoline/ethanol blends
  feedstocks for production of, 137–38
  fuel for ICEVs, 134
e-fuels. *See also* power-to-liquid; power-to-X
  from "artificial photosynthesis", 139–40
  in comparison with batteries for energy storage, 275
  comparisons with existing fossil fuels, 129
  as component in sustainable reduced-driving model, 254
  from concentrated waste CO$_2$, 289
  in the context of geopolitical independence, 255–56, 256f
  in the context of model global automotive economies, 232–34, 233f, 238f–39
  in the context of whole mobility economies, 217, 218
  cost to customers, 255, 293–95
  energy-intensiveness of production of, 252
  environmental impact compared with battery technology, 239–41
  global production potential, 175–76
  non-alcoholic ~, 141
  pilot production plant in Chile, 290
  political support for, 251
  practical economic considerations of, 265
  principles of, 129

public opinion and reporting on, 217
scale-up considerations, 251–52
from wasted regenerative electricity potential, 176
ectotherms
energy consumption of, 155–56
electrolysis (to produce H_2)
in alkaline solution, 271–72
energy requirement, 194–95
for Fischer-Tropsch synthesis, 142b
in RSOC cell, 282
electron transport chain (ETC)
in chloroplasts, 35
in energy/material metabolism, 279f, 281f
iron-sulfur clusters in, 42
in mitochondria, 38–40
elemental carbon (EC). See also diamond; graphite; graphene
in context of PMs, 174
in space, 6–8
embodied energy
of electronic devices in general, 230
of eletric vehicles, 229
enantiomers
as exemplified by glucose, 33f
endotherms
evolutionary advantage of, 156
thermal efficiency of, 155–56, 157–58
energy balances
in the context of synthetic portable fuels, 129–33
epigenetic modification
in adaptation to changes, 298
in context of metabolism and disease, 82
hypothesized mechanism of ketosis, 50
ethanol
agriculture considerations, 139
California fuel policy, 131
EU fuel policy, 137–38
energy balances of production, 135–36, 137, 138–39
environmental impact in ICE technology, 237f
extent of potential gasoline replacement, 134–35
feedstocks/production methods for bioethanol, 134, 135, 136–38, 139
in gasoline blends, 134
in methanol blends, 130
in Open Fuel Standard Act, 131
political decision-making on, 138
eukaryotes
evolution of photosynthesis, 32, 35

evolution of respiratory pathway/energy metabolism, 38–39
via symbiosis with alpha-proteobacterium, 37–41
Euro emissions norms
emissions standard, 165

F_1F_0-ATPase
in chloroplasts, 36
in mitochondria, 38–39, 279f, 281f
fatty acids
biological synthesis of, 154–55
in central metabolism of energy/materials, 281f
energy comparisons with carbohydrates, 54f
energy content, 51–54b, 55
energy metabolism and storage in plants, 279f
evolutionary advantages of fats, 55, 57
interconversion with carbohydrates, 155
in membrane phospholipids, 122f
as respiratory (energy) substrate, 44–45
saturated versus partly unsaturated, 55–56f
in triglycerides, 46f, 51–52
fingerprint radiation
absorption of IR radiation by moleulces, 87–89, 91b
energy transfers via CO_2 in the atmosphere, 99–100
physical cause of, 89
from vibrational modes in CO_2, 92–94f
Fischer-Tropsch (process)
aviation fuel via, 144, 147
chemical industrial process, 142b
conversion of waste biomass into fuel, 147b
DME via, 143
e-diesel via, 131, 141, 255
fuel prices, 293
methanol production via, 132–33, 255
fluid bed combustion
waste incineration via, 184–85
fullerenes. See Buckminster fullerenes

gasoline
chemical composition of, 123f
consumption of, 191t
cost to the consumer, 294t
energy density of, 130
from Fischer-Tropsch process, 142b
replacement with e-methanol, 140–41, 290
gasoline particulate filter
introduction of, 165–66

genetic modification (GM)
  public debate about, 179
glycemic index (GI)
  carbohydrate metabolism and health, 47–48
glycogen
  animal carbohydrate storage, 48
  chemical structure of, 46f
  in energy metabolism and storage, 48, 52–54
  energy value of, 52–54, 55
  in fried meat flavor, 73
  polymerization from glucose, 52–54, 154f
graphene
  chemical nature of, 20f
  properties and applications, 21
greenhouse effect. *See also* radiative forcing
  discovery, 97–98
  notional atmospheric layers in, 106–7
guanine
  from central energy/material metabolism, 281f

Haber-Bosch process
  role in fertilizer production, 70
heterocyclic amines (HCA)
  exposure and evolutionary tolerance to, 76, 77–78
  flavor of, 73
  from meat searing, 73
  resulting from paleolithic meat cooking, 72f
hydrocarbons
  advantages in cement manufacture, 194–96
  as alkanes, 15
  as aviation fuel, 144–46
  "biological hydrocarbons" in energy storage, 82
  biological hydrocarbon analogs, 57
  biological parallel with hydrocarbon fuels, 59
  biologically relevant physical characteristics, 46f
  in certain meteorites, 123–24
  chemical nature of, 121–22
  compared with other energy storage methods, 272
  as components of the petrochemical industry, 191t
  cost considerations of, 294t, 294–95
  distinction from carbohydrates, 52
  energy densities, 148f, 149–50
    compared with Li-ion batteries, 198
  fatty acids likened to, 51–52
  in geological cycles, 113–14
  physical and combustion characteristics, 147
  production in AGB stars, 8–9
  relationship of oxidative energy to CO_2 emissions, 151–52
  via Fischer-Tropsch process, 142b
hydrogencarbonate (HCO_3^-)
  in ancient terrestrial ocean, 30–31
  involvement in CarbFix technology, 210
  mechanism of ocean pH fall, 110
hydrous ethanol
  product of distillation, 136

infrared (IR radiation)
  absorption by CO_2, 86, 91b, 94f
  in black-body radiation, 88b–89
  energy transitions upon molecular collisions, 104f
  as fingerprint, 89, 91
  nature of, 87
  spectra of CO_2 and H_2O, 96f
internal combustion engine (ICE)
  as analysed in Fraunhofer study, 218
  CO_2 capture from, 159
  in comparison of car-driving with air travel, 198
  efficiency/energy recovery, 158–60, 181–82
  emissions and treatment of, 130, 151, 163–69, 196
  environmental impact (and comparison with other technologies), 141, 181–82, 197–98, 202–3, 219, 235, 246–48
  fuel consumption, 250–51
  further development of, 160–63
  as jet engines, 162–63
  maintenance/life-span, 242
  manufacture and recycling, 243–44, 257
  political dimensions, 167–68
  in public opinon and debate, 179, 217, 258
  use of DME fuel in, 143–44
  use of ethanol fuel in, 130–31, 132, 136, 146, 275
  use of methanol fuel in, 130, 133
Internet of Things (IoT)
  CO_2 footprint of, 241, 297
interstellar medium (ISM)
  material flux into, 8
IR: infrared (radiation). *See* infrared (radiation)
IR-absorptive capacity
  of CO_2 and other molecules, 86, 87–89

Jurassic (period)
  geology, 109

kerogen
  in meteorites, 123–24

nature and occurrence of, 121–22
non-terrestrial ~, in oil formation, 123
origin of, 121, 124
terrestrial ~, in oil formation, 121
kerosene
chemical composition of, 123f
energy density of, 147
from Fischer-Tropsch process, 142b
ketogenesis/ketosis/ketogenic diet
definition, 49
evolutionary significance of, 57, 60
health-promoting effects of, 49–50, 60
in life-span extension, 59
mechanism in energy metabolism, 59
ketone body
chemical nature of, 49
energy metabolism of, 57, 58
history surrounding discovery, 49–50
physiological significance of, 50, 79–80
significance of chirality in, 58
Krebs cycle. See TCA cycle

leightweighting
of cars, 182, 219
Li-ion (battery)
energy of manufacture, 228–29
environmental impact of, 231
recycling and sustainability considerations, 243–44, 257–58
technical details and qualities of, 243, 270–71, 276t
use in vehicles, 231
lignin
in coal formation, 118
in context of lignite, 116
in lignocellulose, 48
lignite
formation and composition of, 116
microscopic appearance of, 115f
lignocellulose
as fermentation substrate for biofuel, 137, 175–76
nature of, 48
lipids
appearance in context of alkaline ocean vents, 27
chemical structure of, 46f, 122f
as constituents of living organisms, 45–47, 122f
in evolution of proto-cells, 28f
fats and oils, 46f
in first protein-carrying membranes, 29–30
in the formation of oil and gas, 120–22

formation of primitive membranes from, 27
as sites of proto-biological energy metabolism, 29, 30
lithium hydroxide (LiOH or LiOH·H$_2$O)
chemistry of in batteries, 270–71
energy of manufacture, 229–30
production facility, Brandenburg, 240–41
sources, 241
Lupo 3L
production, technology, fuel consumption, 248–49

meat
consumption trends, 44, 62–63, 68, 70–71
cooking, 71–73
GHG footprint of, 47, 64b
health-related qualifications and concerns, 62, 70–71, 74
in human nutrition, 44–45
necessary reduction in consumption, 70–71, 262, 263
resource use in production of, 63–65, 64b
methane. See also natural gas
in cement manufacture, 194–95
cost of synthetic, 293
in diamond synthesis, 19–20
emissions from agriculture, 47, 48
emissions from waste, 184–85, 288–89
energy density of, 148f, 152t
as the first (C$_1$) alkane, 123f
hybrid sp orbitals in, 15, 16f
as an industrial raw material, 186
in Miller-Urey experiment, 23
origin from super novae, 6
from RSOC cell, 282–84, 286
in steel and cement manufacture, 150
synthetic~ for energy storage, 267, 272, 273–74, 276t, 290
as an underestimated GHG, 69, 106–8, 184–85, 218
as vehicle fuel, 237f
methanogens
in ruminant digestion, 48
methanol and. See alcohols; power-to-liquid; power-to-x
e-methanol from regenerative electricity, 139–41, 290
energy balance of synthesis, 275
as fuel for ICEs, 141
milk (and ~products)
energy consumption and equivalences, 65–66f

mitochondrion
  biochemistry within, 38b
  in cell respiration, 38b
  central role in reversible energy<>material metabolism, 281f
  comparison with chloroplasts, 35
  D-beta-hydroxybutyrate metabolism by, 58–59
  evolution of, 35
  in health-promoting ketogenesis, 59
  interplay with chloroplasts in plant central metabolism, 279f
  oxphos in, 34–37
  structure of, 38b

natural gas (predominantly methane)
  Chemical composition, 123f
  energy density of (methane), 152t
  in flex-fuel cars, 131–32
  from kerogen, 121–22
  methanol production from, 131
negative feedback
  as component of new economies, 300–1
nitrogen (N_2)
  concentration in the atmosphere, 92t
  incorporation into carbon compounds in space, 11, 11f
  as source and decay product of ^{14}C, 7
nitrogen oxides (NO_x)
  from candle flames, 175
  influence on the methane cycle, 107–8
  from jet engines, 168–69
  permissible limits, 174
  reduction via ICEV technologies, 164, 165
nitrous oxide (N_2O)
  concentration in the atmosphere, 93t
  as GHG, 69, 218
nuclear energy/power
  history of, 291
  persistence/re-emergence of, 181, 301
  in public debate, 180
nuclear fusion. See also nucleosynthesis
  during the Big Bang, 4
  creation of elements in stars, 5–6
  of hydrogen in stars, 1
nucleosynthesis (also Big Bang nucleosynthesis and stellar nucleosynthesis)
  in Big Bang, 4
  in Stars, 5

ocean
  in the global water cycle, 90–92
  mass of carbon cycling through, 109–10, 113–14, 126
  pH (acidity) of, 110–13
  as potential location for CO_2 sinking, 212–13
  proportion of life in, 108–9, 108f
  role of Southern ~ in ice ages, 99
oxidative phosphorylation (oxphos)
  in mitochondria, 34–37, 38–39
oxycombustion
  in context of CO_2 capture, 208
  economy-promoting measure, 265
oxygen catastrophe
  role in cellular evolution, 37–40
ozone
  concentration in atmosphere, 93t
  role in methane cycle, 107–8
  tolerance toward, 77–78

P450 cytochromes
  in detoxification, 74–75
particulate matter (PM, e.g., PM2.5)
  changing distribution of sources, 168
  in emissions from aviation, 168
  entry routes into the body, 170f
  healthcare implications, 169–74
  from non-ICE sources, 166–67
  removal of from combustion processes, 164–66
  risk perception of, 174–75
  sources of PM2.5s, 171–72
  types and definition, 164–66
PE: polyethylene
  chemical structure of, 188f
  in materials, 188
periodic system (Mendeleev's Periodic System of Elements)
  principle of, 2b
permanent (atmospheric) gases
  concentrations of, 92t
PET: polyethylene terephthalate
  chemical structure of, 183f
  in materials, 183
Peto's Paradox
  cancer incidence in animals, 76
petroleum
  abiogenic origin of, 123–24
  biogenic origin of, 120–23
  from biological lipids, 121
  from kerogen, 121–22
  porphyrin found in, 121f
  products from, 128, 191t
photovoltaic (PV) energy
  E-methanol from PV energy, 139–41
  hydrogen production via, 271–72, 290

polyaromatic hydrocarbons (PAHs)
  associated with PM, 172–74
  produced in meat cooking, 72f, 73
  role in oxidative damage, 76
  in wood and tobacco smoke, 77
polypropylene (PP)
  chemical structure of, 188f
  use in cars, 182
  use in construction, 188
polyurethane (PU)
  in construction, 189
polyvinyl chloride (PVC)
  in car manufacture, 182
  chemical structure of, 188f
  in construction, 188, 189
power-to-liquid (PtL). *See also* e-fuels
  principle of, 141
power-to-X (PtX). *See also* e-fuels
  comparative emissions from, 236
  methane/methanol from RSOC cell, 273–75
  as non-fossil methane, 272
  parallels with biology, 282–84
  principle of, 141
  process efficiency of, 272, 273–74
  for storing peak renewable energy, 273
precautionary principle (PP)
  risk assessment/response, 187
pre-combustion
  in carbon capture, 208
  principle and applications, 159
pro-carcinogens
  in metabolism of PAHs and HCAs, 77–78
purines
  formation in alkaline vent context, 31
  formation in Miller-Urey experiment, 25
  formation via mitochondrial metabolism, 281f
pyrimidines
  formation in alkaline vent context, 31
  formation in Miller-Urey experiment, 25
  formation via mitochondrial metabolism, 281f
pyrolysis
  feedstock production (general), 190
  in oil formaion, 120
  as source of synthesis gas, 142b
  of (waste) biomass, 147b, 271–72
  of waste plastic, 182–83
pyruvate (pyruvic acid)
  in context of cell respiration, 38b
  formation in alkaline vent context, 30–31
  in mitochondrial energy/material metabolism, 281f, 285–86

radiative forcing. *See also* greenhouse effect
  balance with radiative forcing, 107
  caused by methane, 106–7
  caused by water vapor, 92
  damping of, 107
radicals (chemical)
  in context of Oxygen Catastrophe, 37–40
  in damage to organisms, 21
  neutralization via C=C bonds, 16–17
  OH (hydroxy) reaction with atmospheric methane, 107–8
  from PAHs, 173
  produced in energy metabolism, 60
  superoxide, 16–17
Rankine cycle
  in ICE energy recovery, 158–59
rare earth metals
  in EV technology, 230
  from stellar fusion reactions, 6
reactive oxygen species (ROS). *See also* radicals
  organismal protection against, 76
  from PM, 169
  resulting from trans-unsaturated chemistry, 56f
ReCiPe
  principle and components of, 234–35
recycling/recyclability
  batteries, 167–68, 243–44, 251–52, 267, 274–75
  cars and car components, 182–83, 227, 274–75
  comparison ICEV/BEV, 243–44
  in context of energy/material economies, 177
  in context of rock deposits, 83
  cost of, 293, 295
  in (e-)fuel economies, 157–58, 176, 177
  necessary economies of, broad, 129, 186–87, 190, 257–58, 264, 301
  for CO_2, 215
  plastics and other synthetics, 183–88
  in power-to-x, thermal/material, 272, 273–74, 287
  principle inherent to life, 18–19, 42–43, 79–80, 127, 157–58, 304
  steel, 192, 202
  thermal ~, 196
red giant (star)
  fusion producing heavier elements, 5–6
RNA: Ribonucleic acid
  chemical constituents of, 12f
  evolutionary emergence of, 31
  fraction in organisms, 45–47
  nitrogenous bases in, 46f

road transport
  challenges of electrification, 140–41, 167–68
  emissions, various, 165, 166–67, 171–72
  emissions, comparison with livestock products, 65–71
  European Commission policies on, 245
  proportion of CO_2 emissions, 253

saline aquifers. *See also* brines
  for seqestration of CO_2, 210, 212
saturated fatty acids
  biological energy in, 55
  chemical structure of, 54*f*
selective catalytic reduction system (SCR)
  application in ships, 165
spodumene
  battery manufacturing energy, 229–30
  trend in use in battery economy, 241
  water use, 240–41
steam reforming. *See* pyrolysis
steel (manufacturing)
  challenge of de-carbonizing, 192–91
  emissions from, general, 172–73
  manufacturing method/chemistry of, 193*f*
  using hydrogen/methane, 150
sub-bituminous coal
  origin/qualities of, 115*f*, 117
sugars. *See also* carbohydrates
  in biological combustion (glycolysis), 34–37
  from C3 and C4 plant metabolism, 7
  chemical structure, 46*f*
  chirality in, 58–59
  as component in ATP, DNA, RNA, 35
  as disaccharides, 47–48
  as feedstock for fermentation processes, 137
  formation in alkaline vent environment, 31–32
  found in meteorites, 11
  isomers of glucose, 33*f*
  from the Miller-Urey experiment, 25
  in nutrition and physiology, 47–48
  polymerization into starches for storage, 153–55
  production via photosynthesis, 32–34, 36
  structure (of ribose), 11
  as substrates in material metabolism, 286
synthesis gas/syngas
  in DME synthesis, 143
  in Fischer-Tropsch process, 142*b*
  in integrated energy/material economies, 215
  microbial fermentation of, 139
  organic waste as source of, 132–33, 147*b*
  from pre-combustion, 208
  in RSOC context, 282

TCA cycle/Krebs cycle
  in cell respiration, 38*b*, 57, 279*f*
  in energy/material metabolism, 281*f*
Tertiary (period)
  lignite formation during, 116
three Rs (in animal research)
  in models of human cancer, 74
thymine
  from energy/material metabolism, 281*f*
tipping point
  as result of positive-feedback systems, 113
  resulting from increasing CO_2 concentrations, 205
toxic oxygen radicals. *See also* radicals; reactive oxygen species
  cellular defense against, 60
trans-unsaturated fatty acids
  chemical reactivity of, 56*f*
  chemical structure of, 56*f*
triglycerides
  biological fats and oils, 51–52
  chemical structure of, 46*f*
  energy content and release, 53*b*
  in fat metabolism, 52
triisobutane
  as potential aviation fuel, 149
troposphere
  breakdown of methane in, 107–8
  half-life of CO_2 in, 105–6

ultrafine particles (UFPs). *See also* particulate matter
  from coal-fired power stations, 174–75
  in subways, 167
ultraviolet (UV radiation)
  absoption by molecules, 86
  conversion to IR during greenhouse effect, 87
  in DME synthesis, 143
  effect on electrons, 86
  resistance of plastics to, 182
unsaturated (fat/fatty acid)
  chemical structure of, 56*f*, 122*f*
  cis-/trans- conformations of, 56*f*
  energy content of, 55
  fatty acids, 54*f*
uracil
  from central energy/material metabolism, 281*f*
  in RNA, 46*f*

variable (atmospheric) gases
  list of, 93t
  nature of, 90–92

water vapor
  IR absorption and role as a GHG, 92, 97–99
  participation in global cycles, 90–92
  relationship to CO_2's IR absorption spectrum, 93–96f, 100–1
  role in energy transfers in the atmosphere, 95–96, 102, 105

wind energy
  application and efficiency, 265, 290
  availability of, 266, 267, 274
  in e-fuel production, 141, 273–74, 290

wood
  chemical nature of, 48
  chemical signatures of in coal, 118
  CO gas from, 254–55
  in construction, 188, 206
  in lignite formation, 116
  as substrate for biofuels, 137, 139
  sustainability considerations of wood-burning, 150–53
  use of waste ~ in steel industry, 192–94

wood smoke
  evolutionary exposure/adaptaion to, 78, 79
  principal components of, 72f

X-ray (radiation)
  absorption by atoms, 86

yeast
  in fermentative production of ethanol, 135–36

zeolites
  in absorption of ICE combustion products, 160